Questioning Science in East Asian Contexts

Science and Religion in East Asia

VOLUME 1

The titles published in this series are listed at *brill.com/srea*

Questioning Science in East Asian Contexts

Essays on Science, Confucianism, and the Comparative History of Science

By

Yung Sik Kim

BRILL

LEIDEN | BOSTON

The publication of this book was supported in part by a grant from Templeton Religion Trust. The opinions expressed in this book are those of the author and do not necessarily reflect the views of Templeton Religion Trust.

Library of Congress Cataloging-in-Publication Data

Kim, Yung Sik, 1947- author.
[Essays. Selections]
Questioning science in East Asian contexts : essays on science, Confucianism, and the comparative history of science / by Yung Sik Kim.
pages cm. -- (Science and religion, ISSN 2214-8329 ; volume 1)
Includes bibliographical references and index.
ISBN 978-90-04-26509-7 (hardback : acid-free paper) -- ISBN 978-90-04-26531-8 (e-book) 1. Science--East Asia--History. 2. Science--History--Cross-cultural studies. 3. East Asia--Civilization--Confucian influences. I. Title.
Q127.E26K56 2014
509.5--dc23

2013049492

This publication has been typeset in the multilingual 'Brill' typeface. With over 5,100 characters covering Latin, IPA, Greek, and Cyrillic, this typeface is especially suitable for use in the humanities.
For more information, please see brill.com/brill-typeface.

ISSN 2214-8329
ISBN 978 90 04 26509 7 (hardback)
ISBN 978 90 04 26531 8 (e-book)

This book is printed on acid-free paper.

Contents

Acknowledgements

The essays in this book were originally written and presented separately, not as parts of a single monograph with an overall plan. Some were written a long time ago, some as long as thirty years ago (Chapters 5 and 7). As a result, the contents, and especially the references cited, do not properly represent the current state of scholarship. There also are overlaps, sometimes differences in emphasis, and even contradictions among the essays. Nevertheless, except for correcting obvious mistakes, unifying the formats for citation and romanization, and updating the citation data for some obsolete references, I have decided to leave the essays in more or less their original form, as I wanted to keep the integrity of the individual essays with their own purposes, motivations, and relevance at the time they were written.

It is impossible to acknowledge all the debts I incurred with my teachers, colleagues, and students over the years that I have been working on these essays. I will mention only those who helped me in the course of preparing and publishing this book. This book is published as the first book of the series "Science and Religion East Asia" by E. J. Brill. I would like to thank Albert Hoffstädt and Patricia Radder at Brill for accepting my proposal to establish the series, and for their commitment to publishing this book. My thanks also go to my fellow series editors, Benjamin Elman, Jongtae Lim, and Dagmar Schäfer, who agreed to launch the series with my own book, to the anonymous referee whose generous comments contributed to the improvement of the manuscript, including its title, to my student, Minchol Shin, who helped me solve all sorts of problems that came up while preparing and polishing the manuscripts, and to Heddi Goodrich, who copyedited the entire manuscript with care and thoroughness. Finally, I would like to thank Templeton Religion Trust for their support in publishing this book through a grant to the "Science and Religion in East Asia" Project at the Science Culture Research Center, Seoul National University.

Yung Sik Kim
September 2013

Introduction: Questioning Science in East Asian Contexts

1 Science and Confucianism

Modern evaluation of the role of Confucianism in the development of science has changed over the course of the twentieth century. Earlier in the century it was uniformly negative: Confucianism was generally viewed as having 'inhibited' scientific development. In fact, extremely negative views were expressed, both by East Asians themselves and by Westerners, about the Chinese scientific tradition itself. Deeply impressed by the strength of the Western powers, which they thought was based on Western science and technology, most early twentieth-century East Asian intellectuals regarded the Chinese scientific tradition as underdeveloped or even non-existent. A leading modern scholar of Chinese philosophy, Fung Yu-lan, even entitled his article "Why China Has No Science."[1] Western scholars, including not only sinologists, but also, for example, Albert Einstein and Lucien Lévy-Bruhl, shared these unfavorable views and mused over the underlying causes. Confucianism, which dominated the East Asian world for centuries, was naturally fingered as the culprit. Later, thanks to the pioneering works of historians of Chinese science, Joseph Needham foremost among them, there emerged some awareness of the remarkable achievements of traditional Chinese science and technology. But the verdict on the role of Confucianism was on the whole negative. With its traditionalism, conservatism, mistrust of mercantilism, etc., Confucianism continued to be seen as the factor undermining the Chinese potential for scientific and technological development, for which the contributing factors were usually sought elsewhere – in Daoism and pseudo-sciences, for example.[2]

Later in the century, however, there arose a shift of opinion. And this shift was brought about by a growing popular awareness of the current situation, in particular, of the predicament of modern science and the development of the

1 Einstein's letter to J. E. Switzer, quoted in *Science since Babylon*, ed. Derek J. de Solla Price (New Haven: Yale University Press, 1961), 15; L. Lévy-Bruhl, *Les functions mentales dans les sociétés inférieures*, 2nd ed. (Paris: Alcan, 1928), 380; Fung Yu-lan, "Why China Has No Science: An Interpretation of the History and Consequences of Chinese Philosophy," *The International Journal of Ethics* 32 (1922): 237-263.

2 Such a view is most pronounced in Joseph Needham's work, and is found in various places in *Science and Civilisation in China* (Cambridge: Cambridge University Press, 1954).

 | DOI 10.1163/9789004265318-002

East Asian economy. Firstly, people came to realize that the development of modern science and technology could not continue indefinitely but had limits and, furthermore, that progress had created problems that could not be solved by science and technology themselves. This realization made people look for alternatives and solutions in the ideas and practices of non-Western cultures, including Confucianism. Secondly, there was the recognition that several East Asian economies had flourished over the past few decades. It was inevitable that this awareness led many people to think, even assume, that Confucianism, the cornerstone of the region for two thousand years, must have played a positive role in this success. As the result of this insight, and of the conclusions people have drawn from it, a new trend has become widespread, one that sees various Confucian ideas, virtues, assumptions, beliefs and tendencies as important factors underlying the success of the East Asian economies. There have been numerous discussions on the relevance and congruence of various Confucian elements for the modern economic and industrial development in the region, including scientific and technological development.[3]

While I am not well read in the relevant recent literature, I find myself in basic agreement with the belief that some elements of Confucianism, which has dominated the region for so long, must have played a positive part in the recent economic development of the region.[4] Yet we cannot extend these congruences to the role of Confucianism for the scientific development in East Asia.[5] It should also be noted that choice and discussions of the topic have been dictated by modern circumstances and modern concerns. That is, more often than not, people first point out the success of modern East Asian economies and then search for its source in the Confucianism prevailing in the region for so long; in doing so, they often adopt modern, and frequently Western, perspectives. It would be more reasonable, as well as fruitful, if

3 For some examples of the more sophisticated discussion of these congruences, see the papers gathered in Tu Wei-ming, ed., *Confucian Traditions in East Asian Modernity: Moral Education and Economic culture in Japan and the Four Mini-Dragons* (Cambridge, Mass.: Harvard University Press, 1996).

4 The situation is strikingly similar to that surrounding the discussion of the role of Christianity in the scientific development of the West, whereby it is assumed that, as a dominant element in the West, the Christian religion must have played an important role in the development of Western science.

5 A similar extension from economic to scientific development was carried out in the case of the West by Robert K. Merton, who applied Max Weber's insight on the relation between the Protestant ethic and the rise of capitalism to his own famous thesis on the relation between Puritanism and modern science: Robert K. Merton, "Science, Technology and Society in Seventeenth Century England," *Osiris* 4 (1938): 360-632.

instead one first looked at the main ideas and practices of the past Confucian traditions themselves, and found features in them that are relevant to the Confucian understanding of natural phenomena, as well as to the development of science in the Confucian cultures and societies.

This, then, is the approach taken by the essays in Parts One and Two of this book, in which I have tried to examine the East Asian contexts directly, and to avoid bringing in one-sided or idealized Western perspectives that are neither necessary nor useful, in order to look at both the positive and negative influences of Confucianism on science and technology.

2 Intercultural Comparison

In addition to the main theme, 'Science and Confucianism,' an additional theme will emerge from the essays in this book, namely 'intercultural comparison.' Part Three of the book is devoted to essays that explicitly adopt comparative perspectives or address the issues that arise when making comparisons between different cultures. But in the essays in Parts One and Two as well, while discussing the East Asian contexts of science, I have frequently had intercultural comparisons in mind, and sometimes made actual comparisons.

I am fully aware of the general skepticism among historians of science about the possibility of making meaningful and productive comparisons between sciences of different cultures. The skeptics insist that cross-cultural comparisons should be postponed until the sciences of different cultures are themselves more thoroughly understood. They also argue that one has to be very careful in choosing what one compares. If one compares the wrong things, they say, one tends to distort one's understanding, not only of the things compared but also of the two cultures in general. Even when one comes up with the right sort of subject to compare, many other elements within the two cultures – the so-called cultural 'contexts' – can be so different that the comparison of the particular thing may not add to one's understanding of these cultures.

Of course, I agree with this unexceptionable position. But I still feel that we can and should make comparisons between different cultures. A problem frequently encountered in the comparisons that have been made between sciences of traditional China and those of the West is that concepts and factors are often compared in isolation, without much attention to their contexts. Such comparisons tend to stress superficial similarities between things that were in fact quite different; when their contexts are truly examined, one often finds that those things were quite differently perceived or used across the two cultures and played quite different roles. Clearly, one should take context into consideration in making comparisons between ideas and events.

Yet we cannot conclude from this that comparisons of things, concepts, and factors should be avoided altogether, as Lloyd and Sivin, for example, insisted by saying that what should be compared are not "things or concepts but whole processes."[6] I cannot help asking whether it is really so uninteresting or meaningless to compare things and concepts. What then would be the purpose of comparative history? Isn't it comparing X in cultures A and B? Do we want only to understand cultures A and B as a whole? And even if we do, can we understand an entire culture without understanding the individual things and concepts that comprise that culture? Besides, is there even a way to compare the whole processes without looking at the things or concepts that comprise the processes? Should one be silent about the differences in a factor X in different cultures till one has reached a complete understanding of the whole processes? Should we give up wanting to better understand the nature of X by considering its various manifestations across cultures? Should we resist our desire to compare things and concepts, knowing fully well, of course, that their contexts differ from culture to culture?

Another point frequently raised by critics of intercultural comparisons is that in making comparisons, one should not assume "a uniformity in the relevant characteristics across different domains and at different periods" of a given culture.[7] This point is also in principle unexceptionable, but is sometimes difficult, and even undesirable, to stick to in practice. Without denying the theoretical validity of the point, and thus fully accepting the inevitable presence of variations and differences in a given culture that cannot be given justice in simplified accounts, I still believe that one can in practice make interesting cross-cultural comparisons of some general features of the sciences of different cultures.

Another problem lies in the hidden assumption frequently underlying intercultural comparisons, in particular comparisons between sciences in China and the West, namely the assumption that the factors which were significant in Western scientific development are the factors which must have also affected the development of the sciences of China. This has often led those who study traditional Chinese science to look at such concepts as time, causal-

6 Geoffrey Lloyd and Nathan Sivin, *The Way and the Word: Science and Medicine in Early China and Greece* (New Haven: Yale University Press, 2002), 9.

7 Lloyd has called this "anti-generalisation point." G. E. R. Lloyd, *Adversaries and Authorities: Investigations into Ancient Greek and Chinese Science* (Cambridge: Cambridge University Press, 1996), 3.

ity and laws of nature, such methodological elements as mathematics, logic, proof and experiment, and such social and cultural factors as capitalism, urbanization, merchant class, slavery, bureaucracy, university, religion, individualism, toleration, and so on, as if these concepts and factors which were important in Western scientific development would have played the same roles in the development of science in China. This assumption is clearly mistaken. But I cannot agree that one should therefore stop making intercultural comparisons of these sorts of things altogether. For the problem does not lie in making comparisons but rather in the assumptions underlying them. I believe that it is possible to make comparisons of these things without making such assumptions. And obviously, if the objects of comparison are chosen for their importance in the Chinese context rather than for their significance in Western development, they will result in comparisons that are more fruitful for understanding what really happened in the history of Chinese science.

Yet another observation can be made concerning intercultural comparisons of such factors. Nearly all cultural, intellectual, and social elements in traditional China – religion, law, merchant class, causality, bureaucracy, etc. – were so drastically different from their counterparts in the West that some scholars have gone so far as to wonder whether such elements can be said to have existed at all in China. This naturally leads to the conclusion that it is pointless to study the effect of these elements in Chinese scientific development. However, this attitude, which has stifled many potential cross-cultural comparisons (and I think it was fortunate that some were stifled in such a manner), nonetheless presupposes that these elements should be the same in every culture, or that the way these elements existed in the West are the norms to be followed by every culture. Of course, this is wrong. For why can't there be different forms of religion, law, merchant class, etc. in different cultures?

These, then, are the questions that either are addressed explicitly or lay hidden in the discussions of most of the essays in this book.

3 Asking 'Big Questions'

Most of my essays in this book deal with general problems of a broad scope, and usually ask questions, often 'big' questions. And this reflects my feeling that we should not be afraid of finding, asking, and at times attempting to answer some general, interesting questions that have no simple or single answers, or may have no answers at all. Of course, I know fully well that this feeling goes against the general trend of current scholarship in the study of traditional East Asian science, to which most of my essays belong. Thus, in presenting my essays I am offering a plea to reconsider an aspect, a consequence, of the trend that prevails in the current scholarship which worries me deeply.

I refer to this trend as 'neo-Scholasticism,' the expression which I have coined to echo the 'Scholasticism' of the European Middle Ages. It refers to the following: the powerful inclination for – and emphasis upon – close, careful, meticulous, rigorous studies of minute details, of often extremely technical content, while avoiding considerations of more general, broad, interesting, big problems; and the increasingly strong trend of specialization, sometimes over-specialization. The trend makes the researcher avoid asking (not to mention answering) or speculating about general questions of broad scope, their reason being that "one has to know more," or that "one has to know completely," before asking such questions. What are frequently found in this sort of 'neo-Scholastic' research are reports, and not judgments; describing concrete things and events, and not explaining them or not even making comparisons between them. Thus, in the historical writings reflecting this trend, works of analysis are favored far more than those of synthesis; consideration of values, or of relevance to the present situation, is safely left out. The result is something like the Baconian inductive method applied to historical research.

We can find sources of this trend in the past ideas of both East Asia and the West. On the one hand, the trend can be traced back to the research ideology of nineteenth-century reformed German universities, combined with Rankean historicism; on the other hand, it also goes back to the ideals of evidential research (*kaozheng* 考證) of mid-Qing scholarship. The trend is widespread in every field of modern scholarship; in fact, it is the main characteristic of today's academic scholarship. It is so prevalent that it is impossible to break free from this trend in our own field when it remains rampant in other fields. However, this trend is especially strong in the field of the history of science. And compared with many sub-fields of the history of science, it is strongest in the study of traditional East Asian science.

This trend of 'neo-Scholasticism' has discouraged researchers in the field from asking many possibly interesting questions. A prominent example of a question suppressed in such a manner is the so-called 'why not' question – such as 'Why did the Scientific Revolution not take place in China?' – discussed in one of the essays in Part Three. The 'neo-Scholastic' trend has also made researchers in the field steer clear from making statements about general trends, or from dealing with the questions of periodization, and of developmental patterns of science in East Asia as a whole, or in China, Korea or Japan.[8] The trend has also discouraged comparisons of East Asian sciences with Western sciences. I have pointed out that the problems are not in making comparisons themselves, but in the assumptions underlying the comparisons, and quick, easy conclusions based on such assumptions, and argued that it is possible to make comparisons without making such assumptions. And if we are aware of the dangers and are careful in making comparisons, our comparisons will enhance our understanding of both cultures.

Nevertheless, the awareness of the dangers of this sort of comparison has made historians reluctant to consider the roles of certain social and cultural factors – religion, bureaucracy, etc. – in the development of science in East Asia, because they fear that considerations of such factors are likely to plunge them into such comparisons with the accompanying pitfalls. At times, some of the words designating these factors are criticized for not being accurate enough to represent the multifarious, and changing, historical reality. Thus, when those of us in the field begin to use such words as 'science' or 'religion' in our discussion of what happened in the East Asian past, or even such words as 'Confucianism' and 'Daoism,' we are immediately stopped and asked to clarify what we mean by these words before beginning our discussion. Sometimes we are not even allowed to use such words as 'China,' 'East Asia' or 'the West,' because, we are reminded, these words stand for different regions and periods. Clearly, such interventions are necessary and useful. They benefit us by making us consider the complex and multi-layered meanings of the words and their pertinence more carefully. But I often wonder whether it would not be more beneficial to sometimes allow these words to be used in our discussion first before delving into an analysis of the words themselves. Indeed, it is very likely that the discussion itself would throw light back upon the multifarious meanings and relationships between the terms used anyway.

Of course, asking big, general questions has its dangers, as well as its costs. For example, the answers, however promising and insightful they may appear at the beginning, often on closer analyses turn out to present many difficulties: over-simplifications and distortions of realities that are much more complex and nuanced. But everything worthwhile comes with a price. The existence of the hazards should not stifle us or keep us from allowing ourselves to ask the big, interesting questions once in a while. Furthermore, asking these questions

8 I might mention even the question of moral judgment. For while I myself am inclined to believe the value-neutrality of my work as a historian, I cannot be sure in absolute terms that I am right, and even less sure whether I can impose value-neutrality as an ideal for all historians.

facilitates closer analyses of many problems, which otherwise would not have taken place. And that can be the real contributions of these questions and answers. Completely avoiding the big questions, while looking only at solid details, may very well help us accumulate more information, but not necessarily more knowledge, or more understanding.

The 'neo-Scholastic' trend also tends to make the scholar in the field seek, for his or her research, only those problems that have not been studied by others; and having chosen a topic, there is the tendency to look thoroughly through all the available materials related to the topic; and there is a reluctance to leave out any information, even irrelevant data. We are criticized, and our writing rejected, if we fail to treat any detailed aspect of the topic discussed or leave out any text, archive source, or even secondary material, which is even slightly related to the topic. Often the criticisms are unfair, because had the details and sources we overlooked or barely dealt with been treated fully, they still would not have undermined our arguments, general findings, or the evidence behind them. Of course, it would still be possible to criticize us for neglecting something potentially relevant. I do not know how to respond to such criticisms.

Perhaps I need to make clear at this point that I am not advocating laxer scholarly standards. In particular, I am not saying that in asking and answering those big questions it is all right to compromise the high standard of scholarship achieved through the solid research of our colleagues. Rather, I am suggesting that with such a high standard achieved – with such solid research results into many concrete facts and details – we can now feel more confident to base ourselves upon those results to go a step further, to ask more general, exciting questions. Nor am I advocating that everyone should ask big questions or engage in such work. The core of scholarship should, and in any case will, continue to be made up of solid, careful, meticulous research. That will continue to be its foundation. But some of us should keep asking the bigger questions, at least once in a while. We occasionally need to move from the particular to the general – from individuals, concrete problems, and detailed changes to overall developments, general patterns, and underlying causes. I myself have frequently warned my students against the danger of posing the big questions, which they should be very careful in asking. I might have even told some of them that one cannot be too careful. But if I did indeed say that, I was wrong. Because one can indeed be too careful. I think that our careful 'abstention' from asking such questions has in some ways gone too far, and has thus led to research which is sometimes too laborious, uninteresting, and even irrelevant.[9]

9 There have been earlier criticisms of this trend. For example, Lawrence Stone, in a lecture at the American Historical Association meeting in 1982, voiced his disapproval. He saw

Finally, perhaps the most serious problem arising from this kind of 'neo-Scholastic' trend is that if we do not deal with the big questions arising from our work, others will. My favorite quotation along this line comes from the historian of Islamic science, A. I. Sabra.

> As for the argument that "we do not yet know enough to ask the big questions," my answer is this: it is only by attempting to formulate appropriate questions that can be fruitfully examined in light of what we now know that we make it possible for others to come up with deeper and more probing questions in the future. We do not know much (that is for certain), but the day when we know "enough" will never come. On the other hand, by altogether abandoning all programs of full-fledged historical research, we only tempt others to fill the vacuum with easy and useless essentialist generalizations.[10]

I hope that the essays in this book have indeed addressed "appropriate questions that can be fruitfully examined in light of what we now know," and that my discussions of the questions will "make it possible for others to come up with deeper and more probing questions in the future."

---

the current practice of choosing Ph.D. dissertation subjects, and carrying out dissertation research, as one major cause. He said that it was as though Ph.D. candidates – and their supervisors – were engaged in a competition to select the least interesting topic to work on. He even went as far as to propose abolishing the requirement of writing dissertations for Ph.D. degrees. I am not going to advocate such an extreme position, about which I do not think Lawrence Stone himself was truly serious anyhow. But I believe what he pointed out was a real danger. It was already serious thirty years ago, and has not lessened.

10 A. I. Sabra, "Situating Arabic Science: Locality versus Essence," *Isis* 87 (1996): 664.

PART 1
Science and Confucian Scholars

CHAPTER 1

Zhu Xi on Nature and Science*

1 Introduction: Zhu Xi's Knowledge about the Natural World

If we look in the works of neo-Confucian scholar Zhu Xi 朱熹 (1130-1200) for something that would appear as 'science' or 'scientific' knowledge to us, we can find the following three kinds. 1) First, the most widespread were the isolated bits of natural knowledge in his numerous writings and conversations. 2) Then we must look at what we might call 'natural philosophy': his discussions on *li* 理, *qi* 氣, yin-yang 陰陽, the five phases (*wuxing* 五行), etc. 3) Finally, and most conspicuously, he had numerous discussions of various branches of specialized knowledge, such as astronomy (*tianwen lifa* 天文曆法), harmonics (*lü* 律), music (*yue* 樂), geography (*dili* 地理), 'images and numbers' (*xiangshu* 象數), and so on. As a sum, these various kinds of Zhu Xi's natural knowledge showed an impressive breadth, and sometimes even depth. Yet they did not form a single coherent whole for Zhu Xi. Indeed, there was no general notion, comparable to the modern idea of 'science,' that covered all the above. Instead, each of them was a part of a larger genre, none of which was confined to the natural realm.[1]

Thus, the first of the above three categories, that is, his numerous items of natural knowledge, which came up in his various writings and records of conversations, focused mainly on moral and social issues and by no means centered on the natural world. Frequently, such knowledge of the natural world

* Originally published as "Chu Hsi on Nature and Science," *Essays on Zhuzi's Scholarship*, vol.1 (Taipei: Center for Chinese Studies, 2002): 131-160.

1 Much of the content of this paper is based upon my findings discussed in more detail in my book, *The Natural Philosophy of Chu Hsi (1130-1200)* (Philadelphia: American Philosophical Society, 2000). The two basic sources for Zhu Xi's sayings and writings are *Zhuzi yulei* 朱子語類 (compiled in 1270) and *Zhuwengong wenji* 朱文公文集 (compiled in 1534). The following abbreviations are used for references to these sources: WJ60.17a refers to *Huian xiansheng Zhuwengong wenji* 晦庵先生朱文公文集, *Sibu beiyao* 四部備要 (hereafter abridged as SBBY) edition, the 60th *juan* 卷, p. 17a; YL49.2a3 refers to *Zhuzi yulei* (1270 edition, reprinted in 1473), the 49th *juan*, the 3rd passage beginning on p. 2a; YL15.9b0 refers to the 15th *juan*, the passage that begins in the previous page (p. 9a) and is continued on p. 9b. The following approximate formula can be used to convert the page in the 1473 edition (p) into the page in the modern punctuated edition (Zhonghua shuju 中華書局, Beijing, 1986) (q): $q = (q^0 - 1) + 1.2p$, where "q^0" is the first page of the relevant *juan* in the Zhonghua shuju edition.

 | DOI 10.1163/9789004265318-003

entered the discussion when he was commenting upon certain natural phenomena referred to in the classics, which were supposedly written by – or at least contained the intentions of – the sages (*shengren* 聖人). In such cases, Zhu Xi discussed natural objects and phenomena in some detail, and his interest in them appeared quite strong. Nevertheless, his actual concern was with elucidating the sages' intentions. Similarly, Zhu Xi's discussion of *qi*, yin-yang, etc. never did form a separate genre that could be called 'natural philosophy'; it was simply part of philosophy in general. The last aspect of Zhu Xi's natural knowledge, on the other hand, can be seen as part of a broader genre of specialized knowledge that also included law, criminal justice, taxation, military affairs, official duties, etc. Of course, each of the above branches of astronomy, harmonics, geography, etc. was itself a separate, independent area, with its own specialist tradition. But when these areas of specialized knowledge were mentioned together, there was never any clear distinction between what belonged to the natural realm and what did not.

This shows that for Zhu Xi 'science' was not a separate, independent category. The above categories of separate bits of knowledge were part of the greater philosophical system constructed by Zhu Xi, referred to as 'neo-Confucian synthesis,' which included *everything*. One has the feeling that scientific knowledge became part of his system as the system came to embrace everything. Thus, for Zhu Xi, there were no clearly delineated terms which would stand for 'science' and 'nature,' and which were distinguishable from what is 'non-science' and 'non-nature.' Furthermore, 'science' and 'nature,' for him, did not neatly correspond to each other. 'Science,' even if we could find something in him that we could call by that name, was not a field devoted exclusively to the study of the natural world. And the natural world was not something to be studied only 'scientifically.'

Bearing all of this in mind, in this paper I look at the above-mentioned various aspects of Zhu Xi's knowledge about the natural world, and discuss some of their key features. I begin, in Section 2, by outlining Zhu Xi's notion of the natural world and his characteristic attitude toward natural phenomena, viewing the objects and phenomena of the natural world as 'natural' – as obvious and matter-of-fact. Section 3 then shows how this attitude can be related to some key aspects of the basic ideas and assumptions of Zhu Xi's natural philosophy, and suggests how some general features of the conceptual schemes he used in discussing things and events of the natural world might have conditioned his perception and understanding of the natural world. Section 4 follows with the discussion of Zhu Xi's attitude to specialized scientific branches. Finally, Section 5 concludes with some speculations about the possible influence Zhu Xi may have had – in terms of his natural knowledge and attitude to nature and science – on the subsequent development of science in China.

2 Nature as 'Natural': The Natural World of Zhu Xi

For Zhu Xi, the world consisted of the three basic constituents: heaven and earth (*tiandi* 天地), the myriad things (*wanwu* 萬物), and man (*ren* 人). These three stood in various relations to one another.

To begin with, heaven (or heaven and earth) produces men and the myriad things. Thus produced by heaven and earth, men and things live – or exist – between heaven and earth. They receive the *qi* and the mind (*xin* 心) of heaven and earth as their *qi* and minds. Of all that is produced by heaven and earth, man is most numinous, for he is endowed with the *qi* that is most correct, clear, complete, and so on. Man, therefore, forms a triad with heaven and earth, and complements the activities of heaven and earth. Zhu Xi also spoke of many aspects in which men and things, especially men, are parallel to heaven and earth. Sometimes the parallelism changed into an identity, giving rise to the notion that man is one with the whole world of heaven and earth, with everything in it.[2]

The world made of these three constituents covered everything, not merely the things that can be characterized as 'physical' or 'material.' For Zhu Xi, there was no boundary separating what was 'natural' from what was not. Objects and phenomena involving life and mind, and even morality, were included in his world of heaven and earth, the myriad things, and man; there was no clear distinction between 'natural' and 'non-natural' realms in that world. This lack of distinction was evident also in such basic concepts as *qi*, yin-yang, and the five phases, which were endowed with characteristics of life, mind, and morality, as well as matter.

One consequence of this was that, for Zhu Xi, the 'natural' world existed in harmony with the 'non-natural' – human and social – world. There could be no possibility of tension between the morally neutral 'natural' world and the human world that is governed by morality. On the contrary, and in part owing to the above-mentioned idea of parallelism – and even identity – between man and heaven and earth, the 'natural' world, referred to as 'heaven and earth,' was frequently invested with moral qualities. This, then, is what lay at

2 Kim, *Natural Philosophy of Chu Hsi*, Chap. 7.

the basis of the idea of 'cosmic basis of morality,' the notion that a moral order underlies the natural world, providing a kind of basis for morality.[3]

In sum, Zhu Xi's 'natural' world was an integral part of the larger world that also contained human and social realms, with no boundary between them. Nothing was excluded from this world envisaged by Zhu Xi.

This natural world – what exists in it, what happens in it – was 'natural' for Zhu Xi. Most of the objects and phenomena in the natural world were, to him, obvious. He took them for granted, in a matter-of-fact manner, and did not feel any need to explain them. In fact, they were so 'natural' and obvious to Zhu Xi that he frequently alluded to common and familiar natural phenomena in the course of discussing moral and social problems, by adducing analogies between obvious natural phenomena and moral and social problems, which were considered more problematic. Only rarely did he mention such common natural phenomena for their own sake.

For example, Zhu Xi spoke of the fact that once a cart has started to move, no great exertion of force is needed to keep it moving; he argued that, in study also, a great exertion of effort is needed only at the beginning, after which it becomes easy.[4] Similarly, to explain that when impurities enter the mind it loses 'sincerity' (*cheng* 誠) and falls into self-deception, he used the analogy that when gold is mixed with a small amount of silver the whole bulk of gold loses its worth as gold.[5] On 'the method of reading books' (*dushu fa* 讀書法), especially on the need for 'careful reading,' two different analogies argued for the same point:

> Generally, in reading books, it is necessary to read carefully. When read carefully, the essences [of the books] will spontaneously become familiar; after the essences have become familiar, the *li* will spontaneously be seen. It is like eating a piece of fruit. When one first bites it, one does not know the taste; one simply eats it. One has to chew it into tiny and soft pieces, then the taste spontaneously comes out, and one begins to know whether this is sweet or bitter. ... [It is also like] a gardener watering gardens. One who is good at watering waters each tree, following its vegeta-

3 E.g., Thomas A. Metzger, *Escape from Predicament: Neo-Confucianism and China's Evolving Political Culture* (New York: Columbia University Press, 1977), Secs. 3i and 3p; Peter K. Bol, "Chu Hsi's Redefinition of Literati Learning," in *Neo-Confucian Education: The Formative Stage*, ed. William Theodore de Bary and John W. Chaffee (Berkeley: University of California Press, 1989), 151-185.

4 E.g., YL31.8a0,1, 8b0,1,2, 9a1, 9b2, 78.35b1.

5 E.g., YL16.19b0, 59.36a2.

> bles and fruits. After a while the watering becomes sufficient, then water and soil are mixed and the things [i.e. vegetables, fruits and trees] get nourishment and grow naturally. One who is not good at watering administers it hurriedly...[6]

It is not impossible to read into these examples something about Zhu Xi's views on natural phenomena – the tendency of moving objects, the properties of the mixture of metals, the perception of tastes, and the nourishment of plants. But his real concern lay elsewhere: he was arguing for the necessity of a big effort at the beginning stage of study, for the importance of sincerity and purity of the mind, and for the necessity of careful reading, by drawing analogies to the natural phenomena he was mentioning. In none of the examples were the natural phenomena themselves what Zhu Xi was really interested in.

Many concrete natural phenomena and objects arose in this context in Zhu Xi's discussions. Thus, his frequent references to effects of medicines and to clarity of water were usually intended for discussions, by way of analogy, on human nature (*xing* 性) and the state of mind (*xin* 心). He resorted to the fixed sequence of the seasons to make the point that the constant human virtues cannot be changed. Even the retrograde motion of planets came up in a similar context, i.e. to illustrate that man's mind, normally compassionate, can sometimes become cruel.[7] He went as far as inventing natural phenomena to support his points concerning analogous human problems. For example, he said that a tree will die if it stops growing for a single day, in order to back up his point that one should keep studying every day.[8]

The most famous example of Zhu Xi's discussion of natural phenomena came up in a similar context. L. Carrington Goodrich in 1942 drew attention to Zhu Xi's following comment on the shells found on mountains, and considered this as proof that Zhu Xi understood the nature of fossils.[9]

---

6 YL10.6a1. See also YL80.19b1.

7 YL57.12a1.

8 E.g., YL72.24a2,3. For more on this and other examples, see Kim, *Natural Philosophy of Chu Hsi*, Chap. 10. John E. Murdoch has mentioned instances in medieval Europe in which certain theological points appear to have been used as excuses to discuss such natural phenomena as light, motion and celestial spheres, cases which appear to be exactly the opposite to what Zhu Xi was doing: Murdoch, "From Social into Intellectual Factors: An Aspect of the Unitary Character of Late Medieval Learning," in *The Cultural Context of Medieval Learning*, ed. John E. Murdoch and Edith D. Sylla (Boston: Riedel, 1975), 278-279.

9 L. Carrington Goodrich, "Early Mentions of Fossil Fishes," *Isis* 34 (1942): 25.

> I have seen on high mountains shells of conches and oysters, some of them in the rocks. These rocks were soils in earlier days, and the conches and oysters are the things [living] inside water. [What happened is that] what was low has changed and become high, and what was soft has changed and become hard...(常見高山有螺蚌殼, 或生石中. 此石卽舊日之土, 螺蚌卽水中之物. 下者却變而爲高, 柔者變而爲剛...)[10]

This passage, and similar passages by Shen Gua 沈括 (1031-1095),[11] led to some very enthusiastic evaluations, or over-evaluations, of Zhu Xi's – and Shen Gua's – understanding of the nature of fossils, which was achieved, as it was often noted, centuries before the West. The explanation was that in the absence of the obstacles like the belief in the creation of the world only several thousand years earlier, which precluded the correct understanding of the nature of fossils in the West, these Song 宋 Chinese thinkers could see the implications of such common observations, and recognized "that some of the life-like forms discovered in the rocks were in fact remains of ancient animals."[12]

It should be noted, however, that the assertion in the above quotation was made while Zhu Xi was discussing and illustrating the cosmogony of the *Huainanzi* 淮南子 and the theory of yin-yang cyclical repetition, which together formed the basis of his own cosmology.[13] He mentions the transformation of seawater into mountains as a detail of the *Huainanzi* cosmogony, according to which the primordial *qi* in the beginning of the world rotated, and the sediments of the *qi* were precipitated at the center to form the earth. But as the last sentence in the quotation indicates, Zhu Xi was also using the change to illustrate the cyclical alternation of yin and yang characteristics: what is low (the bottom of the sea) becoming high (mountain); what is soft (mud) becoming hard (stone). Thus clearly, what he was discussing was not a paleontologi-

---

10 YL94.3.

11 Shen Gua, *Mengxi bitan* 夢溪筆談 (reprinted in 1975, Beijing: Wenwu chubanshe 文物出版社). *juan* 卷 21, nos. 373, 374.

12 Joseph Needham, *Science and Civilisation in China* vol. 3 (Cambridge: Cambridge University Press, 1954-), 611. For more on this kind of explanation, see Yung Sik Kim, "'Fossils', 'Organic World-view,' 'The Earth's Movements,' etc.: Problems of Judging East Asian Scientific Achievements from Western Perspectives," in *Historical Perspectives on East Asian Science, Technology and Medicine* , ed. A. K. C. Chan, G. K. Clancey, and H Loy (Singapore: World Scientific Publishers, 2002), 14-16.

13 *Huainanzi* (SBBY edition.), 3.1a. For more on this, see Kim, *Natural Philosophy of Chu Hsi*, Sec. 9.1.

cal theory or the nature of fossils, but simply an illustration of the yin-yang scheme and traditional cosmogony.[14]

3 Basic Ideas and Conceptual Schemes

Several key aspects of Zhu Xi's basic ideas and assumptions made him view natural phenomena as so obviously, matter-of-factly, 'natural.'

For instance, qualities and activities of *qi* were considered innate by Zhu Xi, and thus once certain phenomena had been attributed to certain qualities and activities of *qi*, they were deemed sufficiently accounted for, without any need to look for external causes or hidden mechanisms.[15] Zhu Xi's account of the formation of the earth, for example, will illustrate the point.

> In the beginning of heaven and earth there was only the *qi* of yin and yang. This *qi* moved and turned around continuously. When the turning became very rapid, a large quantity of the sediments of *qi* was compressed. And as there was no outlet [for the sediments], these consolidated to form the earth in the center.[16]

He referred to the rapid rotation of *qi* as responsible for the formation of the earth, but he never paid attention to the cause of such rotation. It was almost as if rotation were the natural activity for *qi*.

Zhu Xi's concept of *li* also had an effect on his view of natural phenomena as 'natural' and obvious. The *li* of an object or phenomenon, to him, was merely something because of which the object exists or the phenomenon takes place

14 It might be added in passing that a similar idea is found in the pre-Socratic philosopher Xenophanes (6th century BC), who, according to Hippolytus (3rd century AD), "thinks that a mixture of the earth with the sea is going on, and that in time the earth is dissolved by the moist. He says that he has demonstrations of the following kind: shells are found inland, and in the mountains, and in the quarries in Syracuse he says that an impression of a fish and of seaweed has been found, while an impression of a bay-leaf was found in Paros in the depth of the rock, and in Malta flat shapes of all marine objects. These, he says, were produced when everything was long ago covered with mud, and the impression was dried in the mud. All mankind is destroyed whenever the earth is carried down into the sea and becomes mud; then there is another beginning of coming-to-be, and this foundation happens for all the worlds. [See G. S. Kirk and J. E. Raven, eds., *The Presocratic Philosophers* (Cambridge University Press, 1957), 177.]"

15 For more detailed discussion, see Kim, *Natural Philosophy of Chu Hsi*, Chap. 3.

16 YL1.4b3.

as it actually does: when and only when there is *li* for it, does it exist or take place. Thus, *li* was not conceptually simpler or more fundamental than the object or phenomenon itself. *Li* referred to a given object or phenomenon as a whole in its totality; it was not something that could be used in the explanation or analysis of the object or phenomenon in simpler terms. When *li* was mentioned, it was merely invoked to assure the existence or occurrence of the object or phenomenon. Nor was the content of *li* analyzed; it is to be grasped as a whole. Thus, when Zhu Xi noted regularities in the natural world, he was concerned only with their existence, but not with concrete details, of those regularities which he sometimes referred to as *li*.[17]

The dichotomy of what is 'above physical form' (*xing er shang* 形而上) versus what is 'below physical form' (*xing er xia* 形而下) also facilitated ready acceptance of natural phenomena. Abstract and sublime concepts without manifest 'physical forms' (*xing* 形) – the Way (*dao* 道), *li*, mind and human nature, for example – belong to the former, while concrete things with tangible physical forms are examples of the latter. Naturally, what is without physical form was difficult to understand and was thought to be important and worthy of further consideration, whereas what has physical form and is visible was easy to understand, and was considered obvious and even trivial. Since most common natural phenomena are accompanied by tangible qualities and physical effects and are 'below physical form,' they were thought to be self-evident and were simply accepted in the way they were perceived; no further investigation was attempted beyond the surface of the phenomenal realities of empirical data.[18]

The perennial Confucian emphasis on the reality of the external world also seems to have reinforced the readiness to accept commonly observed natural phenomena. Confucians considered their acceptance of the reality of the world to be what distinguished them from Daoists and Buddhists.[19] They were

17 Kim, *Natural Philosophy of Chu Hsi*, Chap. 2.

18 Typical was Zhu Xi's comment on the famous dialogue between Cheng Yi 程頤 (1033-1107) and Shao Yong 邵雍 (1011-77) on the question of where thunder comes from: after Shao Yong posed the question, Cheng Yi answered, "Thunder comes from where it comes from," to which Zhu Xi commented, "Why must one know where it comes from?" (YL100.11a0) The dialogue is recorded in *Henan Chengshi yishu* 河南程氏遺書 [Surviving Works of the Chengs of Henan] *juan* 卷 21a: *Er Cheng ji* 二程集 (modern punctuated ed., Beijing: Zhonghua shuju 中華書局, 1981), 270.

19 Zhu Xi said, for example: "The difference between Confucians' and Buddhists' sayings about [human] nature is simply that Buddhists speak of the void [whereas] Confucians speak of reality, and that Buddhists speak of nothingness [whereas] Confucians speak of existence. (YL126.7b2)"

not actively engaged in discussions of the concepts like 'void' (*kong* 空, *xu* 虛) and 'nothingness' (*wu* 無), which were too easily associated with Daoists and Buddhists, who encouraged people to concentrate on introspection without paying attention to the actual world. This made Confucians like Zhu Xi simply accept natural phenomena rather than engage themselves in abstract, theoretical discussions about them.[20]

The conceptual schemes – yin-yang, the five phases, etc. – that Zhu Xi used in discussing things and events of the natural world conditioned his perception and understanding of them in other ways as well. For example, because of the lack of distinction between 'natural' and 'non-natural' realms, the concepts of *qi*, yin-yang, the five phases, and so on, were endowed with characteristics of life, mind, and morality as well as matter, and covered everything in the world: they were not confined to what we would call 'natural' world.[21]

Also, as most of these basic concepts form sets of categories, they are associated with various sets of characteristics. Different characteristics associated with a given category are connected to one another, thus giving rise to a network of mutual associations. This kind of association was a key mode of 'explanation' in Zhu Xi's discussion of natural phenomena, and indeed, such categorical and associative character was a universal feature of the traditional Chinese discourse about the natural world, which many commentators have noted and referred to as 'correlative thinking' and 'the system of correspondence.'[22]

We can see many examples showing this feature of Zhu Xi's thought. For one, the four seasons and the four cosmic qualities – *yuan* 元, *heng* 亨, *li* 利, and *zhen* 貞 – formed a network of four-fold sets of mutually associated characteristics, through their respective five-phase associations with weather, compass directions, and the constant human virtues. They were further associated with four of the 'seven sentiments (*qiqing* 七情)' – love, joy, hatred and desire – and with various other four-fold sets of human characteristics. In his list of the four-fold sets, Zhu Xi could even include the four parts of a day – day, night, dusk and dawn – and the four legs of a fire stove.[23] Some sets of characteristics

20 Kim, *Natural Philosophy of Chu Hsi*, Sec. 13.5.

21 Ibid., Chaps. 3-5.

22 E.g., Needham, *Science and Civilisation*, vol. 2; Manfred Porkert, *The Theoretical Foundations of Chinese Medicine: Systems of Correspondence* (Cambridge, MA: MIT Press, 1974); John B. Henderson, *The Development and Decline of Chinese Cosmology* (New York: Columbia University Press, 1984); A. C. Graham, *Yin-Yang and the Nature of Correlative Thinking* (Singapore: The Institute of East Asian Philosophies, 1986).

23 For these and other sets of five-phase associations, see Kim, *Natural Philosophy of Chu Hsi*, Chap. 4. On the four cosmic qualities, see ibid., Sec. 5.4.

were associated directly with each other, without bringing in, as intermediaries, associations with the basic categories like yin-yang and the five phases. Thus, the five musical notes (*wusheng* 五聲) were directly associated with a group of five different things: the *gong* 宫 note with rulers, *shang* 商 with ministers, *jiao* 角 with people, *zhi* 徵 with events, and *wu* 羽 with things.[24] The twelve musical pitches (*shi er lü* 十二律) were associated with the twelve months.[25] Zhu Xi also mentioned associations of man's perceptual organs with visceral organs, the eyes with the liver and the ears with the kidneys, for example.[26]

These associations appear quite arbitrary to us today. At times Zhu Xi himself seems to have felt a need to justify some of those associations; but more often, he simply stated the associations without any explanation.[27] All he did for a certain set of terms was to associate them with a known set of categories as though such an association constituted a sufficient explanation. For example, he explained a cryptic passage of the *Xunzi* 荀子, "Clear brightness (*qingming* 清明) has inner shadow and turbid brightness (*zhuoming* 濁明) has outside shadow," merely by associating the former with metal and water and the latter with fire and the sun.[28] At times he merely mentioned a few distinguishing characteristics for certain sets of related terms without actually discussing them, not even associating them with known categories. His discussion of different kinds of changes is a good example, because all he did in numerous passages dealing with them was to distinguish the sudden change, '*bian*' 變 or '*shen*' 神, from the gradual one, '*hua*' 化.[29]

These categorical and associative basic concepts are also cyclical: they come in cycles that repeat fixed sequences. Not only the concepts like yin-yang and the five phases, but characteristics associated with them also reflect the cyclical character.[30] Various yin-yang characteristics follow each other continuously, forming cycles that have no beginning or end: movement and rest, contracting and expanding, vanishing and growing, going and coming, opening and closing, day and night, life and death, hot and cold, and so on. Many sets of five-phase characteristics – the four seasons, the four cosmic qualities, the life cycle of plants, for example – also repeat their fixed sequences end-

24 E.g., YL78.36b0, 92.10b0.

25 E.g., YL87.11b2.

26 E.g., YL53.10a1.

27 Kim, *Natural Philosophy of Chu Hsi*, Sec. 4.4.

28 YL1.8b4. The original passage is from the "Jiebi" 解蔽 chapter of the *Xunzi*.

29 Kim, *Natural Philosophy of Chu Hsi*, Sec. 8.3.

30 Ibid., Secs. 4.2, 4.2.

lessly. Zhu Xi even said that regret, good luck, parsimony, and bad luck, which are associated with the four seasons, "circulate like spring, summer, autumn and winter."[31] Thus, for him, such endless cyclical repetition was a universal feature of natural phenomena, which is not surprising because the cyclical nature of many natural phenomena – the movement of the luminaries in the sky, the change of the seasons, the tides, plant life cycles, and even the bending and stretching movements of the inchworm – must have been obvious even to a most casual observer. In fact, cyclical repetition became another key feature of traditional Chinese perceptions of natural phenomena.[32]

4 Zhu Xi's Attitude toward Specialized Subjects

Zhu Xi had a strong interest in various branches of specialized knowledge involving natural phenomena, such as calendrical astronomy, harmonics, geography, and medicine, and in some of them his understanding reached a fairly high level of sophistication. His interest in these technical subjects became particularly strong late in his life. Much of his discussion on calendrical astronomy and other scientific subjects took place in the late 1190's, when he and his school suffered vicious political persecution. In the last few years before his death, he spent a great deal of effort trying to master the content of the *Cantongqi* 參同契, a classic of inner alchemy.

Zhu Xi's attitude toward specialized knowledge was different for different branches. The subjects that Zhu Xi discussed most frequently were calendrical astronomy (*li* 曆), harmonics (*lü* 律 or *lülü* 律呂) and geography (*dili* 地理). These subjects, for him, were clearly part of the Confucian tradition. On the other hand, toward the practices connected with these subjects, such as astrology (*tianwen* 天文 or *zhanxing* 占星), music (*yue* 樂) and geomancy (*fengshui* 風水, literally, 'winds and waters'), his attitude varied. He frequently discussed music, but had little to say about astrology and geomancy. It must have been that he did not fully accept the latter activities, whereas music, a part of the Confucian rituals (*li* 禮), was important for him.

Another subject discussed frequently by Zhu Xi was the so-called 'images and numbers' (*xiangshu* 象數). Consisting mainly of numerological speculations involving simple numbers and the *Yijing* 易經 diagrams, it was based

31 YL74.9b0.

32 It is in reference to such cyclical repetition that Needham has characterized "Chinese physics" by the concept of "wave" as opposed to "particle": e.g., *Science and Civilisation*, vol. 4, part I, 3-14.

essentially on the *Yijing* texts and commentaries; Zhu Xi referred to it also as 'the study of the *yi*' (*yixue* 易學). He considered the subject to be fully worthy of the attention of Confucian scholars, though various post-Han influences, including much that could be called 'Daoist,' had infiltrated it. The study of 'images and numbers' was also applied to other practices, such as divination (*zhan* 占 or *bu* 卜) and alchemy (*dan* 丹 or *liandan* 煉丹). Zhu Xi did not refrain from discussing either of these activities; indeed, he wrote quite extensively about various aspects of both practices. Then, there were various other techniques linked with these practices, whose experts were called 'the masters of the Way' (*daoshi* 道士). Zhu Xi had less to say, however, and perhaps had rather low opinions, about these subjects, except for the technique of 'nourishing life' (*yangsheng* 養生), the so-called 'inner alchemy' (*neidan* 內丹), in which he had a considerable interest.

Medicine (*yi* 醫) did not command Zhu Xi's interest quite as much as the above four subjects, that is, calendrical astronomy, harmonics, geography and 'images and numbers.' In his comments on Zixia's 子夏 saying in the *Analects* 論語 – "Even 'the small ways' (*xiaodao* 小道) must have something worth 'looking at' (*guan* 觀). But if pursued too far, one may become bogged down. For this reason great men do not 'do' (*wei* 爲) them." – Zhu Xi mentioned medicine among his examples of "the small ways," which also included such fields as agriculture (*nong* 農), horticulture (*pu* 圃), divination and techniques (*gong* 工), but not the above four subjects.[33] And for him, medical specialists (*yijia* 醫家) belonged to the same category as Daoists, Buddhists, diviners, technicians and 'nourishing-life specialists' (*yangshengjia* 養生家). It is understandable, then, that while frequently mentioning various drugs and other remedies, Zhu Xi rarely touched upon the technical content of medical knowledge. Nor did he speak much of the related subject, 'materia medica' (*bencao* 本草); and although he did record descriptions of many plant and animal species, most of these were in his commentaries on the *Book of Poetry* (*Shijing* 詩經) and the *Songs of Chu* (*Chuci* 楚辭).

As for the remaining specialized branches of mathematics (*suan* 算, literally, 'computation'), agriculture and techniques, Zhu Xi had even less to say. But he could not ignore them altogether, for he must have faced problems involving knowledge of these subjects, agriculture in particular, in performing duties as a local official.

It has to be noted, however, that, along with these subjects and activities, Zhu Xi also spoke of, and wrote about, many other specialized topics such as protocols for rites and ceremonies, burial methods, military strategies and

33 E.g., YL49.2a3. The *Analects* passage comes from 19.4.

transportation vehicles, methods of calligraphy and painting, laws and criminal justice, land, tax and finance, civil service administration and other institutions. For him, these latter subjects, which take up large portions of his conversations and writings, were no different from the above-mentioned ones in that they were also specialized, practical branches of knowledge with their own experts. And then, there were the traditional 'six arts' (*liuyi* 六藝), the six basic skills which Zhu Xi considered essential for the education of children (*xiaoxue* 小學): rites, music, archery, charioteering, calligraphy, and computation.[34]

Zhu Xi's basic position on these specialized subjects in general was that they should not be ignored but should also be studied. He said, for example: "As for things like harmonics, calendars, criminal justice, laws, astronomy and geography, armies and official positions, all [these] must be understood."[35] This attitude was closely linked with his doctrine of *gewu* 格物, which, interpreted to mean 'investigating the *li* of things,' tended to emphasize studying all concrete events and things in all areas of human concern.[36] He said repeatedly that every thing or event in the world has its *li* and should be studied and understood.[37] He interpreted certain expressions from the classics, the *Analects* in particular, in this spirit of the *gewu* doctrine. For example, he used the emphasis in the *Analects* on 'broad study' (*boxue* 博學) in supporting his insistence upon studying and understanding everything.[38] Confucius's phrase, "study down below and attain up above" (*xiaxue shangda* 下學上達), was used also in the same vein by Zhu Xi, who emphasized that one should start with concrete things that are clearly manifest and easy to understand.[39]

There were other reasons that Zhu Xi was interested in the specialized branches of knowledge. For instance, many of them were associated with important philosophical terms and concepts. The importance of the concept of 'heaven' (*tian* 天), for example, made calendrical astronomy – the subject that deals with physical heaven – significant to him. Geography and geomancy, likewise, could be seen to be connected with the other half of the term 'heaven and earth.' The value of music as part of Confucian rituals also lent importance to the related subject of harmonics. Similarly, the importance of the *Yijing*, and

34 E.g., YL7.1a1, 1b4.

35 YL117.22b0.

36 On Zhu Xi's ideas of the *gewu* doctrine, see Kim, *Natural Philosophy of Chu Hsi*, Chap. 2.

37 E.g., YL15.4b2, 18.22b0, 34.33b0, 116.13b0, 117.12b0.

38 YL117.22b0. The expression '*boxue*' appears, e.g., in the *Analects*, 12.5, 19.6.

39 E.g., YL44.19b0-21b2; WC47.13a. See also YL117.21a3, where Zhu Xi criticized those who "study down below today and wish to attain up above tomorrow." Confucius's phrase appears in the *Analects*, 14.37.

the ideas and diagrams in this classic text, could be translated into the importance of the subject of 'images and numbers,' as well as divination and alchemy which applied them. Alchemy, especially in the form of 'inner alchemy,' could be correlated to the concept of 'the Way' (*dao* 道), because it was among the techniques practiced by those who seek the Way, the so-called 'masters of the Way'. Investigating these subjects would clearly help one understand the *li* of the ideas and concepts associated with them.

Moreover, these subjects were present to some degree in the texts widely studied by scholars, particularly the standard commentaries to the classics and the official dynastic histories. The latter almost always included treatises on astronomy, calendars, harmonics, geography, as well as on rites and music. Zhu Xi studied the relevant portions of these commentaries and treatises, and his understanding of them reached a considerable level, to the point where he could make his own judgment as to which of these texts were best suited to the study of a particular subject or issue.[40] There were also treatises on specialized subjects written by scholars themselves.

It should not be ignored that knowledge of some of these subjects was actually needed by Confucian scholars in order to perform their official duties. Admittedly, traditional Chinese civil service did include offices devoted to specialized branches and filled by specialists. But generalist officials like Zhu Xi could also face tasks involving specialized knowledge and, in any case, had to manage and supervise the specialist officials who were usually of lower official status. This need was at least in part the reason why Zhu Xi included the subjects in his proposal for examination, classifying them under the category of 'current tasks' (*shiwu* 時務).[41]

Yet, in spite of his emphasis on the need to study and understand specialized subjects, Zhu Xi did not hide his feeling that there were more important subjects, namely moral and philosophical matters. In fact, he said repeatedly that one must understand 'the basis' (*ben* 本), or what is 'great,' before moving on to 'small' matters.[42] In other words,

40 Indeed, this was what he had to do, for, having asserted the importance of the specialized subjects and the necessity to study them, he had to decide which texts were the best or correct ones to be studied, just as he did for the moral and social philosophies. For this aspect of Zhu Xi's "program of learning" in general, see, e.g., William Theodore de Bary, "Chu Hsi's Aims as an Educator," in *Neo-Confucian Education*, ed. de Bary and Chaffee (op. cit., note 2), 186-218; Daniel K. Gardner, *Learning to Be a Sage* (Berkeley: University of California Press, 1990), 35ff.

41 WC69.22a.

42 E.g., YL64.5b1, 84.3b0, 116.13b0.

> if one does not understand this basis first and merely desires to take up [particular] events and to understand them, then even if one understands many curiosities [they will] only add to much confusion and disorder, and only to much overbearing and parsimony.[43]

It must have been for this reason that, despite the fact that Zhu Xi studied specialized subjects and attained various levels of knowledge in them, his grasp never equaled that of the specialists. He even admitted that it was not necessary to try to reach a complete understanding of all the details of these subjects.[44] Indeed, people with adequate knowledge of these subjects were rare. On Zhu Xi's proposal that topics like calendrical astronomy, geography, music and harmonics should be included as subjects of the civil examination, a disciple even feared that "there may be no [qualified] examining officer after all."[45]

It was natural, then, that Zhu Xi did not hold a very high opinion of specialists. For him, they were merely technical experts in specialized areas which he did not master himself. Sometimes he even appeared confident that he could have mastered them if only he had tried. Such low opinions were reflected in his tendency to criticize his contemporary specialists whose understanding did not reach the high standard that he believed to have existed in the golden ages of ancient sages.

Finally, Zhu Xi's attitude towards specialists was not the same for all areas. For subjects like calendrical astronomy and harmonics, which he valued more and thus studied more in depth and was more well versed in, he did not quite accept the expertise of specialists. On the other hand, for other subjects that he did not know as much about, he tended to accept the expertise of the specialists; then again, he did not really value what they did. In fact, at times he did not hide his feelings of disdain and distrust toward specialists of certain techniques, those called 'the masters of the Way,' 'the recipe masters' (*fangshi* 方士) and 'the yin-yang specialists' (*yinyangjia* 陰陽家).

---

43 YL84.5a0. See also YL57.7a1, 84.3b0.

44 For example, to his remark, quoted earlier, on the need to understand specialized subjects, he added: "Although one may not be able to see through their essences and subtleties, one should nevertheless know the general outlines ... " YL117.22b0.

45 YL109.8b0. Zhu Xi's response to this was that one "should first order the examining officers to study them."

5 Concluding Speculations: Zhu Xi and the Scientific Development in China

As Zhu Xi had such an overwhelming influence on Chinese thought in the subsequent period, it was inevitable that there have been many speculations about his influence on the scientific development in China. I shall conclude this paper with my thoughts about such speculations.

Some scholars have found positive aspects in Zhu Xi's discussion of, and attitude to, natural phenomena and scientific subjects. For example, some have identified a scientific 'method' and 'spirit' in Zhu Xi, especially in his *gewu* endeavor. Hu Shih, for example, believed that this constituted a 'scientific spirit.' He saw in Zhu Xi's *gewu* doctrine "a set of principles on the spirit, the method, and the procedure of investigation and research," and went as far as to characterize it as "the method of hypothesis and verification by evidence." Zhu Xi's problem, according to Hu Shih, was that this scientific spirit was directed exclusively to textual studies, and not to nature.[46]

It is doubtful as to whether we can go this far in discerning a scientific spirit in Zhu Xi's *gewu* doctrine. Nor is it fair to fault Zhu Xi for not doing what the European "new scientists" did in the Scientific Revolution, that is, employing their 'new' method in the study of nature. But apart from these problems, there is a sense in which Hu Shih's basic position itself is mistaken. For, in spite of the emphasis on investigating many concrete things, the *gewu* endeavor did not primarily involve intellectual procedures. In fact, man's understanding of the *li* of things, achieved as the result of *gewu*, was considered a kind of 'resonance' between the mind's *li* and the things' *li*. This was so because the *li* of heaven (*tianli* 天理) resides both in man's mind (as the mind's *li*) and in things and events (as their *li*). Thus, when a man has reached the *li* of a thing or an event, it was described usually as 'seeing' the *li* rather than as 'knowing' it. In other words, what he has gained is not so much knowledge of the *li* as an insight into it.[47] Also, for such resonance to take place, the mind needs to be in a certain state – 'empty' (*xu* 虛), 'bright' (*ming* 明), and 'tranquil' (*jing* 靜). These are the original states of man's mind, manifesting the *li* of heaven to its fullest, unhindered by human desires (*renyu* 人慾). In such a state, the mind spontaneously

46 Hu Shih, "The Scientific Spirit and Method in Chinese Philosophy," in *The Chinese Mind*, ed. Charles A. Moore (Honolulu: The University of Hawaii Press, 1967), 104-131, originally presented at the Third East-West Philosophers' Conference, held in Honolulu in 1959. See especially pages 116-118.

47 Gardner, *Learning to Be a Sage*, 46-53.

sees the *li* in things and events, which are nothing but manifestations of the *li* of heaven contained in the mind itself.

Moreover, gaining insight into the many *li* of individual things and events was not the real aim of the *gewu* endeavor, the ultimate purpose of which was to reach the *li* of heaven via the many individual *li*. The key step in the *gewu*, then, lay in moving from those individual *li* to reach the one *li* of heaven. Yet the connection between the many individual *li* and the one *li* of heaven was not easily traceable. It was never clear exactly how the grasp of many individual *li* can lead to the apprehension of the one *li* of heaven.[48]

All seemed to agree, however, that the step must involve something more than a purely intellectual process. In Zhu Xi's words, one needs 'laborious efforts' (*gongfu* 工夫) and 'nourishing' (*yang* 養) in addition to 'knowing' and 'understanding.' The mental state described by the term 'reverence' (*jing* 敬) was important, for when a man is reverent, his mind is 'bright,' 'transparent,' and 'alive,' all *li* are in his mind and the *li* of heaven becomes 'brilliant.' Thus, moral and intellectual endeavors of a scholar converged in his search for *li* – the many individual *li* and the one *li* of heaven – through *gewu*. And in this convergence, the moral side was clearly the more important. Of course, the intellectual aspect could not be ignored altogether, but, on the whole, the intellectual elements of the *gewu* endeavor were fused into its ultimately moral aims. It was to uphold morality and to avoid errors that one investigates things.[49]

---

48 Kim, *Natural Philosophy of Chu Hsi*, Chap. 2.

49 Moreover, Zhu Xi did not show much interest in methodological problems of knowledge. For example, he did not engage himself in discussions of how one can obtain knowledge or how one can be sure that what one knows is sound, the key questions in Western epistemology. When he did discuss 'method,' it was almost exclusively with the method of study, the so-called "method of reading books" (*dushufa* 讀書法) in particular. And there the target was the *li* contained in books. The problem for him was how to get at the *li* from the information recorded in books. How to know, or understand, what is recorded in books was not much of a problem for him. His was a 'common sense' epistemology, and at its basis lay the belief that man's mind in its original state can spontaneously grasp the ways and *li* of things and events. Nor was the method of discourse, i.e. the method of presenting what one knows, of much concern for Zhu Xi. He showed little concern with rigorous logical methods, for example. It was simply to be taken for granted that when one sees something or reads something in a book, one knows it, and that when one knows something, one can talk or write about it. What mattered to him was not how to obtain knowledge, or how to present it to others, but rather how to set one's mind upon the knowledge, or how to act upon it. This was indeed the direction that the *gewu* doctrine took, as we have seen. See Kim, *Natural Philosophy of Chu Hsi*, Sec.14.4.

While Hu Shih focused upon Zhu Xi's *gewu* doctrine and saw scientific method and spirit in it, Joseph Needham looked at the general character of Zhu Xi's worldview and called it 'organic.' This characterization is not unreasonable given that Needham used the word 'organic' in the sense opposite to 'mechanical,' and Zhu Xi's worldview is clearly more 'organic' than 'mechanical.' And, as Needham further suggested, Zhu Xi's ideas might indeed have inspired Leibniz's philosophy of monad, and then the organic philosophy of Whitehead, leading Needham to draw similarities between Zhu Xi's ideas and modern science.[50] But what we have here is nothing more than a possible, though interesting, case of intercultural influence and congruence. It would be wrong to proceed from this to conclude either 1) that Zhu Xi's worldview somehow had the character of modern science which has overcome, or shed itself of, the predominantly mechanical character of Newtonian classical science, or 2) that it is a desirable worldview for people to have in the modern world plagued by many problems created by 'mechanical' science.

To be sure, it is true that modern science has a relatively more 'organic' character compared to classical science. Yet this 'organic' character of modern science is on a completely different level from that of Zhu Xi's worldview. Although modern science does not depict the natural world in such a completely mechanical way as classical science did, it has not forsaken its mechanical character to such an extent that it has become similar to Zhu Xi's worldview. Modern science remains essentially mechanical, though less so than classical science.

To characterize Zhu Xi's worldview as 'organic' is also misleading if it is interpreted as a systematic worldview in which everything finds its proper place in a harmonized whole. For, in fact, Zhu Xi's worldview was rather particularistic. He basically treated most natural phenomena, and problems involving them, as particulars; he did not generalize from them, as he took each phenomenon or problem as it came along, without much concern to correlate it with his views about the rest of the natural world.[51] In this respect, the medieval European worldview was far more organic than Zhu Xi's. This can be seen clearly if we look at the way Zhu Xi dealt with issues and phenomena related to the idea of 'weight' and compare it with how medieval European scholastics dealt with them. Zhu Xi noted that water tends to move downward, which is similar to the views of the medieval scholastics in that whether an object falls or rises is determined by what material substance it is made of. However, Zhu Xi's was not a general idea applicable to all heavy objects. To flow

50 Needham, *Science and Civilisation*, vol. 2, 505.

51 Kim, *Natural Philosophy of Chu Hsi*, Sec. 14.3.

downward, for him, was an innate tendency of a particular substance – water – and not of all heavy objects. It is even unclear whether he realized that the downward flow of water is a particular case of the fall of heavy objects. Furthermore, for the medieval scholastics, the fall of heavy objects and the rise of light objects were tied with their conceptions of the structure of the world: heavy things fall down as they tend to their natural place, the center of the earth, which, for them, was the 'center of everything.' Thus, the issue of the weight of the earth, i.e. the reason why such a heavy thing would remain suspended in the middle of the sky without falling down, never arose for the scholastics. The concept of weight was built into their world-picture itself; the earth, made of the heaviest of the four elements, lies at its natural place, the center of the world, and thus does not move. But Zhu Xi, for whom the problem of weight was separate from that of the structure of the world, had to deal with the problem of the earth's weight. Whereas the scholastics' world-picture included the particular empirical fact of weight as an integral – and thus 'organic' – part of it, Zhu Xi's had to take account of the notion of weight as an additional and independent fact.[52]

Of course, there have also been negative judgments on Zhu Xi's influence. This is inevitable because Chinese science is frequently considered to have begun to stagnate just when the Zhu Xi orthodoxy became dominant.

One source of the problem could be the categorical and associative conceptual schemes of Zhu Xi's natural philosophy: yin-yang, the five phases, etc. Needham, for example, believed that the sixty-four hexagram system of the *Yijing* was a 'hindrance' to the development of scientific ideas, while the yin-yang and the five-phase schemes 'helped' it. Graham, on the other hand, thought that the five-phase associations, unlike the yin-yang characteristics, lacked objectivity and thus were not paid much attention, perhaps because they were felt to be less useful in practice.[53]

Yet Zhu Xi's knowledge of the natural world was not completely conditioned by these conceptual schemes. It was never Zhu Xi's intention to use these schemes for constructing a coherent system of natural knowledge covering all the particulars; they were not the sole means for explaining natural phenomena. Thus, the presence of the schemes should not lead one to ask, or to answer, such questions as to whether the yin-yang – or the five-phase, or the *Yijing* diagram – schemes 'helped' or 'hindered' the understanding of natural phe-

52 Ibid., Sec. 13.2. For a brief account of the medieval Western world-picture, see Edward Grant, *Physical Science in the Middle Ages* (New York: Wiley, 1971), Chaps. 4-5.

53 E.g., Needham, *Science and Civilisation*, vol. 2, 304; A. C. Graham, *Two Chinese Philosophers: Ch'eng Ming-tao and Ch'eng Yi-ch'üan* (London: Lund Humphries, 1958), 33.

nomena, for alternative explanations, where available, could always be adopted. Indeed, the presence of categorical schemes could not possibly hinder, inhibit, or stand in the way of, the adoption of other explanations.

Consider the example of the yin-yang scheme. Zhu Xi used it in explaining numerous natural phenomena. The idea of the yin-yang cyclical alternation dominated his discussions of such phenomena as lunar eclipses, seasonal characteristics and sea tides. But he was not forced to use the idea of yin-yang alternation to explain all such phenomena. When he could adduce a more concrete explanation, he could adopt it. His explanation of the phenomenon of the moon's phases is a good example: he chose to account for it in terms of the relative positions of the sun and the moon even though the yin-yang alternation could have easily explained it. Thus, what 'prevented' Zhu Xi from reaching the 'correct' explanation of phenomena like lunar eclipses and the tides must lie elsewhere – perhaps in the difficulties that the correct explanation posed, or in the lack of additional knowledge necessary for it.[54]

Similar things can be said about Zhu Xi's use of the five-phase scheme. Consider his discussion of the five regular and the five intermediary colors, for instance. Zhu Xi had a rather long discussion in which he spoke of the production of 'the five intermediary colors' (*wujianse* 五間色) by mixing various pairs of 'the five regular colors' (*wuzhengse* 五正色):

> Blue, red, yellow, white and black are the regular colors of the five [compass] directions. Green, vermilion-red (*hong* 紅), emerald-blue (*bi* 碧), violet (*zi* 紫) and horse-brown (*liu* 騮) are the intermediary colors of the five directions. Now, with blue of Wood [one can] conquer yellow of Earth. [Thus,] combining blue and yellow produces green and becomes the intermediary color of east. With white of Metal [one can] conquer blue of Wood; combining blue and white produces emerald blue and becomes the intermediary color of west. With red of Fire [one can] conquer white of Metal; combining red and white produces vermilion-red and becomes the intermediary color of south. With black of Water [one can] conquer red of Fire; combining red and black produces violet and becomes the intermediary color of north. With yellow of Earth [one can]

54 The 'correct' understanding of the tidal phenomena, for example, required knowledge of gravitation and of the earth's rotation, which Zhu Xi could never have imagined. In the absence of such knowledge, some medieval Europeans also resorted to the idea of 'microcosmic-macrocosmic respiration.'

> conquer black of Water; combining yellow and black produces horse-brown and becomes the intermediary color of the center.[55]

However, Zhu Xi was not interested in the actual processes by which the colors are mixed to produce new ones. What prompted him to discuss this was the matter of which were the proper colors for gentlemen to wear.[56] It is natural, then, that he should be interested in classifying the colors – according to whether they are 'regular' or 'intermediary,' and according to their five-phase associations. Thus, the five-phase scheme did not play the role of a coherent theoretical framework covering all phenomena involving color. It was merely that Zhu Xi on this particular occasion was concerned with colors for gentlemen's clothing, an issue for which certain five-phase associations were well suited, that he brought in the five-phase scheme. Moreover, the five-phase concepts and associations were ambiguous enough to accommodate any particular color phenomenon and any interpretation of it in the scheme.

On the whole, it is difficult, and perhaps impossible, to judge whether Zhu Xi's influence was positive or negative. One would be hard-pressed to come up with a single unequivocal judgment on all the various aspects in which Zhu Xi's knowledge, attitude, method, style, etc., influenced the way Chinese in the subsequent period dealt with objects and phenomena in the natural world.

For example, even the great range – breadth and inclusiveness – of his natural knowledge cannot be unambiguously considered to have been beneficial. In fact, paradoxically, the very breadth of Zhu Xi's intellectual scope may have been a factor in the subsequent narrowing of interest among scholars. His followers could have felt that everything they might wish to know about the natural world had already been included in the corpus of their master. It was almost as if, once Zhu Xi had produced a corpus of knowledge that covered everything, those who came after him did not feel the need to worry about problems other than their principal concern – morality and self-cultivation – everything else, including knowledge of the natural world, was already there in Zhu Xi after all.[57]

---

55 WC32.12a.

56 The idea that some of the intermediary colors could be produced in the manner described by Zhu Xi – violet from red and black, horse-brown from yellow and black, for example – also bespeaks his lack of interest in the actual process of mixing colors.

57 Yet, in view of what we have seen in previous sections, what happened turns out to be not so 'paradoxical' after all. We have noted, for example, that most natural phenomena were obvious and unproblematic for Zhu Xi, and were simply accepted by him. We have also noted some characteristics of his basic ideas and assumptions that could have facilitated this attitude. It should come as no surprise then that Zhu Xi's followers went on to ignore

Finally, there is a more fundamental question to ask, perhaps the ultimate question. To judge the role of Zhu Xi in the development of science in China is essentially to ask what would have happened to science in China if Zhu Xi had not existed. But would it make any sense at all to speak of China without Zhu Xi? Could 'China without Zhu Xi' still be considered China?

---

natural phenomena; they were simply following in the steps of their master, who was not very concerned with gaining detailed knowledge about objects and phenomena of the natural world that he took for granted.

CHAPTER 2

'Analogical Extension' (*leitui*) in Zhu Xi's Methodology of 'Investigation of Things' (*gewu*) and 'Extension of Knowledge' (*zhizhi*)*

The 'investigation of things' (*gewu* 格物) and 'extension of knowledge' (*zhizhi* 致知) were the key elements of neo-Confucian learning and self-cultivation. Their purpose was to reach the *li* 理 – first, the many *li* of individual things and events, and then ultimately, through them, the one universal *li*. There were extensive debates among neo-Confucian thinkers about the detailed methods involved in various stages of these endeavors. One method in particular was deemed important, namely 'analogical extension' (*leitui* 類推), so called because it was the method of 'extending' (*tui* 推) the knowledge of a thing or an event to that of other things or events belonging to the same 'kind' (*lei* 類).[1]

---

* Originally published as "'Analogical Extension' (*leitui*) in Zhu Xi's Methodology of 'Investigation of Things' (*gewu*) and 'Extension of Knowledge' (*zhizhi*)," *Journal of Song-Yuan Studies* 34 (2004): 41-57.

1 While Morohashi Tetsuji 諸橋轍次, in *Dai Kanwa jiten* 大漢和辭典 (Tokyo, 1957-1960), 12:43636:43 (page 299), identifies passages from *Huainantzi* 淮南子 and *Hanshu* 漢書 as instances of the earliest use of the term *leitui*, the idea constituting this method had been available before Han times: See, e.g., the passages from *Mencius* and *Mozi* 墨子 discussed in D. C. Lau, "On Mencius' Use of the Method of Analogy in Argument," in his *Mencius* (Baltimore, 1970), 235-63 [originally published in *Asia Major*, new series, 10 (1963)].

Previous studies on Zhu Xi's methodology have focused heavily on the methods of self-cultivation and of reading books, and have usually neglected the *leitui* method. For some works containing more than passing remarks about it, see, e.g., Zhang Liwen 張立文, *Zhu Xi sixiang yanjiu* 朱熹思想研究 (Beijing: Zhongguo shehui kexue chubanshe 中國社會科學出版社, 1981), 400-402; Chen Lai 陳來, *Zhu Xi zhexue yanjiu* 朱熹哲學研究, 2nd ed. (Shanghai: Huadong shifan daxue chubanshe 華東師範大學出版社, 1987), 309-314. On Zhu Xi's theory of *gewu*, see, e.g., D. C. Lau, "A Note on *Ke Wu*," *Bulletin of the School of Oriental and African Studies* 30 (1967): 353-57; Qian Mu 錢穆, *Zhuzi xin xuean* 朱子新學案 5 vols. (Taipei: Sanmin shuju 臺北三民書局, 1971), vol. 2, 504-50; Hoyt C. Tillman, "The Idea and the Reality of the 'Thing' during the Sung," *Bulletin of Sung & Yuan Studies*, no. 14 (1978): 68-82. On Zhu Xi's idea about "the method of reading books," see, e.g., Daniel K. Gardner, *Chu Hsi: Learning to Be a Sage* (Berkeley: University of California Press, 1990), 42-54.

My translation of the character *tui* 推 as 'to extend' poses a problem because I stick to the standard translation of the term *zhizhi* 致知 as 'extension of knowledge.' To minimize confu-

 | DOI 10.1163/9789004265318-004

In this paper, I shall take up the example of Zhu Xi 朱熹 (1130-1200), the author of the 'neo-Confucian synthesis' that dominated Chinese learning for centuries, and examine what he thought about this method of 'analogical extension.' And while doing so, I shall try to put Zhu Xi's ideas about the method in the context of his theories of 'investigation of things' and 'extension of knowledge.' In particular, I shall show that the method of 'analogical extension,' by making it possible to extend knowledge of one thing to that of other things, played a key role in the Zhu Xi's efforts of 'investigation of things' and 'extension of knowledge,' the ultimate purpose of which was to achieve a 'thorough penetration' (*guantong* 貫通). I shall begin, in the first section, by recapping some relevant aspects of Zhu Xi's thought about *li* and 'investigation of things.'

1 *Gewu*: *Li* and the Mind

Li, the target of Zhu Xi's endeavor of 'investigation of things,' had a variety of characteristics with meanings of various shades and dimensions.[2] An aspect of the concept of *li* emphasized by Zhu Xi, however, was that there are many individual *li* for individual objects (a tree, a bird, a teapot) and phenomena (a flood, a child laughing, a rock falling) as well as the one *li* for the sum of these objects and phenomena: every object and phenomenon of the world has its own individual *li*, and all individual *li* are manifestations of the one *li*. He praised Cheng Yi's 程頤 (1033-1107) famous expression, "*Li* is one and its mani-

sion, in this essay I have indicated in parentheses where I have chosen the word 'to extend' as the translation of the character *zhi* 致, which has various other meanings such as 'to reach,' 'to achieve,' and 'to cause to come.'

The two basic sources for Zhu Xi's sayings and writings are *Zhuzi yulei* 朱子語類 (compiled in 1270) and *Zhuwengong wenji* 朱文公文集 (compiled in 1534). The following abbreviations are used for references to these sources: WJ60.17a refers to *Huian xiansheng Zhuwengong wenji* 晦庵先生朱文公文集, *Sibu beiyao* 四部備要 (hereafter abridged as SBBY) edition, the 60th *juan* 卷, page 17a; YL49.2a3 refers to *Zhuzi yulei* (1270 edition, reprinted in 1473), the 49th *juan*, the 3rd passage beginning on page 2a; YL15.9b0 refers to the 15th *juan*, the passage that begins on the previous page (p. 9a) and is continued on page 9b.

2 For discussions of various aspects of the concept of *li*, in Zhu Xi as well as other neo-Confucian thinkers, see, e.g., Allen Wittenborn, "*Li* Revisited and Other Explorations," *Bulletin of Sung-Yuan Studies*, no. 17 (1981): 32-48; Cheng Chung-ying, "Chu Hsi's Methodology and Theory of Understanding," in *Chu Hsi and Neo-Confucianism*, ed. Wing-tsit Chan (Honolulu: University of Hawaii Press, 1986), 169-196; Willard J. Peterson, "Another Look at *Li*," *Bulletin of Sung-Yuan Studies*, 18 (1986): 13-32.

festations are many" (*liyi fenshu* 理一分殊), and said: "Speaking of heaven and earth and the myriad things together, [they] simply are the one *li*; when it comes to [the *li*] in man, then again each man has [his own] *li*."[3] Zhu Xi himself expressed the same idea in various other ways: "All things have [their] *li*; as a whole [they are] simply the one *li*;" [4] "Heaven and earth, yin 陰 and yang 陽, life and death, day and night, and *gui* 鬼 and *shen* 神 – [all these] are simply the one *li*."[5]

The 'one *li*' is thus a kind of 'universal' *li* that encompasses all the individual *li* of things and events. Zhu Xi identified the one *li* with 'the heavenly *li*' (*tianli* 天理),[6] which is what original human nature (*xing* 性) manifests in the form of such ethical virtues as humaneness (*ren* 仁) and righteousness (*yi* 義). It is because the heavenly *li* is obstructed and blocked by human desires (*renyu* 人欲) that man loses those virtues and displays vices. This dichotomy of heavenly *li* versus human desires, which can be traced back to ancient times,[7] became the basis of Zhu Xi's moral philosophy, and thus, the state of mind free from human desires was, for him, the ultimate goal of man's self-cultivation.[8] When that goal is attained, man's mind manifests the heavenly *li* to its fullest.[9]

It is not only original human nature that manifests the heavenly *li*; every thing and event in the world has its *li*, which is but a manifestation of the one universal, heavenly *li*. Therefore, man can aim at the heavenly *li* through the method of *gewu*, because the term – usually translated as 'investigation of things' – meant, for Zhu Xi, reaching the *li* in 'things.'[10] Zhu Xi said, "*gewu* is

---

3 YL1.1b6. Cheng Yi's saying appears, e.g., in *Cuiyen* 粹言, *juan* 1 [*ErChengji* 二程集 (Beijing: Zhonghua shuju 中華書局 edition, 1981, hereafter abridged as ECJ), 1202-1203]; *Yichuan wenji* 伊川文集, *juan* 5 (ECJ, 609), where Zhang Zai's 張載 (1020-1077) discussion of the idea is praised.

4 YL94.8b0.

5 YL25.13b1.

6 YL40.8a1. See also YL41.7a0.

7 E.g., *Liji* 禮記: *Liji zhushu* 禮記註疏 (reprinted in 1977, Taipei: Xinwenfeng 新文豊), 37.10a-10b: "A man who is transformed into a thing is one who destroys the heavenly *li* and exploits human desires."

8 Zhu Xi's discussions of this dichotomy can be found, e.g., in YL13.2a6-3b2, 31.14b2, 40.7a0, 41.5b1, 7a0, 20b2, 68.10a2, 90.7a5-7b0; WJ39.20a, 73.42b.

9 Zhu Xi even said that "the entire body of the Sage [i.e. Confucius] is the heavenly *li*." YL58.1a1. See also YL44.18b1.

10 E.g., *Sishu jizhu* 四書集註 (SBBY ed.) (hereafter abridged as SSJZ), (*Daxue* 大學), 2a. For some brief discussions of what *gewu* meant in Song times, see, e.g., A.C. Graham, *Two Chinese Philosophers: Ch'eng Ming-tao and Ch'eng Yi-ch'üan* (London: Lund Humphries, 1958), 74-82; D. C. Lau, "A Note on *Ke Wu*"; Qian Mu, *Zhuzi xinxuean*, vol. 2, 504-550;

exhaustively investigating the ultimate *li* with respect to everything,"[11] or "One who 'investigates things' simply wants to exhaust the *li* of things."[12] 'Investigation of things,' then, was to get at the *li* of things (*wu* 物). And since every 'thing' has its own *li* which is a manifestation of the single universal *li*, everything is worth investigating.

The heavenly *li* resides in man's mind too, and constitutes the mind's *li*. In its original state, the mind, unhindered by human desires, perfectly manifests the heavenly *li*. And since the individual *li* of things and events are also manifestations of the heavenly *li*, Zhu Xi could go on to say that man's mind in its primordial state contains all 'the myriad *li*.'[13] He went on to equate the *li* in the mind and the *li* in things and events:

> *Li* in things and in my mind are originally one thing. The two do not have even a small gap [between them]. But it is necessary that I should respond to them [i.e. the *li* in things], and that is all. Things and the mind share this *li*.[14]

Thus, according to Zhu Xi, the mind in its original state – 'empty' (*xu* 虛), 'bright' (*ming* 明 or *guang* 光), 'tranquil' (*jing* 靜 or *ning* 寧), 'settled' (*ding* 定) or 'peaceful' (*ping* 平) – can see the *li* in objects and phenomena, which are nothing but manifestations of the heavenly *li* contained in the mind itself. He said: "If one has emptied the mind, one can see the Way and the *li* clearly;"[15] "When the mind is bright, it can spontaneously see that this event has this *li* and this thing has this other *li*;"[16] "When the mind is peaceful and the *qi* 氣 is settled, one can see the Way and the *li*."[17]

It may be noted, before concluding this section, that reaching the *li* of a thing or an event was described by Zhu Xi as 'seeing' (*jian* 見 or *kan* 看) the *li* rather than as 'knowing' it.[18] In other words, what a man has gained as a result

---

Tillman, "The Idea and the Reality of the 'Thing' during the Sung." For Zhu Xi's own account of some earlier thinkers' ideas about *gewu*, see YL18.23b1-24b2.

11 YL15.8a5.

12 YL63.7a0. Daniel K. Gardner has made this aspect explicit by translating the term as "apprehension of the principle in things": *Chu Hsi: Learning to Be a Sage*, x-xi, 52-53, 88n, 117-118 et passim.

13 E.g., YL9.6a3; SSJZ (*Mengzi* 孟子), 7.1a.

14 YL12.17b3.

15 YL11.3a5-3b0.

16 YL12.8b3-9a0.

17 YL11.3a1.

18 E.g., YL11.2a2, 3b5, 12.8b3-9a0.

of the 'investigation of things' was not so much a knowledge of the *li* as an insight into it. In this sense, the moment in which the mind attained the *li* of something represented a kind of 'resonance' between the mind's *li* and the things' *li*.[19] It is no wonder, then, that investigation of things, for Zhu Xi, was not primarily an intellectual process, and that the emphasis of his *gewu* endeavor was overwhelmingly focused on moral and social matters.

2 The Problem of 'Thorough Penetration': From Many Individual *Li* to the One Heavenly *Li*

Understanding the many *li* of individual things and events, however, was not the real aim of Zhu Xi's endeavor of 'investigation of things'; the ultimate purpose of such efforts, for him, was to reach the one universal *li* – the heavenly *li*, which would ensure the moral virtues of man's mind as it is in its original state. Thus, it was necessary for one to make a transition from understanding many individual *li* to understanding the one heavenly *li*; when one has reached such a transition, one is said to have achieved a 'thorough penetration' (*guantong* 貫通).

The necessity of such a transition posed a problem, however, for it was never clear how one could make the transition. An obvious question that presented itself was: how many things should one investigate – and understand the *li* of – before achieving a thorough penetration? This was essentially the question that had been posed to Cheng Yi by a disciple, in the following dialogue:

> Question: In investigating things, should one investigate everything or should one investigate just one thing and know all the myriad *li*?
>
> Answer: How can there be such a complete penetration just like that? If [you are speaking of] penetrating the many *li* by simply investigating one thing, even Yanzi 顏子 would still not dare say [that he could be] like that. One should investigate one item today and investigate another item tomorrow. After having accumulated and familiarized [oneself with] many [of them], suddenly there will spontaneously be a 'thorough penetration.'[20]

19 Gardner, *Chu Hsi: Learning to Be a Sage*, 53.
20 *Henan Chengshi yishu* 河南程氏遺書, *juan* 18 (ECJ, 188).

Yet Cheng Yi's answer was hardly a solution, for while rejecting both of the alternatives posed by the questioner, he voices his own position in an extremely ambiguous manner – that after accumulating and familiarizing oneself with many items, one will 'suddenly' and 'spontaneously' achieve a thorough penetration. It was precisely Cheng Yi's reply on which Zhu Xi's discussion of the problem centered.[21]

Zhu Xi frequently quoted Cheng Yi's answer and praised it for not taking an extreme viewpoint. Zhu Xi said, for example: "Neither did he say that one can penetrate simply after investigating one item, nor did he say that one can begin to penetrate only after exhausting all the *li* of things under heaven."[22] He noted that while Yang Shi 楊時 (1053-1135) took the former position, and Hu Anguo 胡安國 (1074-1138) the latter, "[it was] only [Cheng] Yichuan who said that 'neither should one simply investigate one *li*, nor can one cover exhaustively all the *li* of the myriad things.'"[23]

Yet how one was supposed to then reach the state of thorough penetration was not spelled out by Cheng Yi, nor did Zhu Xi provide any further insight than his northern-Song master. Following Cheng Yi, Zhu Xi emphasized that the penetration comes 'spontaneously' and 'gradually' after spending a long time seeing, experiencing, and studying many things.[24] Typically he said:

> When one encounters an event, one takes up the event and examines its *li* to the very end. After a while there are many [such *li* that one has examined thus], and spontaneously, one can 'penetrate thoroughly.'[25]

> This one item is understood thoroughly and that one item is also understood thoroughly. When many [such items] have been accumulated, then one can 'penetrate thoroughly.'[26]

While making this sort of remark, Zhu Xi recognized that the process is a difficult one: the state of thorough penetration comes spontaneously, but it does so only after a long, painstaking process of study and exertion. Thus, although he refused to accept either of the two alternatives mentioned above, he tended

21 E.g., YL18.2a1-4a1, 6a1-7a0. Zhu Xi also quoted Cheng Hao's 程顥 (1032-85) slightly different saying in, e.g., YL18.4b3-6b0.

22 YL18.2b1. See also YL18.5b1, 6a0, 25a1.

23 YL18.24b2.

24 E.g., YL8.12a11, 9.1b5, 6b2.

25 YL18.5b2.

26 YL44.20a4.

to highlight the necessity of investigating numerous things. For example, he said:

> Although the myriad *li* are simply one *li*, scholars still need to proceed to understand all the thousand ends and hundred beginnings inside the myriad *li*. [As the myriad *li*] come together from all the four sides, [scholars] can spontaneously see that [they] are one *li*. If one does not proceed to understand those myriad *li* but merely keeps [trying] to understand that one *li*, . . . [then it will] simply be empty speculation.[27]

Indeed Zhu Xi insisted that, in investigating things and extending knowledge, one should not exclude any thing or event.[28] Zhu Xi said that "in extending [the daily effort], there must be nothing that *li* does not include or penetrate, so that its extent is the same as the heaven and earth in its breadth and magnitude."[29] If one falls short of this, "one understands what is refined but loses what is coarse; one understands what is large but does not consider what is small. [If one proceeds] like this, one will have gaps and deficiencies in the end."[30] "If one sees one side and does not see another side, one will not be able to penetrate."[31] One should aim at the learning of the Sages (*shengren* 聖人), which has no biases or leanings, unlike the learning of common men (*changren* 常人), which is inclined to one side.[32]

Yet it is obviously not possible to investigate every single thing and event and understand all the *li* in the world. One cannot cover everything simply by continuing the process of investigating one item after another. One cannot achieve it by simply multiplying; one must 'extend' (*zhi* 致) what one has attained. Zhu Xi interpreted Mencius's phrase "to expand and fill [the knowledge]" (*kuo erh chong* 擴而充) in this sense of 'extending' knowledge.[33] It is clear, then, that somewhere in the process one has to cross a gap – the gap between grasping the individual *li* and grasping the one *li*, the result of which is thorough penetration. The essence of the problem, then, was how to cross this gap.

---

27 YL117.12b0.
28 E.g., YL14.21a1, 15.4a1, 7a1, 25b3.
29 YL8.1b1.
30 YL15.4a1.
31 YL15.7a1.
32 E.g., YL8.1b1,2.
33 E.g., YL16.8a1, 18.11b2. The Mencius phrase appears in II.A.6.

3 'Analogical Extension' from What Is Known to What Is Unknown

This is where the method of analogical extension becomes significant. For through this method, one can reach an understanding of the things and phenomena that one does not know without having to investigate them directly, but instead by inferring from one's understanding of what one already knows. Zhu Xi said: "Concerning [the five great virtues], scholars [should] investigate their profound depths, and 'extend by analogy' (*tui lei* 推類) so as to exhaust the rest."[34]

This is possible when those objects and events which one knows and those which one does not know are of the same 'kind' (*lei*). In other words, the extension takes place between similar things belonging to the same category. In this respect, the method of analogical extension is congruent with a key mode of traditional Chinese thought, namely the so-called 'correlative' thinking based on systems of 'correspondences' involving various categorical schemes like yin-yang 陰陽 and the five phases (*wuxing* 五行).[35] In explaining Cheng Yi's theory that "having accumulated and familiarized oneself with many [of them], suddenly there will spontaneously be a thorough penetration," Zhu Xi said:

> Now let us speak about it with ten events. If one can understand seven or eight items, then those two or three items can be penetrated by means of [their being of the same] kind. If all the four sides are understood, then the way and *li* of what is not understood in the middle are also [penetrated] like this. ... When one suddenly comes across an event, one must know which event agrees with this and which event agrees with that, then this event which has just happened will also have something that can be understood.[36]

Of the two characters forming the word *leitui*, however, Zhu Xi's emphasis lay more on the second character, *tui*. What was important for him was the aspect

---

34 SSJZ (Daxue), 3b. Zhu Xi quoted one of the Cheng brothers as even having said that "having investigated one event exhaustively, it can be extended to the others by 'analogy'." YL18.6b1.

35 On "correlative" thinking and the categorical schemes, see, e.g., John B. Henderson, *The Development and Decline of Chinese Cosmology* (New York: Columbia University Press, 1984); A. C. Graham, *Yin-Yang and the Nature of Correlative Thinking* (Singapore: The Institute of East Asian Philosophies, 1986); Yung Sik Kim, *The Natural Philosophy of Chu Hsi* (Philadelphia: American Philosophical Society, 2000), Chaps. 4-5.

36 YL18.15a1.

of 'extension' (*tui*); he did not define what the character *lei* meant. Obviously, such categories as yin-yang and the five phases were among the things that he had in mind, but the extension did not necessarily have to take place between things belonging to those established categories. He seemed to consider two things or events as belonging to the same *lei* if any sort of similarity or analogy could be seen between them. This is evident in his discussions about 'serving' (*shi* 事) parents, rulers and elders, the subject he took up most often in illustrating the method of analogical extension, such as in the following passage.[37]

> As for serving parents, for example, one must exhaustively understand the way (*dao*) to serve them. [One should understand] what one does when [one's service is] accepted by the parents and also what one should do when [it is] not accepted by the parents. If one extends from this to serving rulers, then one knows what to do when [one's service is] accepted by the rulers and also what one should do when [it is] not accepted by the rulers. Extending [from this] to serving elders, it is also like this. Go on extending from this. Everything must be so.[38]

The method of 'analogical extension,' then, was essentially a method of extending from what one knows to what one does not know. When a disciple asked about the method, Zhu Xi responded: "[It] is to go on extending like this from what one has understood."[39] The character *tui* thus came to have the meaning of 'inference' or 'to infer.'[40]

---

37 E.g., YL14.11b4, 16.5a1, 18.6b1, 7b2.

38 YL18.7b2-8a0.

39 YL18.23a1.

40 In fact, the clear emphasis on the aspect of extension in the following *Shiji* 史記 biography of Zou Yan 鄒衍 (ca. 305-240 BC), the principal source of correlative thought, shows that extension from what is known to what is not known was perceived as the characteristic method of correlative thought as early as the time of Zou Yan, or at least by the time of Sima Qian 司馬遷 (ca. 145-90 BC). "[Zou Yan] always examined small objects first and then extended (*tui*) it to large ones until he reached what is without limit. He began with the present and from this went back to the time of the Yellow Emperor, all of which had been recorded by scholars. Then he followed the great events in the rise and fall of the ages, and by examining the omens and systems, extended it far backward to the time when heaven and earth had not yet come into being and it was mysterious and impossible to investigate. He first listed China's famous mountains, great rivers, connecting valleys, birds and beasts, the products of her waters and soils, and based on these extended it to what is beyond the seas and cannot be seen by men. [*Shiji* (Beijing: Zhonghua shuju edition, 1959), Chap. 74, 2344.]"

The term, 'extension of knowledge' (*zhizhi*) is what refers to such extensions. Zhu Xi said, for example:

> What is called 'the extension of knowledge,' is to base oneself on what one has already known, and to infer (*tui* 推) and extend (*zhi* 致) from it, so that one reaches what one has not known and stretches one's reach to the limit. One must reach all the *li* of heaven and earth and the myriad things, and 'penetrate through it by one' (*yi yi guan zhi* 一以貫之). Only then it becomes the 'arrival' (*zhi* 至) of knowledge.[41]

Zhu Xi characterized not only 'the extension of knowledge,' but other key terms designating various stages and aspects of this intellectual endeavor – 'the investigation of things,' and 'the exhaustive investigation of *li*' (*qiongli* 窮理), for example – in the same way. "The investigation of things," according to him, "is merely to understand what has not been understood."[42] "Those who 'investigate the *li* exhaustively' arrive at what they have not known by means of what they have already known, and [they arrive] at what they have not attained by means of what they have already attained."[43]

4 Stages of 'Analogical Extension': From the 'Beginning' to the 'Arrival' of Knowledge

The quotations in the last section illustrate several key aspects of the method of analogical extension, which characterize different stages in the process of extension – from the beginning when it is based on the knowledge one already has to the end when the knowledge is 'arrived at' (*zhi* 至).

First, the extension *is based on what one knows already*. Zhu Xi referred to such prior knowledge, on which one can base the extension, as the 'beginning' or a 'clue' (*duan* 端 or *duanxu* 端緒).[44] He used such expressions as 'in accordance with' (*yin* 因), 'on the basis of' (*ju* 據), and 'from' (*cong* 從), to describe its role.[45] Thus, the investigation of things and extension of knowledge must

---

41 WJ42.16b. The phrase "penetrate through it by one" is from the *Analects*, 4.15.

42 YL18.3a2.

43 YL18.3a1.

44 E.g., YL16.8a1, 18.11b2.

45 E.g., YL15.2a2, 9a1; WJ42.16b.

start from "that which is bright,"[46] "that which is very close to oneself,"[47] and "that which is manifest" (*faming* 發明 or *faxian* 發見).[48] Zhu Xi even said: "The beginning of the exhaustive investigation of *li* is like attacking a solid thing. One must look for cracks that will enable one to enter it."[49] The extension can start like this because, as Zhu Xi noted, there are certain things that every man knows and can be used as 'beginnings' or 'clues': "For example, filial piety for parents, brotherly love for younger brothers, that water must be cold and fire must be hot – it is impossible to say that [there is anyone who] does not know these."[50]

Next, one extends one's knowledge by *inferring* (*tui*) from this previously gained knowledge. It is in this sense that Zhu Xi glossed the character *zhi* (致) in the term *zhizhi* (extension of knowledge) as *tui*: "To extend (*zhi* 致) is to 'infer to the utmost' (*tuiji* 推極)";[51] "The character *zhi* has the meaning of 'inferring and bringing out' (*tuichu* 推出)."[52] While discussing 'extension of knowledge,' Zhu Xi illustrated it with a metaphor in the following manner.

> The [meaning of the] character *zhi* is something like 'to keep inferring and opening (*tuikai* 推開).' It can be compared to [a situation when there are] a few bright places inside a dark room. One should simply proceed from these bright places till one suddenly finds the way out. [Then] one can see everything with great brightness. What people call 'extension of knowledge' is also like this.[53]

Such inferences enable one to *know what one has not known*. And as the result, one's knowledge is extended: from what is 'easy' to what is 'difficult,'[54] from what is 'manifest' (*xian* 顯) and 'commonplace' (*pingyi* 平易) to what is 'hidden and subtle' (*youwei* 幽微),[55] from 'close' to 'far,'[56] from 'coarse' (*cu* 粗) to

---

46 E.g., YL14.13a1, 15.7a1, 8b1.
47 E.g., YL15.2b1.
48 E.g., YL15.5b1.
49 YL15.7a1.
50 YL14.14b3.
51 SSJZ (Daxue), 1b.
52 YL15.8b2.
53 YL15.8b1.
54 E.g., YL11.6b5, 14.1a2.
55 E.g., YL8.11b2.
56 E.g., YL14.14b3, 15.12a3, 18.12a2; WJ64.26a.

'refined' (*jing* 精),[57] from 'low' to 'high,'[58] from 'surface' to 'inside,'[59] from 'shallow' to 'deep,'[60] and from what one 'sees and hears' to what one 'does not see or hear.'[61] And Zhu Xi characterized such extension as 'extending and brightening' (*tuiming* 推明),[62] 'extending and opening' (*tuikai*),[63] and 'extending and broadening' (*tuiguang* 推廣).[64]

One should continue the effort of such extension *till it is completed.* In Zhu Xi's own words, "one must continue to extend (*zhi* 致) one's knowledge to its extreme."[65] "He who extends (*zhi*) his knowledge starts from inside and goes outside to infer till the point when there is nothing left to exhaust; he starts from outside and goes inside to infer till there is nowhere to go."[66] He even said: "When the knowledge has been extended (*zhi*), there is not a single event not exhausted nor a single thing not known."[67]

When the extension is completed in this manner, it is said that *"the knowledge is arrived at"* (*zhi zhi* 知至).[68] And in this state, one's 'intention' (*yi* 意) becomes 'sincere' (*cheng* 誠).[69] This is so because, in the course of extension of knowledge, one "tests [the knowledge] with one's mind and incorporates it into one's body,"[70] which enables one to extend the way (*dao*) and the *li* one has acquired to oneself and to other things.[71] Thus, "when one's knowledge is arrived at, one will have [the ability to] control oneself, and will be able to distinguish and select."[72] This, for Zhu Xi, is the state described by the phrase "to know where to stop" (*zhi qi suozhi* 知其所止), in the *Great Learning* (*Daxue* 大學). He said: "In dealing with the affairs of the world, one must always know where to stop and be free of doubt."[73]

---

57 E.g., YL15.25b3.
58 E.g., YL8.11b1.
59 E.g., YL15.25b3.
60 E.g., YL14.14b3.
61 E.g., YL14.12a3.
62 E.g., YL14.13a1, 15.7a1.
63 E.g., YL15.8b1, 9a1.
64 E.g., YL15.2b1, 8a4, 16.5a1.
65 YL14.14b3. See also YL18.1a4.
66 YL15.16a2.
67 YL15.15b2. See also YL15.9a1.
68 E.g., YL15.12b2, 15.12b3, 15.25b3.
69 E.g., YL15.16a2, 25b3; WJ16b-17a.
70 YL15.15b2.
71 YL18.17a1.
72 YL15.25b3.
73 SSJZ (Daxue), 3b.

When one reaches this state one is said to have achieved a '*thorough penetration*,' the ultimate goal of the endeavor. One should not cease one's efforts till one reaches this state. In fact, Zhu Xi harshly reprimanded a disciple who had asked the question, "Will it still be harmful if one, in one's effort to investigate things, does not achieve a thorough penetration?"[74]

All this is summarized in Zhu Xi's celebrated supplement to Chapter 5 of the *Great Learning*:

> Encountering things under heaven, one must base oneself upon the *li* of what one has already known and exhaust it further so that one may seek to reach the limit. When it comes to [the point when], having practiced this for a long time, one day one suddenly 'penetrates thoroughly' (*guantong*), then one will be able to reach all the inner and outer aspects and the fine and coarse points of numerous things, and the entire substance and the great function of one's mind will be all luminous. Referring to this, it is said that "things have been investigated" and that "knowledge has been arrived at."[75]

5 'Analogical Extension' Based on a Complete, Exhaustive Understanding

For the above kind of extension to be possible, one must know what one knows – or investigate the thing that one investigates, or understand the *li* that one understands – to the utmost. Only when one knows and understands a thing completely can one's knowledge of the thing function as a solid basis from which one can proceed to extend one's knowledge to other things by the method of analogical extension. Thus, Zhu Xi emphasized that in investigating a thing one must not stop at a partial knowledge of the thing but should seek to arrive at a complete knowledge of it. "To investigate a thing," for him, is "to investigate and exhaust that thing,"[76] and "to bring [one's] reach of the *li* of a thing or event to the extreme and to exhaust it to the end."[77] "If you know one or two portions [of a thing], then beginning from what you know about that one or two [portions], you should keep inferring till you reach [all] the ten [portions]. When you have exhausted [all portions of a thing] and there is

74 YL15.6a1.

75 SSJZ (Daxue), 5a.

76 YL15.11b1.

77 YL18.22b0.

nowhere [else] to proceed, then [you have accomplished] the investigation of the thing."[78] Zhu Xi even said that "one should investigate exhaustively more than ninety-nine percent [of a thing]. Only then [the thing can be said to have been] investigated."[79] In the following passage, he illustrated the point by means of actual examples of 'things' – coal, water, and a stove.

> As for [lumps of] coal, for example, there are some that are white and also some that are black. It will be no use if one exhaustively investigates only the black ones without exhaustively investigating the white ones. As for water, although it is cold and wet, it also has many other different aspects. If one simply recognizes the single features of coldness or wetness, it has not been investigated [properly]. ... As for a stove, if one has understood one corner, one should also understand all the [other] three corners, and one should understand the top, bottom and the four sides. Only then [can it be said that] the thing [i.e. the stove] has been investigated. If even a single spot is not understood, it cannot be [said] that the thing has been investigated.[80]

At times, Zhu Xi used the term 'extension of knowledge' to designate the efforts to reach thorough knowledge of a single thing or event. He then used the phrase, 'knowledge is at rest' (*zhi zhi* 知止), to refer to the result of such efforts. He said:

> [Suppose] the way (*dao*) and *li* of one thing has ten portions. If one only sees two or three portions, then one has not seen exhaustively. One must keep extending and must try to see [all] ten portions exhaustively. Only then can it be the investigation of the thing. If one has seen [all] ten portions exhaustively, only then can knowledge be at rest.[81]

He even said that "if one knows a single event thoroughly, it can be said that knowledge [of that one event] is at rest."[82]

Hence, according to Zhu Xi, it was fair to leave out some items from one's investigation, because one can make inferences based on one's prior knowledge of other similar things; however, the single item chosen for investigation

78 YL18.15a1. See also YL15.2a3, 18.1b5.

79 YL15.11a2.

80 YL18.6b1-7a0.

81 YL15.11a1. See also YL15.16a2.

82 YL15.22b0.

should be then investigated thoroughly, for otherwise one cannot know it well enough to base further inferences on it. He said:

> In the extension of knowledge and investigation of things, if, out of ten events, one can investigate and understand nine events thoroughly but cannot understand one event thoroughly, it is not harmful. [On the other hand, in investigating] one event, if one can investigate nine portions and cannot [understand] one portion thoroughly, it is most undesirable.[83]

That "one portion" which one has not understood could become "a root of confusion and impropriety" in the course of the extension.[84]

Again, it is not an easy task to investigate a single thing or a single event, and to understand its *li*, completely and exhaustively. Indeed, it is not always possible to do so for all things and events; there can be some for which such complete investigation and understanding cannot be achieved. For such an instance, Zhu Xi borrowed Cheng Yi's comment: "If the *li* cannot be investigated exhaustively for one event, then investigate exhaustively another event,"[85] and advised that if one cannot achieve a thorough understanding of one thing or an event, one should not stick to it but should move on to others.

> In this case, if, having failed to understand [a thing], one still concentrates oneself on [this] one [thing] and sticks to it, one will become confused. One should set out to investigate another event exhaustively. [If this second event is understood,] one might also be able to understand that [i.e. the earlier event] on the basis of [one's understanding of] this [i.e. the second event].[86]

The idea of extension by analogical extension presents itself again in the last sentence of the passage.

Of course, Zhu Xi duly emphasized that it is only in instances where one is dealing with very difficult things that this position should be adopted, and pointed out that Cheng Yi had made the above remark for such exceptional cases.[87] Zhu Xi, however, warned of the potential danger of abusing this attitude:

---

83 YL15.20a1.

84 YL15.15b2.

85 *Henan Chengshi yishu*, Chap. 15 (ECJ, 157).

86 YL18.7a1.

87 E.g., YL18.5b2-6a0, 7a2-7b0, 7b1, 28b0.

> If, when faced with ordinary events, one [proceeds to] understand a second item while not understanding this one item thoroughly, [and if] one [proceeds to] understand a third item while not understanding the second item thoroughly, then one will not advance far in this way in an entire lifetime.[88]

6 'Analogical Extension' in a Slow, Gradual Process toward a 'Thorough Penetration'

It can be seen from the above discussion that although the 'thorough penetration' is frequently characterized as 'sudden' (*tuoran* 脫然 or *huran* 忽然) and 'spontaneous' (*zi* 自 or *ziran* 自然), in truth, it can only be a slow, gradual process. Admittedly, there has to be a moment in which a thorough penetration occurs. But that moment 'suddenly' and 'spontaneously' comes only after, and as the result of, a long, painstaking process of exertion. Zhu Xi said, for example:

> How under heaven can there be [an instance in which] having understood [only] one *li* completely, a person can understand all the myriad *li* completely? One should also keep accumulating … In study, there should be gradualness; there is no *li* for pressing oneself in haste. Someone once said that, in study, if one applies exhaustive investigation only to one great subject, then all the others will be understood completely. As for me, certainly I would not dare say [that I might be] like this. One must keep working little by little.[89]

While commenting on Cheng Yi's remark that "investigating one item today and another item tomorrow" would 'suddenly' and 'spontaneously' lead to a 'thorough penetration,' Zhu Xi added, "On this matter, one must not desire quickness. If one desires quickness, then one cannot arrive at [a thorough penetration]. One must go on working slowly."[90] Thus, it is that in reading, "there must be an order,"[91] and "one should not press oneself in haste."[92]

---

88 YL18.7b1.
89 YL18.2a2.
90 YL18.4b1.
91 YL11.11b1.
92 YL11.8b4.

Zhu Xi sketched out how such a gradual process can lead to a 'thorough penetration':

> One must understand one [thing or event] after another. After a while, [one will understand] many of them, and gradually one will be able to penetrate thoroughly. [One's understanding of] two [things or events] combines and becomes one. After a while seven or eight [things or events] will combine and become one. Then everything will be penetrated thoroughly at once.[93]

He even compared it to the process to walking:

> One should wait for one's learning to increase day by day. Having investigated one thing today, investigate another thing tomorrow; this way, one's learning cannot [possibly] remain unchanged. It is like the following. When the left foot has advanced one step, the right foot also advances one step; when the right foot has advanced one step, the left foot also advances. By continuing without stopping, [there will] spontaneously [come] a 'thorough penetration.'[94]

For Zhu Xi, 'analogical extension' was the method to be used in this gradual process. He answered the question, "What is analogical extension?" in the following manner:

> It is not necessary to jump higher and look farther. Nor is it [necessary] to roam about aimlessly. It is simply to take up those subjects that are near at hand and are understood and to keep going. It is like having understood this one event thoroughly, one proceeds to infer from [the knowledge of] this event to work on that event, and knows that it [i.e. that event] is also similar. It is like knowing that this lantern has a great deal of light, one then proceeds to infer from [the knowledge about] this lantern to know that that candle is likewise bright. It is like climbing up stairs. Having climbed to the first level, one bases oneself on this first level and advances to the second level. Also, basing oneself on the second level, one advances to the third level. One simply keeps proceeding like this...If, [standing on] the first level, one wants to jump to the third level,

93 YL14.21a1.

94 YL18.3a1.

> the step will have to be too wide, and one will [end up] wasting one's effort.[95]

'Analogical extension,' then, was the key method in Zhu Xi's endeavor of the 'investigation of things' and the 'extension of knowledge,' which would lead to a 'thorough penetration.' And because of the close connection between intellectual and moral endeavors, it was natural that analogical extension was also the method to be used in self-cultivation. Indeed, the passage quoted above goes on to discuss that it is by analogical extension that one moves from 'being filial to parents' (*qin qin* 親親) to 'being benevolent to people' (*ren min* 仁民) and to 'loving things' (*ai wu* 愛物); or from 'cultivating the self' (*xiu shen* 修身) to 'regulating the family' (*ji jia* 齊家) and to 'ordering the country' (*zhi guo* 治國).[96]

95 YL49.4b1.
96 YL49.5a0.

CHAPTER 3

Ideas about the Role of Heaven in Production Techniques in Song Yingxing's *Heaven's Work in Opening Things* (*Tiangong kaiwu*): 'Natural Theology of Industry' in Seventeenth-Century China?*

1 Introduction

I have chosen the title of my essay intentionally to imitate that of Charles Gillispie's classic paper, "The Natural History of Industry."[1] From that short paper, I, like many other historians of science of my generation, learned to think critically about the relationship between science and technology, and about the real nature of written materials on techniques and trades, those in the famous *Encyclopédie*, for example. Much of what Gillispie said in showing 'natural historical' aspects of many French technical writings of the eighteenth century applies to the subject of the present essay, the *Tiangong kaiwu* 天工開物 ("Heaven's Work in Opening Things," published in 1637), a famous seventeenth-century Chinese book of production techniques written by Song Yingxing 宋應星 (1587-1666, ca.), whom Joseph Needham has called "the Diderot of China."[2] In this essay, I will reexamine Song Yingxing's attitude to

* Originally published as "'Natural Theology of Industry' in Seventeenth Century China?: Ideas About the Role of Heaven in Production Techniques in Song Yingxing's *Heaven's Work in Opening Things* (*Tiangong kaiwu*)," in *A Master of Science History: Essays in Honor of Charles Coulston Gillispie*, ed. Jed Z. Buchwald (New York: Springer, 2012), 197-214.

1 Charles C. Gillispie, "The Natural History of Industry," *Isis* 48 (1957): 398-407.

2 The two major studies on the *Tiangong kaiwu* are: Yabuuchi Kiyoshi 藪内清 ed., *Tenkō kaibutsu no kenkyū* 天工開物の研究 [Studies on the *Tiangong kaiwu*] (Tokyo: Kōseisha 恒星社, 1954); Pan Jixing 潘吉星, *Tiangong kaiwu jiaozhu ji yanjiu* 天工開物校注及研究 [Critical Annotations and Studies on the *Tiangong kaiwu*] (Chengdu: Bashu shushe 巴蜀書社, 1989). In addition, Pan Jixing has written an extensive biography of Song Yingxing: Pan Jixing 潘吉星, *Song Yingxing pingzhuan* 宋應星評傳 (Nanjing: Nanjing daxue chubanshe 南京大學出版社, 1990). There have been translations of the *Tiangong kaiwu* into English, modern Chinese, Japanese, and Korean: Sun, E. Z. and Sun, S.-c. trans., *T'ien-kung K'ai-wu: Chinese Technology in the Seventeenth Century* (University Park: Pennsylvania State University Press, 1966); Zhong Guangyan 鍾廣言 ed., trans., *Tiangong kaiwu* 天工開物 (Beijing: Zhonghua shuju 中華書局, 1978); Yabuuchi Kiyoshi 藪内清 trans., *Tenkō kaibutsu* 天工開物 (Tokyo: Heibonsha 平凡社, 1969); Ch'oe Chu, 崔炷 trans. *Ch'ŏn'gong kaemul* 天工開物

 | DOI 10.1163/9789004265318-005

the production techniques and his motivation for writing the book, and call attention to another aspect of the book that has not been noted so far: a 'natural theology of industry.'

Many modern scholars have praised the *Tiangong kaiwu* for dealing with various production techniques relating to agriculture, mining, and other traditional Chinese industries. Yabuuchi Kiyoshi 藪内清, for example, has pointed out that the book is important as it covers all the major industries, and is detailed in discussing the production processes of each industry.[3] The most enthusiastic praise has come from Pan Jixing 潘吉星, who has worked on the book most extensively. He referred to it as a 'systematic,' 'conclusive' treatise of Chinese agriculture and production techniques up to Ming times (1368-1643).[4] There have also been comments on the background and the motivation behind Song Yingxing's writing of this specialized book devoted to production techniques. Most scholars have pointed out the general intellectual climate of the late Ming, which displayed a great interest in practical utility.[5] It has also been noted that the publication of the *Tiangong kaiwu* was in line with the widespread interest during the late Ming in miscellaneous and strange things,

(Seoul: Chŏnt'ong Munhuasa 傳統文化社, 1997). For various editions and reprints of the *Tiangong kaiwu*, see Pan Jixing, *Tiangong kaiwu jiaozhu ji yanjiu*, 131-171. Needham's reference to Song Yingxing appears in Joseph Needham, *Science and Civilisation in China*, vol. 1 (Cambridge: Cambridge University Press, 1954), 13.

3 Yabuuchi Kiyoshi 藪内清, "*Tenkō kaibutsu* ni tsuite" 天工開物について (On the *Tiangong kaiwu*), Yabuuchi Kiyoshi, ed., *Tenkō kaibutsu no kenkyū* 天工開物の研究 (Tokyo: Kōseisha 恒星社, 1954), 1-24, esp. 2.

4 Pan Jixing, *Tiangong kaiwu jiaozhu ji yanjiu*, 21, 92. Pan Jixing went as far as to say that at the time there was no book on technology even in the West that could compare to the *Tiangong kaiwu* in depth and breadth of coverage: ibid, 98-100. He also saw some characteristics of modern science in the book, using such expressions as "experimental science," "mathematization," "critical spirit," and "enlightenment thought": ibid., 76-78.

5 Yabuuchi, "*Tenkō kaibutsu* ni tsuite," 15-16; Pan Jixing, *Tiangong kaiwu jiaozhu ji yanjiu*, 16. Peter Golas has also pointed out such aspects as "stress on concrete accomplishments whether as an official or in a private capacity," "willingness to accept the legitimacy of profit-taking by merchants," and "interest in scientific and technological learning that had practical application," as well as more immediate motivations as "reputation" and "income": Peter Golas, "'Like Obtaining a Great Treasure': The Illustrations in Song Yingxing's *The Exploitation of the Works of Nature*," in *Graphics and Text in the Production of Technical Knowledge in China: The Warp and the Weft*, ed. Francesca Bray, Vera Dorofeeva-Lichtmann, and Georges Métailié (Leiden: Brill, 2007), 574.

which resulted in the publication of many books of miscellanea and encyclopedias.[6]

The meaning of the expression '*tiangong*' 天工 (heaven's work) in the title, and thus the role of heaven in production techniques, has also received some attention. Pan Jixing was of the view that this expression referred to both the natural world and man's work: "'*Tian*' (heaven) of the [term] '*tiangong*' refers to the natural world; '*gong*' (work) refers to man's skill or technique."[7] He then interpreted the whole title – '*tiangong kaiwu*' – as referring to the cooperation between "heaven's work and man's work in developing the myriad things."[8] Of the roles of heaven and man, however, Pan Jixing placed the emphasis on man's role, suggesting a translation of the title: "Exploitation of Products by Artificial Skill from the Nature."[9] Adopted by Joseph Needham in a modified form as "The Exploitation of the Works of Nature," this interpretation has become standard among many scholars.[10] Given the emphasis placed on man's role, not much attention has been paid to Song Yingxing's ideas about the role of heaven in the production techniques he discussed in the *Tiangong kaiwu*.[11]

The purpose of this paper is to throw more light on what Song Yingxing meant by the expression 'heaven's work' (*tiangong*) by examining what he thought about the role of heaven in man's acquisition and use of techniques. His basic idea was that while man plays various roles in producing many things, it is really heaven that lies behind all this. In particular, I will show that heav-

6 E.g., Pan Jixing, *Tiangong kaiwu jiaozhu ji yanjiu*, 16-19. Benjamin A. Elman, *On Their Own Terms: Science in China, 1550-1900* (Cambridge, Mass.: Harvard University Press, 2005), Chap. 1. When such interest died out later among the literati, the book went out of sight in China – to appear in Japan.

7 Pan Jixing, *Tiangong kaiwu jiaozhu ji yanjiu*, 72. In this, Pan Jixing was following the view of Ding Wenjiang 丁文江 (1888-1936), one of the pioneers of the modern studies of the *Tiangong kaiwu*. Dagmar Schäfer's translation, "Heaven, Work, and the Inception of Things," is essentially along the same lines. Dagmar Schäfer, "The Congruence of Knowledge and Action: The *Tiangong kaiwu* and Its Author Song Yingxing," in *Chinese Handicrafts Regulations of the Qing Dynasty: Theory and Application*, ed. Hans-Ulrich Vogel, Christine Moll-Murata and Song Jianze (Munich: Indicum, 2005), 35-60.

8 Pan Jixing, *Tiangong kaiwu jiaozhu ji yanjiu*, 69.

9 Ibid., 73.

10 Peter Golas, for example, has sided with this line of interpretation by emphasizing the importance of "the idea that technology involved human beings employing their skills to make useful the resources provided by nature." See Golas, "'Like Obtaining a Great Treasure,'" 585.

11 Craig Clunas has translated the title as "Heaven's Craft in the Creation of Things", but has not elaborated on it: Craig Clunas, *Superfluous Things: Material Culture and Social Status in Early Modern China* (Honolulu: University of Hawaii Press, 2004), 166.

en's role of taking care of man appears prominently in his discussion of production techniques in the *Tiangong kaiwu*. I believe that with this understanding of heaven's role, it makes better sense to include it in the title of the book, thereby also bringing the title in line with the original use of the expression '*tiangong*' in the *Book of Documents* (*Shujing* 書經): "It is heaven's work; man carries it through." (*Tiangong, ren qi daizhi* 天工, 人其代之). Examination of Song Yingxing's ideas about 'heaven's work' will also shed light on his general ideas about and attitude toward production techniques: the nature and the role of such techniques; their place within the corpus of man's – and scholars' – knowledge; and on the one hand, the relationship between production techniques (and knowledge about them) and other human activities and knowledge, and on the other hand, the relationship between such techniques and natural phenomena.

2 Content of the Book and Song Yingxing's Motivation

Before looking at Song Yingxing's ideas about the role of heaven, I will reexamine the content of the *Tiangong kaiwu* to find more clues as to what he may have thought of the production techniques themselves, and as to what might have been his motivation for writing the book, which, in turn, will provide a context for my discussion on his views about heaven's role in production techniques. I will begin the reexamination with his preface, a natural place where one would expect to find the author's thoughts about the book and its subject matter.

2.1 *The Preface: What Song Yingxing Did and Did Not Intend to Do*

Song Yingxing's preface to the *Tiangong kaiwu*, a work he labored on greatly, is rather brief and casual. He does not seem to have been interested in using the preface to systematically lay out his thoughts about the subjects discussed in the book, or about his own discussions of them. In fact, he did not even explain what he meant by his title, which was hardly transparent.

On the other hand, Song Yingxing's remarks about the situation surrounding various commodities like foods and clothes do reveal his motivation for writing the book. For example, he pointed out that people were interested only in things that were rare while knowing very little about the things that were common. He then noted that people in his time were much more likely to see rare things firsthand than in the past. These remarks reveal an intention on his part to deal with the common things that were frequently seen. Song Yingxing's main interest, however, was not in showing the things themselves, but in showing the processes and instruments used in producing them. He said:

> The royal princes are born and raised deep inside the palace. [Smelling] the good fragrance of the best rice in the royal kitchens, they may wish to look at the plows. [Looking at] the court ladies cutting brocade clothes, they may imagine the looms and the silk. At such times, [if they can] unroll a picture [book] and look at it, it would be like acquiring a heavy jewel.

This echoed the recurring motivation of the Chinese technical books to show elite readers how the everyday commodities providing them with various comforts had been produced.[12]

In his preface, Song Yingxing also made remarks about what he did not intend to do and about the topics he would not be discussing. For example, toward the end of the preface comes his famous remark that "scholars with great ambitions" should throw away the book because the book has nothing to do with "fame and advancement." This remark indicates his view that the topics he had chosen for the book fell outside the curriculum for the civil service examination, the primary concern of many scholars. Song Yingxing also said that originally he had written chapters on astronomy and music, but eliminated them before printing the book. This shows that at first he had not meant to restrict his topics to production techniques; perhaps the book he had in mind was not a book on techniques only but one covering all concrete subjects (including astronomy and music) lying outside the learning for the civil service examination, which was characterized instead by abstruse philosophical discourse.[13] In fact, the reason provided by Song Yingxing himself for his decision to exclude the two topics – that "their ways are too exquisite and are not my affairs" – indicates his general distaste for abstract and abstruse topics and his inclination towards concrete subjects.[14] The preface also shows Song Yingxing's practical bent. Even the order of the topics – with the "five grains" coming first

12 For example, Francesca Bray has spoken of "a general interest among the leisured class of the period in how everyday objects were made." Francesca Bray, "Introduction: The Powers of *Tu*," in Bray, Dorofeeva-Lichtmann, and Métailié, *Graphics and Text*, 30.

13 Dagmar Schäfer has seen "a very Confucian idea of presenting an all-encompassing world-view" in Song Yingxing's original plan. See Schäfer, "The Congruence of Knowledge and Action," 53.

14 Of course, it is also possible that he did not include them because there had already been many texts written about such subjects. Pan Jixing has suggested that Song Yingxing's essays, "Tan tian" 談天 (Discourse of Heaven) and "Lun qi" 論氣 (Discussions on *Qi*), may

and "gold and jade" coming last – reflects his priorities: he valued basic commodities like grains while slighting luxuries.[15]

Thus, on the whole, this book on production techniques seems to have been born of its author's inclination for everyday, practical, and useful topics, and of his distaste for the learning geared towards the civil service examination, which he had failed repeatedly.

2.2 *Practical Interest of Entrepreneurs*

Song Yingxing's selection of the topics in the main part of the *Tiangong kaiwu* was mainly based on the practical utility of the techniques and products. What he wrote about these topics, however, was usually not the kind of information that could be used by actual practitioners of these techniques; rather, his discussions were more frequently about topics that entrepreneurs, or owners, of industries would be interested in.[16]

Thus, Song Yingxing frequently mentioned prices and costs of source materials, products, and tools used for producing them. But more typically, he mentioned the ratios of the quantities of various materials involved in the production processes. For example, he mentioned the ratios of different raw materials required in a given process: the ratio of saltpeter to sulfur in making gunpowder is 9 to 1 (III.15.31b)[17]; that of fat and wax to copper in making bells

---

have originally been written as the contents of the chapters on astronomy and music: Pan Jixing, *Song Yingxing pingzhuan*, 266, 273.

15 Yabuuchi Kiyoshi saw an emphasis on the importance of agriculture in this: Yabuuchi, "*Tenkō kaibutsu* ni tsuite," 5. Indeed, a large part of the content of the *Tiangong kaiwu* is devoted to agricultural topics: not only topics like producing grains and making clothing, but even chapters on dyeing, sugar, etc. contain discussions of agricultural matters.

16 The printers of the second edition (published between the 1650's and the 1680's) seem to have been aware of this aspect of the book, as can be seen in the content of the "advertising sentence" on the title page. See Schäfer, "The Congruence of Knowledge and Action," 42. Peter Golas has given an excellent, if somewhat too modernized, list of the items described in the *Tiangong kaiwu*. "The topics Song regularly discussed about any given technology included raw materials, areas of production, the technical processes involved, essential operations, equipment used, consumption of raw and processed materials, consumption of energy, products and rate of production, special characteristics of the products and their uses. (Golas, "'Like Obtaining a Great Treasure,'" 585.)"

17 Song Yingxing divided the *Tiangong kaiwu* into three parts and 18 chapters, using the same character *juan* 卷 to refer to both the 'parts' and the 'chapters.' In my citation, I put the numbers for the parts (I, II, III) and the chapters (1-18), followed by the page numbers of the original edition (1637), reprinted as volume 1115 of the *Xuxiu Siku quanshu* 續修四庫全書 series (Shanghai: Guji chubanshe 古籍出版社, 1995-1999). 'III.15.31b', for example, refers to Part III, Chapter 15, page 31b.

is 1 to 10 (II.8.18b-19a). He spoke of the quantity of products obtained from a given quantity of source material: the weights of oil obtained from a given quantity of different kinds of seeds (II.12.64a-64b); the weights of flour obtained from a given quantity of different kinds of wheat (I.4.56a). He mentioned different quantities of a product that could be made by using different tools or methods and from different source materials: the quantity of wheat flour produced by different grinding mills (I.4.56b). He noted the extent to which the quantity of raw materials is reduced after undergoing a given process: three tenths of the wrought iron wasted during firing and hammering (II.10.44b); three tenths of silk yarn lost after boiling, and two tenths if the yarn is good (I.2.40b). He also provided information on the number of men required to carry out a particular task, or number of days required to do a given task to conclude that by using a particular machine, cows, or human power alone, etc: the plowing done by two men is equivalent to the work of one ox, and the work done by rotary harrow pulled by two men is equivalent to the work of three oxen (I.1.4a). Such considerations led him to comment upon the economic aspects of the production techniques: the skin of one sable is less than one square foot, and thus more than sixty sables would be needed to make one fur garment (I.2.45b); a household with 100 sheep will earn 100 taels [of silver], because each sheep yields wool for three pairs of socks, and each pair of sheep produces two lambs per year (I.2.47a); although the price of a good-grade flour is two-tenths higher, it is more economical to use if the byproducts of gluten and starch are taken into consideration (I.4.56b).[18]

2.3 *'Natural History of Industry'?: 'Investigation of Things'* (*gewu*) *and 'Test and Verification'* (*shiyan*)

Song Yingxing was usually very brief about the actual production methods. Instead, he frequently discussed the origins, histories, and geographical distributions of the production techniques and the products. He also spoke of various differences in the products: kinds (or species), parts, sizes, names, methods of production, as well as uses and potential damage.

In the course of such discussions Song Yingxing sometimes revealed his general views about production techniques. For example, he was interested in the progress and disappearance of the techniques. He pointed out that firearms techniques progress continuously (III.15.25a), predicting that the gunpowder-making technique in current use would no longer be used after ten years (III.15.35a). He noted that the method of producing 'Japanese silk'

18 Contents like these made Pan Jixing state that Song Yingxing was dealing with 'quantitative relations.' See Pan Jixing, *Tiangong kaiwu jiaozhu ji yanjiu*, 77.

(*woduan* 倭緞) was likely to be lost because it was not used anymore (I.2.41a). He seems to have been of the view that knowledge of techniques cannot be kept secret. Discussing the methods for obtaining grains, he said: "How can this way (*dao* 道) be kept secret forever?" (I.4.53a). These examples reflect not so much an entrepreneurial interest as a scholarly intellectual curiosity about the techniques for producing various things. Indeed, what we find in the *Tiangong kaiwu* are usually not practical instructions to be followed, but rather records of observations, or of descriptions found in books, about what was done and how it was done.[19] One can discern here an aspect of what Charles C. Gillispie called "the natural history of industry."[20]

It is then possible that Song Yingxing thought of the work he had put into composing the book as a result of 'investigation of things' (*gewu* 格物), the basis of Confucian intellectual and moral endeavor.[21] In fact, in his preface, he actually used the term *gewu* in expressing his respect for his friend and sponsor, Tu Shaokui 涂紹煃 (1582-1645), whose "sincere intention moved heaven, and [who with] his numinous mind 'investigated things.'" And commenting on people who mistake clamshell powder for oyster lime, he said that it was "because they did not investigate things." (2.11.54b) Also, he frequently used the expression '*li* of things' (*wuli* 物理), thus invoking the word '*li*' 理 (principle), the object of 'investigation of things,' the ultimate aim of which was to attain the '*li* of heaven' (*tianli* 天理). For example, speaking about the Sichuan people feeding the Zhejiang species silkworms with the leaves of a local tree when mulberry leaves are scarce, he said: "'The *li* of things' are one." (I.2.27a) Speaking of the difficult process of using fragmented jade pieces to fill the cracks in lutes, he said that it was "nearly impossible to understand 'the *li* of things' controlling and obeying each other." (III.18.62a)

---

19 Dagmar Schäfer has noted that Song Yingxing's stance on craft was based on theoretical reflections, not on practical experiences: Schäfer, "The Congruence of Knowledge and Action," 55.

20 Gillispie, "The Natural History of Industry." In fact, in the *Tiangong kaiwu* we frequently find the expression '*bowu*' 博物 (literally, 'broad [study of] things'), which is used in modern Chinese to translate 'natural history'. For example, in the preface Song Yingxing spoke of "the intelligent '*bowuzhe* 博物者' ('those who broadly [study] things')." He ended Chapter 1 on grains by saying, "How can a '*bowuzhe* 博物者' neglect them?" (I.1.22a). See also III.16.46b.

21 On the idea of 'investigation of things' (*gewu*), see Yung Sik Kim, *The Natural Philosophy of Chu Hsi* (*1130-1200*) (Philadelphia: American Philosophical Society, 2000), Chap. 2.

Sometimes Song Yingxing gave the impression that what he was reporting was the result of an examination of the actual procedures.[22] For example, the fact that Song Yingxing stated that he "could not examine in detail" the processes of making so-called 'dragon robes' (*longpao* 龍袍) (II.2.41a) may be taken as an implication that he must have examined the detailed processes of the other techniques, for which he therefore had no such regret. He made a similar comment about the "seventy-two processes necessary to make a porcelain cup": "I have not yet been able to thoroughly cover this in its minute details." (II.7.14a) Sometimes he seemed to be speaking of actual observations. For example, after recording various features about gunpowder, he said: "All of these can be discussed in detail after they have been tested by observation." (III.15.32a)

Song Yingxing even used the expression 'test and verification' (*shiyan* 試驗). For example, after mentioning different kinds of vegetable oils and different yields of oil per a given quantity of different vegetables, he said that he did not "investigate, test and verify" (*qiongjiu shiyan* 窮究試驗) all of them; for some he did test but was not entirely sure of his results (II.12.64b). Concerning many writings on gunpowder and firearms, he said that they "have not necessarily gone through 'test and verification.'" (III.15.31b) Again, these assertions seem to suggest that he did perform 'test and verification' on the other techniques, for which he did not add such comments.

It is difficult to pin down exactly what Song Yingxing meant by the word 'test and verification' (*shiyan*). It is not likely, however, that he had 'experiment' in mind.[23] Perhaps what he had in mind may be something similar to the idea of 'evidential study' (*kaozheng* 考證).[24] Indeed, in his preface, he expressed a regret that he could not verify, by means of books or actual objects, what he wrote from his own observations or from hearsay, and that he could not discuss them with other scholars, implying his belief that, in order to obtain proper

22 Pan Jixing has noted that some items in the book are based on an examination of the actual practice. See Pan Jixing, *Tiangong kaiwu jiaozhu ji yanjiu*, 20.

23 On the other hand, Christopher Cullen has mentioned the possibility that Song Yingxing actually made measurements: Christopher Cullen, "The Science/Technology Interface in Seventeenth-Century China: Song Yingxing on *qi* and the *wu xing*," *Bulletin of the School of Oriental and African Studies* 53 (1990):312.

24 Francesca Bray and Georges Métailié have noted that the method used by Xu Guangqi 徐光啓 (1562-1633) in compiling his agricultural treatise, the *Nongzheng quanshu* 農政全書, was that of evidential studies, rather than scientific experimentation: Francesca Bray and Georges Métailié, "Who Was the Author of the *Nongzheng quanshu*?" in *Statecraft & Intellectual Renewal in Late Ming China: The Cross-Cultural Synthesis of Xu Guangqi (1562-1633)*, ed. Catherine Jami, Peter Engelfriet, and Gregory Blue (Leiden: Brill, 2001), 342-343.

knowledge of the techniques, he should have done so. What Song Yingxing said he should have done is exactly the kind of things that were done by the scholars of evidential studies.[25]

2.4 *Miscellaneous Casual Knowledge and Critical Spirit*

The pages of the *Tiangong kaiwu* also contain a good deal of casual comments, observations, and records of readings and hearsay about all sorts of things, often without any technical significance. In fact, the passages discussed so far in this section are scattered throughout the book in an utter medley with this kind of casual miscellanea. To give some examples, Song Yingxing mentioned bans on the sales of saltpeter and sulfur (III.15.32a) and on the private use of the special tiles for palaces and temples (II.7.2b), and noted that the laws had tightened in order to prevent thefts of and fights over silver (III.14.4b). He spoke of merchants coming from Sichuan to sell silk fabric and buy pepper to take back (I.2.41a). He said that wine-makers must choose yeast sellers who are reliable and well known (III.17.49a). He mentioned the pearl gatherers of an area offering sacrificial services to a 'sea deity' (*haishen* 海神) every third lunar month (III.18.54a), and said that when pearl gatherers come out of the water, they must be wrapped in a warm blanket, for otherwise they may die of cold (III.18.54b). Speaking of the grinding mills turned by oxen, he said that blinders should be placed on the eyes of the oxen to prevent dizziness, and wooden pots should be suspended under their bellies to collect excretions (I.4.56a). While discussing the carts, he even said that the drivers should stop the carts when they come across pedestrians (II.9.40b).

Among such casual remarks are stories of strange things and events that today we would find quite difficult to believe. For example, Song Yingxing noted that while nothing happens when a person kills one or two bees, when more than three bees are killed, an entire group of bees swarm in to attack, calling it a 'bee rebellion' (*fengfan* 蜂反), or that when human urine is mixed with a substance emitted by the bees, the fragrance and sweetness are increased (I.6.79b). He related the story of exquisite red porcelain being produced after a potter jumped into a burning kiln; the story was later elaborated as the tale of a kiln that produced porcelain deer and elephants (II.7.14a). While discussing damage done to rice plants, he spoke of a 'ghost fire' (*guihuo* 鬼火) burning the plants, and provided a long explanation for it (I.1.6a). He also wrote of legendary swords turning into dragons (II.10.44a).

25 On the scholars of 'evidential studies,' see Benjamin A. Elman, *From Philosophy to Philology: Intellectual and Social Aspects of Change in Late Imperial China* (Cambridge, Mass.: Harvard University Press, 1984).

Sometimes Song Yingxing expressed doubts about such reports of mysterious events and strange beliefs. For instance, regarding stories of a 'mercury sea' (*honghai* 澒海) and of mercury being made from a certain type of grass, he called them "unfounded and absurd [stories], which [only] gullible people believe."(III.16.42a). He was especially critical of 'recipe books' (*fangshu* 方書), and 'recipe masters' (*fangshi* 方士). For example, he said that some of the claims the recipe books made about tin and arsenic were "absurdities." (III.14.20a) In an appendix to the section about so-called 'cinnabar silver' (*zhusha yin* 朱砂銀), he spoke of the "dishonest recipe masters deceiving people by [the alchemical techniques of] 'furnace and fire,'" and said that he added the appendix because foolish people were greedy and ignorant, and thus were easy prey for those recipe masters (III.14.7a). In fact, his desire to prevent people from being deceived by tall tales seems to have been part of his motivation to write the *Tiangong kaiwu* in the first place.

Furthermore, Song Yingxing frequently criticized the contents of the *materia medica* (*bencao* 本草). For example, after mentioning methods for obtaining silver, he stated that there were no other ways to make silver, and found it "extremely objectionable that the recipe books and the books of *materia medica* contain unfounded conjectures and notes." (III.14.6a) It is possible that the target of Song Yingxing's criticism was the *Systematic Materia Medica* (*Bencao gangmu* 本草綱目) by Li Shizhen 李時珍 (1518-1593), which was widely circulated among scholars of the time.[26] In fact, the *Tiangong kaiwu* borrowed quotations from the *Systematic Materia Medica* in more than twenty instances, although Song Yingxing did not explicitly mention the title of the book he was mostly criticizing.[27] It may have been that Song Yingxing wanted to present his own book as containing more accurate and reliable information than this celebrated text, which drew on numerous reports about strange things and events he could not accept.[28]

3 The Role of Heaven in Production Techniques, and Elevation of the Status of Knowledge about the Techniques

That heaven played a part in the production techniques is implied in the rhetorical question at the beginning of the preface to the *Tiangong kaiwu*: "How

26 Yabuuchi, "*Tenkō kaibutsu* ni tsuite," 14; Pan Jixing, *Tiangong kaiwu jiaozhu ji yanjiu*, 34.

27 Pan Jixing, *Tiangong kaiwu jiaozhu ji yanjiu*, 80, 103, 186.

28 Carla Nappi, *The Monkey and the Inkpot: Natural History and Its Transformations in Early Modern China* (Cambridge, Massachusetts: Harvard University Press, 2009), Chaps. 3-6.

can [all the numerous things in the world] be [due solely to] man's power?" Song Yingxing went on to make more explicit remarks about heaven's role in the main part of the book. Such comments can be found scattered throughout the book, but mostly in the introductory paragraphs of chapters (which begin with the phrase "Master Song said" [Songzi *yue* 宋子曰]). For example, in the introductory paragraph of Chapter 3 on dyes, he said "Heaven hung down images; the sages followed them and manifested the five colors [of heaven] into the five colors [of man]." (I.3.49a) The introductory paragraph of Chapter 4 on the preparation of grains begins and ends with remarks suggesting the role of heaven: "Heaven produces the five grains to nourish people...Is it possible that the one who made these is not heaven [disguised] in the form of man?" (I.4.53a). Indeed, Song Yingxing's main motivation may have been to show that heaven's role, or 'heaven's work' (*tiangong*), is manifest not only in such abstract ideas as human nature (*xing* 性), *li* 理, etc., but also in tangible objects and phenomena.

At times Song Yingxing used different expressions in referring to heaven's role such as the 'Creator of things' (*zaowu* 造物) instead of 'heaven' (*tian*). After pointing out that "although silk, hemp, fur, and wool all have plain uncolored qualities, [one can] make them precious by dyeing them with different colors," he added: "As for those who say that the 'Creator of things' did not care [enough to give these things different colors], I do not believe them." (I.3.49a) Having noted that silver is not produced within 300 *li* 里 (one *li* being about 400 meters) of a place yielding gold, and vice versa, he went on to say: "The 'feeling' (*qing* 情) of the 'Creator of things' can be seen abundantly [in this] too." (III.14.6a) Sometimes he used the more traditional expression 'creative transformation' (*zaohua* 造化)[29] instead of 'Creator of things.' While speaking about coal, he said: "Between metal and 'earth and stone,' 'creative transformation' also made this lump [of coal] appear." (II.11.55b). After saying that mercury cannot be recovered once it has been converted into vermilion, he wrote: "The skill of so-called 'creative transformation' has already been exhausted." (III.16.42a) After explaining that graphite and saltpeter produced a bright light, he added that this was placed at the end of the book on heaven's work because "this is what 'creative transformation' of the *qian-kun* 乾坤, [i.e. hexagrams corresponding to heaven and earth,] conveniently revealed on the earth's surface." (III.18.66a-66b)

---

29 The expression *'zaohua'* usually referred to the subtle, marvelous, mysterious aspect of some natural phenomena. But sometimes the word had a meaning quite close to that of 'Creator.' See Kim, *The Natural Philosophy of Chu Hsi*, 101.

Song Yingxing also brought in the expression 'the divine' (*shen*). For instance, the introductory paragraph of Chapter 1, in which he spoke of Shen Nong 神農 (Divine Agriculturalist), ends with the following remark: "The character '*shen*' is put before the character '*nong*' (agriculture) and is connected to it. How can [agriculture] be the result of man's power alone?" (I.1.1b). In the introductory paragraph of Chapter 9, he said that the person who built the first cart should be called a "divine man" (*shenren* 神人) (II.9.29a). After explaining the explosive effect of mixing sulfur and saltpeter in terms of the interaction of the essences of yin and yang, he said: "This is a 'divine thing' (*shenwu* 神物) that the *qian-kun* [i.e., heaven and earth] worked out fantastically." (II.11.60b) He also used the expression 'divine transformation' (*shenhua* 神化) in referring to the disappearance of coal once it is burned (II.11.55b).

One thing Song Yingxing aimed at by showing heaven's role in production techniques seems to have been an elevation of the status of knowledge about them.[30] Such intention can be seen too in his frequent citations of classical passages mentioning things related with the techniques he discussed in the *Tiangong kaiwu*. For example, in Chapter 1 on grains, Song Yingxing spoke of the names of grains appearing in the *Book of Poetry* (*Shijing* 詩經) and the *Book of Documents* (I.1.18b, 19a, etc.). Other ancient texts cited in the *Tiangong kaiwu* include the *Book of Changes* (*Yijing* 易經), the *Rites of Zhou* (*Zhouli* 周禮), the *Record of Rites* (*Liji* 禮記), the *Analects* (*Lunyu* 論語), the *Mencius* (*Mengzi* 孟子), the *Laozi* 老子, the *Zhuangzi* 莊子, the *Xunzi* 荀子, the *Hanfeizi* 韓非子, the *Liezi* 列子, the *Record of the Grand Historian* (*Shiji* 史記), the *Han History* (*Hanshu* 漢書), and the *Classic of Mountains and Seas* (*Shanhaijing* 山海經).[31] In fact, many of the chapter titles themselves came from classical passages. Song Yingxing also mentioned the commentaries of Zhu Xi (II.9.40a, II.10.46a, etc.). By making connections with the classics and with Zhu Xi in this manner, Song Yingxing was obviously trying to lift the intellectual status of knowledge about the techniques he was discussing.

---

30 This was what fifteenth- and sixteenth-century European writers on technical arts did as well. See, e.g., Paolo Rossi, *Philosophy, Technology and the Arts in the Early Modern Era* (New York: Harper, 1970), Chap. 1. Song Yingxing's contemporary Wang Zheng 王徵 (1571-1644) also showed a similar tendency in the *Yuanxi qiqi tushuo luzui* 遠西奇器圖說錄最 [Selected Records of the Diagrams and Explanations of the Strange Machines from the Far West]. See Yung Sik Kim, "A Philosophy of Machines and Mechanics in Seventeenth-Century China: Wang Zheng's Characterization and Justification of the Study of Machines and Mechanics in the *Qiqi tushuo*," *East Asian Science, Technology, and Medicine* 31 (2010): 64-95.

31 Pan Jixing has identified many such citations: Pan Jixing, *Tiangong kaiwu jiaozhu ji yanjiu*, 175-183.

4 Heaven and Man: Production Techniques and Natural Phenomena

The actual patterns in which heaven's role is evident in Song Yingxing's discussion of production techniques reflect his views about the relationship between the techniques, natural phenomena, heaven, and man.

4.1 *Man's Role*

Song Yingxing's basic position was that heaven prepares the techniques and transmits them to man; man uses them. To illustrate, when speaking of the weaving technique in Chapter 2, he noted that "the looms of Heaven's Grandchild (*Tiansun* 天孫)[32] transmitted the skill [of weaving] to man. Man saw figured patterns in the original materials, and then embroidered and washed [the material] to get silk [clothes]." (I.2.23a) Later in the chapter, he stated: "With the looms of Heaven's Grandchild, the skill of man has become complete." (I.2.39b)

Of course, Song Yingxing did not deny the role of man in production techniques. At times he even admitted man's role in creating the techniques. When heaven does not perform the work necessary to produce things, man produces them. For example, when it does not rain from the heavens, man has to draw water to irrigate the land (I.1.8b). Also, things are not produced by themselves; man has to perform some tasks in order to get them.[33] In particular, Song Yingxing spoke of the mutual dependence of man and crops: "Men cannot live long [by themselves]; the five grains sustain them. The five grains cannot grow themselves; men grow them." (I.1.1a) He also noted that the oil inside plants does not flow out by itself, but man has to do something – using the power of water, fire, or tools made of wood and stone – to cause it to flow out (II.12.63a).

In Song Yingxing's view, however, man's role is to be found mainly in using the techniques. Man's role is especially apparent in his selection of different methods according to different conditions. For instance, man chooses to take lesser or greater quantities of raw materials, depending on his needs. Thus, speaking about the so-called 'ten-thousand men killer' (*wanrendi* 萬人敵) bomb, Song Yingxing said that the proportions of 'poison gunpowder' (*duhuo*

32 Here Song Yingxing is referring to Zhinü 織女, a mythological female figure associated with weaving, recorded as 'Heaven's Grandchild' in the astronomical treatise (*Tian'guan shu* 天官書) of the *Record of Grand Historian* (*Shiji*).

33 This aspect must have been what made Pan Jixing conclude that the notion that man cannot depend completely on the natural world, but has to do something himself, was 'the leading idea' of the *Tiangong kaiwu*: Pan Jixing, *Tiangong kaiwu jiaozhu ji yanjiu*, 70. This idea can be traced back to the above-mentioned phrase from the *Book of Documents*, "It is heaven's work; man carries it through."

毒火) and 'divine gunpowder' (*shenhuo* 神火) to be mixed are varied flexibly by man (III.15.35a). In making inkstones, the quantities of precious raw materials to be used are also man's decision (III.16.46b). Man chooses the timing of various processes in the production of things. For example, in southern China where there is no frost, the timing of the sugarcane harvest is decided by man (I.6.76a). Man is also able to prevent damage from occurring while using the techniques. In raising silkworms, for example, the task of preventing damage caused by dampness, heat or pressure lies with man (I.2.27b). Also, it is man who chooses among the different methods for deterring birds, rats, and mosquitoes, which spoil silkworms (I.2.29b).

When referring to these roles man plays in the use of techniques, Song Yingxing frequently used the expression 'man's work' (*rengong* 人工), in addition to the expression 'man's skill' (*renqiao* 人巧). For instance, when speaking of the techniques used to produce different ceramic wares suitable for the soils of different localities, he said: "Man's work exhibited differences, and produced good ceramic wares." (II.7.1a) Speaking of tiles produced in various shapes of birds and animals, he said: "All these are made one by one through man's work." (II.7.2a) And after telling of a complicated, ingenious method of extracting pure silver from impure silver, he said: "Bits of both man's work and heaven's work are visible." (III.14.6b)

4.2 *Man, the Sages, and Heaven*

In the last example, the expression 'heaven's work' appears together with 'man's work.' In saying this, what Song Yingxing obviously had in mind was some sort of comparison between man's work and heaven's work. Indeed, he frequently underlined the greatness and excellence of man's work or skill. For example, he said that man was skilled enough to make rice jelly using a thousand different methods (I.6.80b). While speaking of a kind of water mill that can simultaneously perform three different tasks, he said that it was made by "someone whose mind's deliberation leaves out nothing." (I.4.55a) He even used the expression 'divine achievement' (*shengong* 神功) in referring to the techniques of forging metals (II.10.44a).

It was natural then that Song Yingxing, who saw such great, superior qualities in man's work in the preparation and use of production techniques, should attribute the origins of some key techniques to the ancient sages (*shengren* 聖人).[34] An example is when, speaking of dyeing in the introductory paragraph

34 In this Song Yingxing was following the tradition of the "Commentary on the Appended Words" (Xici zhuan 繫辭傳, Chap. B2) of the *Book of Changes*, which attributed the origins of key institutions and techniques to the ancient sages.

of Chapter 3, he said: "Heaven hung down images; the sages followed them and manifested the five colors [of heaven] into the five colors [of man]." (I.3.49a) In the introductory paragraph of Chapter 8 he attributed the origin of mining and casting metals to sages such as the Yellow Emperor (Huangdi 黃帝) and King Yu 禹 (II.8.17a). In the introductory paragraph of Chapter 17, he pointed out that yeasts required in the making of liquors were made by sages like Yandi 炎帝 and the Yellow Emperor (III.17.48a), implying obviously that ordinary men cannot reach the level of Yandi or the Yellow Emperor. In his attribution of the invention of paper-making to the high ancient period instead of Han times (206 BC–AD 220), one can further see his desire to associate it with the sages (II.11.70a).

It is significant that Song Yingxing chose to attribute the origin of the techniques to the sages, and not to ordinary men. The sages, although human, were endowed with superior qualities that were close to heaven – the 'divine', 'creative transformation', and the 'Creator of things.' Indeed, the skill and intelligence involved in various production techniques at such a level seem to have made Song Yingxing suspect that they could not be of a solely human nature. In the introductory paragraphs to many chapters, he implied that there must be nonhuman agents possessing such superior qualities and performing such marvelous feats involving the techniques discussed in the chapters. For example, after speaking of the great advances in weaponry, he said: "Even though man has a skillful mind, how could he possibly reach such a height?" (III.15.25a-25b), implying that perhaps some other agents must have intervened. After mentioning different kinds of metals satisfying different human needs, he wrote: "As for differentiating between good and bad ones, and indicating their weights, who established beforehand [that there would be so many different kinds of metals] and made man depend on them forever?" (III.14.1a). It is possible that here again he was suggesting the possibility of a non-human agent. After mentioning the great variety of metal goods produced by metal casting techniques, he said: "Who can count them all? In short, man's ability cannot reach this [level]." (II.8.17a) Having illustrated various methods of extracting oil from plant seeds, he added: "As for such human skill and intelligence, I do not know what endowed man with them." (II.12.63a) In a discussion on bees collecting honey from flowers, he asked: "Who was in charge of this and grew and spread them [i.e. honey bees] all over the world?" (I.6.74a). After pointing out that every color is available to man, he asked who other than the 'extreme divine' (*zhishen* 至神) can provide this (III.16.40a). His belief that even the sages cannot stop wars implies that they are beyond man's will (III.15.25a).

Sometimes Song Yingxing was more explicit in suggesting that heaven was responsible for such supreme abilities and skills. Speaking of those who made such tools as the mortar and pestle, he said: "Is it possible that the one who made these is not heaven [disguised] in the form of man?" (I.4.54a). After illustrating an exquisite process for obtaining mercury, he remarked: "This is a most sublime and mysterious transformation. Everything is heaven's 'mechanism' (*ji* 機)." (III.16.41b)

Song Yingxing emphasized that heaven's ability apparent in many production techniques is so excellent, mysterious, and abundant that man's skill, however great it may be, cannot compare. For instance, while discussing calcination techniques of stones, he spoke of the examples of changes appearing in nature as the result of heat and fire, such as the colors of alums and the powers of sulfur. He then added that even the great skills of the recipe masters cannot reach even one ten-thousandth of heaven's work (II.11.53a). Thus, in Song Yingxing's view, man cannot completely know heaven's work, or heaven's ability. In fact, after noting that some form of salt is produced everywhere in the world, he wondered: "Who knows the reason why it is so?" (I.5.66a). When speaking of the red fire turning things black and white mercury changing into deep red, he said: "As for the smelting and forging brought about by 'creative transformation,' how could [man's] thought possibly contain them?" (III.16.40a). After mentioning the precious gems gathered by man, he asked: "Can the brilliant essences of heaven and earth stop at these few?" (III.18.53a), hinting again that what men know is only part of the extreme abundance of heaven and earth.

4.3 *Heaven's Role of Taking Care of Man: Natural Theology of Production Techniques?*

One aspect of heaven's great ability that Song Yingxing highlighted was that heaven – or the 'Creator of things' or 'creative transformation' – creates a great variety of things. He frequently spoke of the opposite qualities of the things created by heaven and earth. For example, while discussing various metals, he said: "Generally, when heaven and earth produce things, those things that are bright are the opposite of those that are dark and turbid; those things that are moist are the opposite of those that are withered and harsh. If noble things are here, lowly things are there." (III.18.53a) He ended the introductory paragraph of Chapter 2 on clothes with the following remark: "In general, man and things go together; the noble and the lowly contrast each other. It is really heaven that does this." (I.2.23b) While speaking of coal, as we have seen, he said that "between metal and 'earth and stone,' 'creative transformation' also made this lump [of coal] appear." (II.11.55b)

Song Yingxing noted that such distinctions in the natural world can be found in the human world as well. In fact, the main point of the introductory paragraph of Chapter 3 on colors was that just as there are different colors in nature, so there are also distinctions in the colors of clothes people wear (I.3.49a). In this manner, the distinctions in nature were analogically extended to the human world. He used the expression "its meaning is also like this" to refer to such analogical extension. Sometimes, however, the analogical extension started from the human world, which reflected its pattern upon nature. For example, he noted that in the same way that human order was established owing to the existence of the ten classes among men, "the great earth [also] produced the five metals so that they could be used all under heaven and in posterity." He then added the same words, "Its meaning is also like this," implying that there is a distinction of the noble and the lowly among metals too (III.14.1a). He even said that "the 'Creator of things' prepared different clothing materials to distinguish between noble and lowly men, and between men and animals." (I.2.23a).

Indeed, in those cases where Song Yingxing was illustrating ways in which heaven took care of man, he frequently used the expression 'Creator of things.' Song Yingxing once even said directly that "the 'Creator of things' cared [about man]," after pointing out, in the passage we have already seen, that clothing materials can be dyed in many different colors (I.3.49a). Speaking of fragrant, colorful, and sweet things that are greatly desired by men, he said: "The 'Creator of things' had the idea of making these particularly stand out." (I.6.74a).

One aspect of heaven's concern for man that Song Yingxing emphasized was that heaven makes sure that whatever is needed by man is produced somewhere and somehow. The best example was salt, which is essential for life. Song Yingxing noted that while there are places that do not produce grains or vegetables, there is no place on earth where no salt is made. In the introductory paragraph of Chapter 5 on salt, he first pointed out that while man can go for years without foods with a bitter, sour, sweet, or acrid taste, man cannot live without salt even for ten days, and then said that therefore heaven made salt the source of man's 'life *qi*' (*shengqi* 生氣). He added: "It cleverly comes into being [everywhere] and waits [for man's use]." (I.5.66a) Speaking of different forms of salt, he noted that in places far from the sea where sea salt is rare, other forms of salt are produced for man's use, or transported from other places. In the places where it is difficult to carry salt by boats or carts, "the 'Creator of things' produces [salt somehow] according to the situation." (I.5.66b) Similarly, he noted that in coastal areas where no limestone has been deposited, heaven produces oyster shells instead (II.11.54a). In places where there is no coal, plants are abundant, from which "the subtlety of heaven's mind is seen." (II.11.55b) And as we have seen, after noting that silver is not produced within 300 *li* of a place yielding gold, he said "the feeling of the

'Creator of things' can be seen abundantly [in this] also." (III.14.6a) He even noted that the number of honey-producing bees is lower in places producing sugarcane, the implication clearly being that heaven or the 'Creator of things,' although Song Yingxing did not explicitly name either, is behind this to make sweet things available everywhere (I.6.79a).

The 'natural theological' sentiment is unmistakable in these examples showing Song Yingxing's belief that heaven took care of man. We might ask what the source of this belief was. Would it have been the Christian idea of God? Given the widespread dissemination of Christian doctrines among late Ming thinkers, it is not impossible that Song Yingxing picked up such a natural theological sentiment from this intellectual climate.[35] But any answer to this question will remain speculation until we have come to know a lot more about the details of Song Yingxing's life.

5 Concluding Remarks

Song Yingxing's basic idea about the roles of heaven and man in production techniques was that while man does play a role in producing many things, heaven lies behind all this and is the ultimate source of everything. When man uses the various techniques, he is in reality relying on the work of heaven. Not only was the importance of heaven's role highlighted in the introductory paragraphs to each chapter, but it was also reflected in the title of the book itself: "Heaven's Work in Opening Things."

35 Christopher Cullen has noted that Song Yingxing "certainly had some contact with Jesuit teaching." Cullen, "The Science/Technology Interface in Seventeenth-Century China," 316. Xu Guangtai 徐光台 (Hsu Kuang-t'ai) has suggested a possible connection between Song Yingxing and Xiong Mingyu 熊明遇 (1579-1649), an influential scholar with a favorable attitude towards Catholicism: Xu Guangtai 徐光台 (Hsu Kuang-t'ai), "Mingmo xixue yu Bailutong shuyuan" 明末西學與白鹿洞書院 [The White Deer Grotto and the Western Learning in Late Ming], Gao Feng 高峰, Hu Qing 胡青, Lai Gongou 賴功歐, Ye Cunhong 葉存洪 ed., *Zhongguo shuyuan luntan* 中國書院論壇 (Beijing: Zuojia chubanshe 作家出版社, 2007), 370-384, esp. 378-379. It is also highly likely that Song Yingxing read Matteo Ricci's (1552-1610) *The True Meaning of the Lord of Heaven* (*Tianzhu shiyi* 天主實義), which was widely circulated among the Chinese scholars of the late Ming, and which contained many remarks of a natural theological character. Ricci said, for example: "Considering Heavenly Lord's producing this heaven and earth and these myriad things, there is not a single thing that He did not create for man's use." Zhu Weizheng 朱維錚 ed., *Li Madou Zhongwen zhuyiji* 利瑪竇中文著譯集 [Collection of Matteo Ricci's Writings and Translations in Chinese] (Hong Kong: City University Press, 2001), 69.

This idea of heaven's role in production techniques, however, was not something entirely new. It was essentially in agreement with the traditional Chinese concept of heaven creating everything in the world: man and the ten thousand things. Even the natural theological notion that heaven prepared everything for the sake of man did not go beyond the fundamentally man-centered world-view of the Confucian tradition. What did stand in conflict with Song Yingxing's ideas about heaven's role in production techniques was the established convention in traditional China of associating the supreme level of human affairs with the human sages and that of natural phenomena with the 'divine.' While traditional Chinese generally accepted that the ancient sages were responsible for establishing key institutions and techniques, they usually reserved terms like 'divine' and 'heaven' for mysterious, awe-inspiring natural phenomena.[36] In view of this convention, Song Yingxing's association of 'heaven', 'divine', etc. with human activities of production techniques was a departure from the tradition.

Song Yingxing seems to have done this because he wanted to demonstrate that heaven's role can be seen not only in such abstract and sublime ideas as human nature (*xing*), *li*, and mind (*xin* 心), valued in the learning for the civil service examination. Heaven also plays a role in concrete, practical, and lowly things such as everyday commodities and in the techniques used to produce them.

---

36 Willard J. Peterson, "Making Connections: The 'Commentary on the Attached Verbalizations' in the *Book of Changes*," *Harvard Journal of Asiatic Studies* 42 (1982): 67-116; Kim, *The Natural Philosophy of Chu Hsi*, Chaps. 6-7.

CHAPTER 4

Science and the Confucian Tradition in the Work of Chŏng Yak-yong*

There is no single subject to be identified as 'science' or 'science and technology' for late Chosŏn Confucian scholar Chŏng Yak-yong (丁若鏞 1762-1836). What is usually regarded as such was a collection of many separate things: 1) basic natural philosophical ideas, usually involving such concepts as *li* 理, *qi* 氣, yin and yang 陰陽, the five phases (*wuxing* 五行), etc.; 2) knowledge about heaven and earth (*tiandi* 天地), the myriad things (*wanwu* 萬物), man (*ren* 人), etc., usually on a very general level; 3) specialized knowledge in such subjects as calendrical method (*lifa* 曆法), harmonics (*lü* 律), medicine (*yi* 醫), geography (*dili* 地理), etc. (This genre also includes areas that we do not consider 'scientific': e.g., the bureaucratic system [*guanzhi* 官制], criminal justice [*xingyu* 刑獄], finances and taxes [*caifu* 財賦], the military [*junlü* 軍旅], etc.); 4) knowledge about various techniques; and 5) such borderline ideas and practices as *guishen* (鬼神), occult techniques (*shushu* 術數), etc., which touch upon natural phenomena.

From studies carried out by many scholars so far, we more or less know what Chŏng Yak-yong knew and wrote about these topics, and what his attitudes to them were.[1] But what did 'science' mean for him? What weight, in other words, did he give to his own knowledge and thoughts about these topics? In taking up this difficult question in this paper, however, I will not attempt to answer it. Instead, I will try to put the question in context, both of Chŏng Yak-yong's life and work and of the society he lived in. In doing so, I shall raise a number of related questions, consideration of which will hopefully throw light on our attempts to answer the original question, and will help us better understand Chŏng Yak-yong's thought and his place and significance in Korean history.

* Originally Published as "Science and the Confucian Tradition in the Work of Chong Yagyong," *Dasanhak* 茶山學, no. 5 (2004): 127-168.

1 Pak Sŏng-rae 朴星來, "Chŏng Yak-yong-ŭi kwahak sasang 丁若鏞의 科學思想", Yun Sa-sun, ed. 尹絲淳 編, *Chŏng Yak-yong* 丁若鏞 (Seoul: Koryŏ daehakkyo chulp'anbu 高麗大學校出版部, 1990), 327-350; Kim Yŏng-ho 金泳鎬, "Chŏng Tasan-ŭi kwahakkisul sasang" 丁茶山의 科學技術思想, *Tongyanghak* 東洋學 19 (1989): 277-300; Ko Pyŏng-ik 高柄翊, "Tasan-ŭi chinbo gwan" 茶山의 進步觀 in *Chŏng Yak-yong*, Yun Sa-sun, 313-326.

 | DOI 10.1163/9789004265318-006

1 Chŏng Yak-yong's Scientific and Technical Knowledge

Much is known about Chŏng Yak-yong's knowledge of the natural world and techniques. Most notable is the weight-lifting machine he developed, which was used and said to have proved very efficient in the construction work of the Hwasŏng (華城) castle. Also known are his trials of a smallpox vaccination method, his advocacy for the unification and standardization of weights and measures, the adoption of the decimal system in particular. We are also familiar with his belief in the continuing advancement of techniques, and his proposal of policies for attracting craftsmen ("*lai baigong*" 來百工) and importing advanced techniques. He proposed, for example, to send technical specialists, including officials in the Astronomical Bureau, to Beijing for training. Also well known are his criticism of the five-phase theory,[2] his rejection of the stem-branch (*gan-zhi* 干支) system and the various occult techniques based on it, as well as his objection to superstitious beliefs and practices, including divination, astrology, and the so-called technique of 'winds and waters' (*fengshui* 風水). We also know that he accepted the idea of a spherical earth, and that he accepted the three-way distinction between plants that have knowledge (*zhi* 知), animals that have knowledge but are not numinous (*ling* 靈), and man who both has knowledge and is numinous.[3] His discussions of certain optical phenomena – the bending and deflection of light, the lens, the camera effect, and man's eyesight – are also well known.

Of course, there are items of particular scientific and technical knowledge Chŏng Yak-yong had which are not so well known. For example, he had certain ideas about human reproduction: "comparing it to plants, the father is the seed, while the mother is the soil."[4] He believed nevertheless that man receives his *qi* from his maternal grandfather as well.[5] Regarding reproduction in general, Chŏng Yak-yong was of the view that only those things that have seeds can "keep coming into being ceaselessly." Things like water, for example, do not

---

2 For Chŏng Yak-yong's criticism of the five-phase theory, see Kim Mun-yong 金文鎔, "Chosŏn sidae Yuhakchadŭl-ŭi Ŭmyang ohaeng-ron" 朝鮮 時代 儒學者들의 陰陽五行論 in *Chosŏn yuhak-ŭi chayŏn ch'ŏrhak* 朝鮮 儒學의 自然哲學, ed. Han'guk sasangsa yŏn'guhoe, 韓國思想史硏究會, (Seoul: Yemun sŏwŏn 藝文書院, 1998), 333-339.

3 *Nonŏ kokŭm chu* 論語古今註, Chap. 9, in *Yŏyudang chŏnsŏ* 與猶堂全書 (Seoul: Yŏgang ch'ulp'ansa 驪江出版社, 1985), Part II 第2集, vol. 15 第15卷, 11a-11b (hereafter cited as "*Yŏyudang chŏnsŏ* II.15.11a-11b"). See also *Chungyong kang'ŭi po* 中庸講義補: *Yŏyudang chŏnsŏ* II.4.47a.

4 "Sang chungssi" 上仲氏: *Yŏyudang chŏnsŏ* I.20.19a. He goes on to say that it is the "seed" that determines the species.

5 Ibid. I.20.22b.

have seeds, and thus cannot keep coming into being.[6] He was also of the view that "what heaven made is all circular, and what man made is all square."[7] He had explanations for the sea tides. He rejected the popular theory that views the tides as "the respiration of heaven and earth," and instead attributed it to movements of the moon. He believed that the strength of the tides is greater near the equator, and provided an explanation for this.[8] He also believed that salt water could turn into fresh water, which is the reason why inland water stays fresh despite the presence of holes inside the earth connecting the sea-water to the inland water.[9] His long discussion with King Chŏngjo 正祖 (r. 1776-1800) shows that he had some interest in, and knowledge about, the names of rare animal species.[10] He also knew of various methods of dyeing cloth, and said that he had actually tried the methods himself.[11] This, of course, does not exhaust all the scientific and technical knowledge found in his corpus.

2 The Confucian Tradition for Chŏng Yak-yong

To place this aspect of Chŏng Yak-yong's knowledge into proper perspective, we should keep in mind that his main intellectual concern aligned with the Confucian tradition. As was the case with most Confucian scholars of the Chosŏn 朝鮮 period (1392-1910), the neo-Confucian philosophical system of Zhu Xi 朱熹 (1130-1200) was the basis for his entire thought.

To be sure, Chŏng Yak-yong's views deviate from Zhu Xi's on some important philosophical points, as has been discussed and emphasized by many modern scholars.[12] To mention only a few salient examples, Chŏng Yak-yong rejected the identification of 'heaven' (*tian* 天) and 'human nature' (*xing* 性) with *li*

6 *Chungyong kang'ŭi po*: *Yŏyudang chŏnsŏ* II.4.52a.

7 "Mun tongsŏ nambuk" 問東西南北: *Yŏyudang chŏnsŏ* I.9.2b.

8 "Haejo tae" 海潮對: *Yŏyudang chŏnsŏ* I.22-20b-21b. Of course, his explanations of tides would not find acceptance today.

9 *Chungyong kang'ŭi po: Yŏyudang chŏnsŏ* II.4.52a.

10 Ibid. II.4.35a-b. The animal names are mostly those mentioned in the *Classic of Poetry* (*Shijing* 詩經).

11 "Sang chungssi": *Yŏyudang chŏnsŏ* I.20.20b.

12 E.g. Han Hyŏng-cho 韓亨祚. *Chu Hi-esŏ Chŏng Yak-yong-ŭro: Chosŏn yuhak-ŭi ch'ŏrhak-chŏk p'aerŏdaim yŏn'gu* 朱熹에서 丁若鏞으로 – 朝鮮 儒學의 哲學的 패러다임 研究 (Seoul: Segyesa 世界社, 1996).

理.[13] In particular, he had the idea of 'heaven' as a personal deity, which led him to use the word 'high heaven' (*shangtian* 上天) and such expressions as 'to serve heaven' (*shi tian* 事天). The separation of heaven from *li* led him to believe that morality is not innate to man but is something to be acquired through great effort.[14] He also rejected Zhu Xi's view of *guishen* based on *li* and *qi*,[15] saying flatly that "*guishen* cannot be discussed in terms of *li* and *qi*."[16] He rejected Cheng Yi's (程頤 1033-1107) characterization of *guishen* as 'traces of creative transformation' (*zhaohua zhi ji* 造化之跡) and Zhang Zai's (張載 1020-1077) as the 'innate ability of the two *qi*' (*erqi zhi linagneng* 二氣之良能), which had formed the basis of Zhu Xi's ideas about *guishen*.[17] He went as far as to implicitly identify *guishen* with heaven.[18] On the relationship between *li* 理 and things (*wu* 物) as well, Chŏng Yak-yong differed from Zhu Xi. He separated *li* from things, and rejected Zhu Xi's assertion that "when *li* is exhausted, things are exhausted," because he believed that *li* was not something that could be exhausted.[19] He maintained a clear distinction between man and animals, and criticized Zhu Xi for "speaking about man and things together."[20] According to Chŏng Yak-yong, unlike other creatures that continue to be born owing to "the *li* of coming into being continuously" (*shengsheng zhi li* 生生之理) endowed to them at the beginning of the world, each man receives his own 'numinous brightness' (*lingming* 靈明) at the time of his birth.[21] This led him to refute the idea that "the myriad things are one body" (*wanwu yiti* 萬物一體) and to reject the belief that morality can be derived from the natural world.[22]

Yet, in spite of all this, Chŏng Yak-yong stayed within Zhu Xi's neo-Confucian framework. His was simply a modification and adjustment to Zhu Xi's sys-

---

13 Chŏng Yak-yong himself mentioned three main defects of the Confucians of his time: *Taehak kang'ŭi* 大學講義: *Yŏyudang chŏnsŏ* II.2.40a.

14 E.g., Kŭm Chang-t'ae 琴章泰, *Tasan Chŏng Yak-yong* 茶山 丁若鏞 (Seoul: Sŏnggyun'gwan daehakkyo ch'ulp'anbu 成均館大學校出版部, 1999), 290.

15 On Chŏng Yak-yong's views of *guishen*, see Kim Hyŏn 金炫, "Chosŏn yuhak-esŏŭi kwisin kaenyŏm" 朝鮮 儒學에서의 鬼神 概念 in Han'guk Sasangsa yŏn'gu-hoe, *Chosŏn yuhak-ŭi chayŏn ch'ŏrhak*, 349-418, especially 403-413. For Zhu Xi's ideas of *guishen*, see Yung Sik Kim, *The Natural Philosophy of Chu Hsi (1130-1200)* (Philadelphia: American Philosophical Society, 2000), Chap. 6.

16 *Chungyong kang'ŭi po: Yŏyudang chŏnsŏ* II.4.23a.

17 E.g., Ibid. II.4.20b-22b.

18 Ibid. II.4.21a.

19 Ibid. II.4.49b-50a.

20 Ibid. II.4.4a.

21 Ibid. II.4.2b.

22 E.g., Kim Yŏng-ho, "Chŏng Tasan-ŭi kwahakkisul sasang," 281-282.

tem; he considered his work not as a departure from, but as a recovery of, the true Confucian Way. Chŏng Yak-yong never wavered in his respect for Zhu Xi and in fact gave him only the highest praises. He said, for example, that Zhu Xi's sayings could not be doubted because they had come from "the essential and subtle (*jingwei* 精微) parts of his life experience."[23] He characterized Zhu Xi's criticisms and corrections of earlier scholars and texts as showing 'a great insight' (*da yanmu* 大眼目).[24] While commenting on a phrase in the *Doctrine of the Mean* (*Zhongyong* 中庸), he said that it was Zhu Xi who had corrected "the extremely shallow explanation" in the standard commentary by Zheng Xuan 鄭玄.[25] At times, when speaking of past achievements, the only two men that he deemed worthy of mention were Confucius and Zhu Xi. For example, the two great achievements in the *Yijing* 易經 tradition that Chŏng Yak-yong brought up were the composition of the Wings (*yi* 翼) by Confucius and the commentary to the *Cantongqi* 參同契 by Zhu Xi.[26] He said that "since the Han, *Yijing* studies have largely been completed by Master Zhu. Many excellent words and the supreme *li* are found in [his *Zhouyi*] *benyi* 周易本義."[27] Moreover, as for the *Rites of Zhou* (*Zhou li* 周禮), he said that "only Zhu Xi both understood and trusted [the *Rites of Zhou*]," whereas earlier scholars either had not understood it or did not trust it.[28]

Chŏng Yak-yong repeatedly said that Zhu Xi was free from the shortcomings of current scholarship, or that he rectified such defects.[29] Such was his respect for the Southern Song master that even while rejecting Zhu Xi's remarks, he sometimes tried to explain what Zhu Xi had really meant, or why Zhu Xi had said such things. Thus, after rejecting Zhu Xi's view that "the natures of both man and things are the same," Chŏng Yak-yong provided an explanation: "What Master Zhu meant when he said that 'the natures of both man and things are same' was to say that what both man and things originally received from heaven are the same."[30]

It is natural, then, that Chŏng Yak-yong expressed his dissatisfaction with the widespread criticism of Zhu Xi. For example, he criticized the Qing scholar Mao Qiling 毛奇齡 (1623-1716) for "collecting the leftover words of the ancients

23 *Chungyong kang'ŭi po: Yŏyudang chŏnsŏ* II.4.23b.

24 "Tap Chungssi" 答仲氏: *Yŏyudang chŏnsŏ* I.20.18b, 19a.

25 *Chungyong kang'ŭi po: Yŏyudang chŏnsŏ* II.4.49a. The phrase in question is "*siti* 四體" in the *Zhongyong*, Chap.4.

26 "Ohak ron" 五學論: *Yŏyudang chŏnsŏ* I.11.23b.

27 *Yŏkhak soon* 易學緒言: *Yŏyudang chŏnsŏ* II.46.14a.

28 "Tap Chungssi": *Yŏyudang chŏnsŏ* I.20.15b.

29 E.g., "Ohak ron": *Yŏyudang chŏns* I.11.20a, 20b.

30 *Chungyong kang'ŭi po: Yŏyudang chŏnsŏ* II.4.47b.

to block the great achievements of Mater Zhu,"[31] and the Tokugawa Confucian Ogyū Sorai 荻生徂來 (1666-1728) for having "the malady of a bad temper," as the latter "always desired to oppose Zhu Xi's theory."[32]

Indeed, Zhu Xi seemed to have been a model for Chŏng Yak-yong. In thinking and speaking about numerous things, he constantly had Zhu Xi in mind. Talking about his own illnesses and regrets, for example, Chŏng Yak-yong said, "This was what Master Zhu regretted in his later years. How can I not be concerned?"[33]

This makes one wonder how Chŏng Yak-yong's views could have possibly deviated from Zhu Xi on such key issues. It is significant in this connection, however, that those ideas of Chŏng Yak-yong's that differed from Zhu Xi's were of the kind that could still be accommodated within the broad scope of Zhu Xi's ideas. Indeed, Chŏng Yak-yong's views rarely diverged so greatly from Zhu Xi's that they could not possibly be embraced by Zhu Xi's very broad and flexible framework, at least in principle. Take the most frequently discussed example of Chŏng Yak-yong's identification of heaven as the 'Lord on High' (*shangdi* 上帝), and not as *li* 理. His concept of the 'Lord on High' did not bear much resemblance to that of the Christian deity but rather was almost exclusively moralistic, wholly in line with Confucian morality. Indeed, it is not impossible to subsume the notion of the 'Lord on High' within the concept of *li*.[34] Moreover, although it is true that the notion of a personal deity did exist in ancient Confucianism, and although it is likely that Chŏng Yak-yong saw its resemblance to the Western religious counterpart and synthesized the two,[35] it would be unfair to deny the presence of this religious aspect of ancient Confucianism in Zhu Xi.[36] Furthermore, despite the fact that Chŏng Yak-yong saw a separation between man and things, and criticized Zhu Xi for merging the two, as we have seen, his ideas about non-human entities were not entirely

31 "Che Mo Taega chamo yŏkkwae tosŏl" 題毛大可子母易卦圖說: *Yŏyudang chŏnsŏ* I.14.41a.

32 *Nonŏ kokŭm chu: Yŏyudang chŏnsŏ* II.9.5a.

33 "Tap Chungssi": *Yŏyudang chŏnsŏ* I.20.24a.

34 See Sŏng T'ae-yong's 成泰鏞 comments in a roundtable discussion: Choadam 座談, "Tasan, Chujahak, kŭrigo Sŏhak" 茶山, 朱子學, 그리고 西學, *Tasanhak*, no. 2 (2001): 271.

35 E.g., Han Hyŏng-cho 韓亨祚, "Tasan-kwa Sŏhak: Chosŏn Chujahak-ŭi yŏnsok-kwa tanjŏl" 茶山과 西學: 朝鮮 朱子學의 連續과 斷絶, *Tasanhak* 茶山學, no.2 (2001): 142; Don Baker, "A Different Thread: Orthodoxy, Heterodoxy, and Catholicism in a Confucian World," in *Culture and the State in Late Chosŏn Korea*, ed. Jahyun Kim Haboush and Martina Deuchler (Cambridge, Mass.: Harvard University Press, 1999), 199-230.

36 E.g., Julia Ching, *The Religious Thought of Chu Hsi* (Oxford: Oxford University Press, 2000).

free from the human-centeredness that characterized traditional Confucian thinking. He said, for example, that "of the sun, the moon, and the stars...and the plants and the animals,...there is nothing by which man is not shone upon, warmed, nourished, or served."[37] Similarly, although Chŏng Yak-yong challenged the view, generally attributed to Zhu Xi, that one has to investigate all things of the world to attain a proper understanding of *li*,[38] such criticisms were also echoed by Zhu Xi himself, who, for example, endorsed Cheng Yi's remark that it was actually not necessary to investigate everything, for "after one has accumulated and familiarized oneself with many [things], suddenly and spontaneously there will be a 'thorough penetration.'"[39] Even Chŏng Yak-yong's emphasis on practical utility should not be considered a move away from the Zhu Xi tradition either, for this was an essential, and persistent, element of Zhu Xi Confucianism.

Nor did Chŏng Yak-yong's criticism of Zhu Xi necessarily lead him to a more modern, objective attitude toward knowledge; in some respects, he became more intuitionistic, or more inclined towards 'the learning of the mind' (*xinxue* 心學), than Zhu Xi had been, as is evident in his rejection – as we have just seen – of the necessity to exhaustively investigate everything in the world. Also, in spite of his general distaste for all things superstitious and mystical, Chŏng Yak-yong was not able to completely divest himself of such characteristics. For example, his separation of heaven and man was not as thorough as he claimed. He once said, for example, that "although [the Way of heaven and the Way of man] each has its own meaning, the ways are not two."[40] Thus, one has to "know heaven" first to be able to "choose what is good."[41] While discussing ideas like *guishen* and the 'bright *shen* of heaven' (*tian zhi mingshen* 天之明神), he even said that those things are "clear evidence of the fact that the myr-

37 *Nonŏ kokŭm chu: Yŏyudang chŏnsŏ* II.15.14b.

38 For example, Chŏng Yak-yong said that "the things of the world are extremely abundant and broad. Even if one covers them skillfully, one cannot exhaust their numbers, and even if one knows broadly about things, one cannot penetrate their *li*." *Taehak kong'ŭi: Yŏyudang chŏnsŏ* II.1.20a. On Chŏng Yak-yong's criticism of Zhu Xi's ideas of *gewu*, see Kim Nak-chin 金洛眞, "Chosŏn yuhakchadŭl-ŭi kyŏngmul ch'iji-ron: Chayŏn insik pangbŏp-kwa kwallyŏnhayŏ" 朝鮮 儒學者들의 格物致知論 – 自然 認識 方法과 關聯하여, in Han'guk sasangsa yŏn'guhoe, *Chosŏn yuhak-ŭi chayŏn ch'ŏrhak*, 98-100. For Zhu Xi's ideas of *gewu*, see Kim, *Natural Philosophy of Chu Hsi*, 21-25.

39 *Henan Chengshi yishu* 河南程氏遺書, Chap. 18 (*Er Cheng ji* 二程集, 188). Zhu Xi quoted this remark repeatedly in conversations compiled in *juan* 18 of *Zhuzi yulei* 朱子語類.

40 *Chungyong kang'ŭi po: Yŏyudang chŏnsŏ* II.4.17a.

41 Ibid. II.4.44a.

iad things have one root," an idea in direct opposition to his previously mentioned rejection of the idea that "the myriad things are one body."[42]

In many respects, then, Chŏng Yak-yong was a conservative thinker, eager to preserve the Confucian tradition. He differed from such unconventional thinkers as Hong Tae-yong 洪大容 (1731-1783) and Pak Chi-won 朴趾源 (1737-1805) of a generation earlier. Many things that were traditionally important were important to Chŏng Yak-yong as well. For example, he showed great concern with the details of ceremonial protocols. In his *Lectures on the Doctrine of the Mean* (*Chungyong kang'ŭi* 中庸講義), he divided the 19th chapter of the *Doctrine of the Mean* into many separate sections and discussed the details of various rituals and ceremonies at length.[43] His letters to his elder brother, Chŏng Yak-chŏn 丁若銓 (1758-1816), also contained detailed discussions of funeral and sacrificial services.[44] Nor was he free from many conventional, secular concerns; he was especially interested in official positions. His letters written while in exile frequently express his worries that his family might be eventually excluded from official service.[45] He expressed regrets about the life of his maternal great grandfather, Yun Tu-sŏ (尹斗緒 1668-1715), who, in spite of sagely talents, had to spend most of his life without an official position, and did not attempt to hide the feeling that he and his brothers, who had inherited such excellent qualities from Yun Tu-sŏ, should have prospered.[46] He also harbored a disdain for local functionaries, a long tradition among Confucian scholars.[47]

3 Science and the Natural World for Chŏng Yak-yong

What, then, did science mean for this fundamentally conservative thinker? The discussion in the preceding section suggests an obvious, and simple, answer, namely that science and the natural world were not important to Chŏng Yak-yong. And although he did not explicitly say what he thought about science and techniques, we can guess about them from his attitudes towards statecraft studies and specialized subjects in general.

42 Ibid. II.4.58b.

43 *Chungyong kang'ŭi po: Yŏyudang chŏnsŏ* II.4.25b-34b.

44 "Tap Chungssi": *Yŏyudang chŏnsŏ* I.20.16b-18.

45 E.g., "Ki ia" 寄二兒: *Yŏyudang chŏnsŏ* I.21.3a-5a; "Ki yang'a" 寄兩兒: *Yŏyudang chŏnsŏ* I.21.13a-13b.

46 "Sang Chungssi": *Yŏyudang chŏnsŏ* I.20.22b.

47 E.g., *Hŭmhŭm sinsŏ* 欽欽新書 "Sŏ" 序: *Yŏyudang chŏnsŏ* V.30.1b; "Hyangri ron" 鄉吏論: *Yŏyudang chŏnsŏ* I.11.17b-19a.

It is generally agreed that Chŏng Yak-yong based his statecraft studies on classical studies.[48] His comment on his own major statecraft work – the so-called '*ilp'yo isŏ*' 一表二書 – in which he called the classical studies of the Four Books and the Six Classics (*sishu liujing* 四書六經) the 'basis' (*ben* 本) and statecraft studies the 'details' (*mo* 末),[49] clearly shows how he ranked statecraft studies. It is not hard to imagine that his views about the sciences may have been similar. Indeed, his criticism of Yen Ruoju 閻若璩 (1636-1704) was based on the fact that the Qing scholar 'confused' beginners by mixing such subjects as criminal administration, calendrical method and geography, into his discussion, making it difficult for them to see the correct *li* 理.[50] His views are even more apparent in a letter to Chŏng Yak-chŏn, in which he made the following remarks concerning a book on marine animals that his brother had been preparing.

> The fundamental principle of learning is that writing books becomes useful only after one has first established the general principles. As for this Way, 'base it upon' (*ben zhi* 本之) filial piety and brotherly love; 'polish it by' (*wen zhi* 文之) rites and music; 'combine it with' (*jian zhi* 兼之) censure and investigation, finances and taxes, armies, and criminal justice; 'supplement it with' (*wei zhi* 緯之) agriculture, horticulture, medicine, calendrical astronomy, mathematics, and the manual techniques. [Then] its whole virtue might be at hand.[51]

It is clear from these lines that, to Chŏng Yak-yong, such scientific subjects as agriculture, horticulture, medicine, calendrical astronomy, mathematics, and the manual techniques, which he felt should merely 'supplement' the Way, must have been of an even lower status than statecraft topics such as censure and investigation, finances and taxes, armies, and criminal justice, which he felt should be 'combined' with the Way. Also, he begins his preface to the *Ch'onbyŏng hokch'i* 村病或治 by saying that he had read the medical books which his son had sent him only because "while in exile there were absolutely no [other] books."[52]

48 E.g., Kŭm Chang-t'ae 琴章泰, *Tasan sirhak t'amgu* 茶山實學探究 (Seoul: Sohaksa 小學社, 2001), Chap. 6.

49 "Chach'an myojimŏng" 自撰墓地銘: *Yŏyudang chŏnsŏ* I.16.18a.

50 "Yŏhaegŏ" 與海居: *Yŏyudang chŏnsŏ poyu* 與猶堂全書補遺, II, p. 197

51 "Sang Chungssi": *Yŏyudang chŏnsŏ* I.20.19b-20a.

52 "Ch'onbyŏng hokch'i sŏ" 村病或治序 : *Yŏyudang chŏnsŏ* I.13.10b

Of the scientific and technical subjects, music and harmonics (*lü* 律) was an exception in that Chŏng Yak-yong had a great interest in it. He spoke of the importance of music in edifying people: "when music decayed, punishments became harsh, battles became frequent, resentments rose, and fraud flourished."[53] Yet the significance of music was due to its traditional connection with the rites (*li* 禮), often combined into one word, 'rites and music' (*liyue* 禮樂). In fact, he even tried to reconstruct the lost text of the *Classic of Music* (*Yuejing* 樂經), one of the 'Six Classics' that had existed in ancient times, and composed the *Aksŏ kojon* 樂書孤存. He never mentioned music in the same breath as other specialized subjects. This, then, is yet another illustration of the importance of the classical Confucian tradition in Chŏng Yak-yong's way of thinking.

The impression one gets from the above paragraphs is clearly different from the judgments of many modern scholars who have been enthusiastic about Chŏng Yak-yong's views about, and achievements in, science, and have seen these as a sign of his departure from the tradition. The most enthusiastic evaluation has come from Kim Yŏng-ho, who has said that Chŏng Yak-yong "took the principles of the natural world as the object of rational scientific investigation,"[54] and that he "prepared an intellectual soil, in which science and technology took root, and which carved a path for their progress."[55] These statements are not supported by the actual sources on Chŏng Yak-yong. Indeed, one would be hard-pressed to explain why Chŏng Yak-yong should have wanted to do what Kim Yŏng-ho has credited him as having done. Kŭm Chang-t'ae's judgment that Chŏng Yak-yong achieved new knowledge in such scientific and technical areas as agriculture, instruments, and medicine, and that he "raised the level of such military techniques as shipbuilding and castle construction,"[56] are also basically overestimations. Most of what Kŭm Chang-t'ae has mentioned are not scientific ideas, but rather techniques pursued by Chŏng Yak-yong for their practical utility. Of course, Chŏng Yak-yong did have some knowledge of Western science that had been in circulation among Korean intellectuals of his time. But no one has shown how important such Western scientific knowledge might have been for his thought as a whole.

It should be noted that by Chŏng Yak-yong's time Western astronomy had become less problematic, after nearly two centuries since it had first entered Korean learned circles. Western scientific knowledge centering on Western

---

53 "Ak ron 2" 樂論二: *Yŏyudang chŏnsŏ* I.11.9a.

54 Kim Yŏng-ho, "Chŏng Tasan-ŭi kwahakkisul sasang," 282.

55 Ibid., 279.

56 Kŭm Chang-t'ae, *Tasan Chŏng Yak-yong*, 97-98.

astronomy, which had proved to yield more accurate calendrical computations, had attained a certain degree of legitimacy, and even authority. To accept and use such Western scientific knowledge was nothing remarkable for a scholar like Chŏng Yak-yong. This can be seen from the ease and matter-of-factness with which he mentioned the idea of a spherical earth. For example, his explanation of why oceans do not overflow while rivers constantly flow into them was based on the idea of a round earth.[57] He even spoke of 'the correct *li*' (*zhengli* 正理) of the spherical earth.[58]

It is also worthwhile to recall that a certain degree of interest in science was nothing new, but existed already in the Zhu Xi tradition.[59] An entry in Chŏng Yak-yong's *Chronological Biography* (*Nianpu* 年譜) is significant in this respect. It records that at 7 *sui* 歲 he was good at numbers, which prompted his father to predict that the young boy would excel in the calendrical method and mathematics.[60] As it turned out, Chŏng Yak-yong did not pursue in much depth those subjects for which he had shown childhood potential. Nonetheless, the fact that his father said such a thing about his son in that manner is evidence that it was possible for scholars of the time to pursue, and to excel in, such subjects, though Chŏng Yak-yong did not. Thus, the Confucian scholars of East Asia did not have to be awakened or inspired by the West to take an interest in scientific knowledge. Indeed, there were some Chosŏn scholars for whom knowledge of the natural world was important, despite basing their philosophy on the Zhu Xi neo-Confucian system. These scholars attempted to incorporate natural knowledge into their Confucian framework based on classical studies. A prominent example is Sŏ Myŏng-ŭng 徐命膺 (1716-1787), who had a considerable knowledge of both traditional and Western astronomy, and integrated this knowledge into a cosmological system based on the *Yijing*.[61] Chŏng

57 *Chungyong kang'ŭi po: Yŏyudang chŏnsŏ* II.4.51b-52a.

58 "Sang Chungssi": *Yŏyudang chŏnsŏ* I.20.22a.

59 Kim, *The Natural Philosophy of Chu Hsi*, especially Chap. 12.

60 Chŏng Kyu-yŏng 丁奎英, *Saam sŏnsaeng yŏnbo* 俟菴先生年譜 (Seoul: Chŏngmunsa 正文社, 1984), 2-3.

61 Pak Kwon-su 朴權壽, "Sŏ Myŏng-ŭng-ŭi yokhakchŏk ch'ŏnmun'gwan" 徐命膺의 易學的 天文觀, *Han'guk Kwahaksa Hakhoeji* 韓國科學史學會誌 20 (1998), 57-101; Moon Joong-yang, "Traditional Cosmology Associated with the *I-ching* and Anti-Cosmological Discourses in 18th-Century Korea," *Seoul Journal of Korean Studies* 12 (1999): 196-208; Yung Sik Kim, "Western Science, Cosmological Ideas, and the *Yijing* Studies in Seventeenth and Eighteenth-Century Korea," *Seoul Journal of Korean Studies* 14 (2001): 322-326.

Che-tu (鄭齊斗 1649-1736), even earlier than Sŏ Myŏng-ŭng, had also tried to incorporate Western scientific knowledge into his classical studies.[62]

Chŏng Yak-yong's interest in science and the natural world does not seem to have reached the level of these earlier scholars. His learning was essentially based on classical studies,[63] in which science and the natural world did not carry much weight. Admittedly, he was informed by Western scientific knowledge, and he accepted some parts of it. But he did not adopt the whole system of Western science, or try to incorporate it into his own philosophical system as the earlier thinkers like Sŏ Myŏng-ŭng and Chŏng Che-tu had.

Nor did Chŏng Yak-yong have a very firm grasp of scientific knowledge. He was far from the relatively high standard reached by such earlier Chosŏn scholars as Yi Ik 李瀷 (1681-1763), Hong Tae-yong, and Hwang Yun-sŏk 黃胤錫 (1729-1791). For example, in his essay "Explanation of the Diagram of the Sphere of the Earth" (Chigu tosŏl 地毬圖說), Chŏng Yak-yong used the idea of the spherical earth to explain the variation of the elevation of the North Star from the horizon at varying latitudes, namely that it becomes higher as one goes northward and lower as one goes southward. However, his discussion was basically of the same type as the traditional explanations based on the idea of a flat earth, and revealed the fact that he interpreted the 'degree' (*du* 度) as linear, and not angular, distance, and that he believed the distance of the North Star from the earth's surface to be more or less of the order of the earth's diameter.[64] In the "Discussion on the Outermost Heaven" (Chongdongch'ŏn pyŏn 宗動天辨), he rejected the leftward rotation of the outermost heavenly sphere (*zongdongtian* 宗動天) and maintained that all the seven planets rotate rightward. But he was not aware that this then would have required him to explain the observed daily leftward rotation of the fixed stars. The earth's rightward rotation on its own axis would have explained it, but the possibility of such an explanation does not seem to have occurred to him.[65] His discussions of opti-

62 Ku Man-ok 具萬玉, "Chosŏn hugi ch'ŏnche unhaengron-ŭi pyŏnhwa" 朝鮮後期 天體運行論의 變化, *Sirhak sasang yŏn'gu* 實學思想研究 17/18 (2000): 361-363.

63 Indeed, his level of accomplishment in classical studies may have been the most ambitious and most extensive among all the Confucian scholars since Zhu Xi. On the range of Chŏng Yak-yong's classical learning, see, e.g., Kim Mun-sik 金文植, *Chosŏn hugi kyŏnghak sasang yŏn'gu* 朝鮮後期經學思想研究 (Seoul: Ilchogak 一潮閣, 1996), 182-193.

64 "Chigu tosŏl" 地毬圖說: *Yŏyudang chŏnsŏ* I.10.9a-11a

65 "Chongdongch'ŏn pyŏn" 宗動天辨: *Yŏyudang chŏnsŏ* I.12.15a-16a. In fact, he did not agree with Chŏng Yak-chŏn, who said that the change of the position of a comet was 'clear evidence' of the earth's rotation. ("Sang Chungssi": *Yŏyudang chŏnsŏ* I.20.24b) Instead, Chŏng Yak-yong said that matters like the rotation of the heavenly sphere "is something that cannot be known. And what cannot be known should not be discussed." While refusing to

cal phenomena such as the workings of convex lenses and the eyes, the camera effect, and the deflection of light have been praised for their accuracy. Yet close examination of his essays show that they were based on a casual – and rather superficial – understanding of these curious but widely known and often discussed phenomena, about which he must have heard from people like Yi Pyŏk (李檗 1754-1786).[66]

4 Western Learning and Christianity for Chŏng Yak-yong

The question once more arises: What did science mean for Chŏng Yak-yong? Before tackling this question, however, we must consider a few related questions.

First, what did Western learning in general mean for Chŏng Yak-yong? While there is no question that he did accept some Western knowledge, we need to ask ourselves why he would have wanted to accept Western knowledge in the first place – that is, unless we are to assume that any Chosŏn scholar of the time would have welcomed whatever Western knowledge was available to him. The most enthusiastic answer has come from Kŭm Chang-t'ae, who has seen "a remarkable creative fusion of Western and Confucian learning" in Chŏng Yak-yong's *Lectures on the Doctrine of the Mean*,[67] and claimed that "Western learning offered Tasan [Chŏng Yak-yong] a logically coherent framework of thought that enabled him to reexamine the orthodox worldview of Zhu Xi and to go beyond it."[68] To answer the same question, Chŏng Ok-cha looks at the so-called

---

discuss Zhuangzi's 莊子 cryptic phrases, "Is it true that heaven moves? Is it true that the earth stays still?" (天其運乎, 地其處乎), he even said that "in speaking of the *li* of heaven and earth, there is something that cannot be known thoroughly. Those who speak of it recklessly should not be trusted blindly." ("Chongdongch'ŏn pyŏn": *Yŏyudang chŏnsŏ* I.12.16a) Nor was he firm in his rejection of the idea of the earth's rotation. Indeed, he made remarks that can be interpreted as implying a rotating earth: "In my view, the earth is [something like] a great carriage of the myriad things…On the south and the north, there originally were the fixed poles, whose rotation is similar to [that of] a wheel." ("Chiri ch'aek" 地理策: *Yŏyudang chŏnsŏ* I.8.2a)

66 "Aech'e ch'ulhwa tosŏl" 靉靆出火圖說: *Yŏyudang chŏnsŏ* I.10.5b-7b; "Ch'ilsil kwanhwasŏl" 漆室觀畵說: *Yŏyudang chŏnsŏ* I.10.7b; "Wan puchŏng sŏl" 碗浮靑說: *Yŏyudang chŏnsŏ* I.10.8a; *Ŭiryŏng* 醫零, "Kŭnsi ron" 近視論, in Kim Tae-wŏn 金大原, "Chŏng Yak-yong-ŭi *Ŭiryŏng*. 1" 丁若鏞의 『醫零』. 1, *Han'guk Kwahaksa Hakhoeji* 韓國科學史學會誌 15 (1993): 245-246. Cf. Pak Sŏng-rae, "Chŏng Yak-yong-ŭi kwahak sasang," 348.

67 Kŭm Chang-t'ae, *Tasan Chŏng Yak-yong*, 190.

68 Kŭm Chang-t'ae, "Tasan Kyŏnghak-ŭi t'al-Chujahakchŏk Segyegwan" 茶山 經學의 脫朱子學的 世界觀, *Tasanhak* 茶山學 1 (2000): 37.

'Sŏngho Left Wing' (Sŏngho *choap'a* 星湖左派) circle, from whom Chŏng Yak-yong derived much of his learning. According to her, this circle of scholars turned to Western learning in their attempt to seek "an intellectual breakthrough" while remaining excluded from power.[69] This may very well have been the case for many in the 'Sŏngho Left Wing' circle who were enthusiastic about Western learning, and even for Yi Ik himself. Nevertheless, the situation seems to have been different for Chŏng Yak-yong, who rarely showed an outright endorsement for Western learning.

Without a doubt, the new political situation surrounding the advent of Christianity in Chosŏn may have masked Chŏng Yak-yong's true sentiment towards Western learning, which may have been more sympathetic than he was willing to reveal. But it should also be noted that it was not expected that a man like him would welcome everything from the West. Living as he did in late eighteenth- and early nineteenth-century Korea, he was under no pressure to accept the overall superiority of Western culture. He had the flexibility to believe that selective adoption of Western culture and supplementary reforms, while maintaining the basic ingredients of tradition, would solve the problems facing him and his country. That was what he thought he was doing, and indeed did, to a considerable extent. Thus, whereas many among his close circles fully embraced Western learning, and even Christianity, Chŏng Yak-yong did not do so. He did not forsake Zhu Xi's philosophical system as his basis; in his heart he remained a member of the mainstream of Chosŏn yangban 兩班 society, which was rooted in that system. Nor should we assume that everything new in his thinking came from the West. For example, although it is possible that many of his ideas regarding social and institutional reforms had something to do with Western learning, it is equally possible that he had read them from earlier reformist thinkers like Yu Hyŏng-wŏn (柳馨遠 1622-1673) and Yi Ik.

There is another related question that has a bearing on our attempt to understand the meaning of science for Chŏng Yak-yong: What did Christianity mean for him? Although the extent of his conversion to, and later denunciation of, Christianity is still being debated, it seems clear that Christianity influenced his idea of 'heaven,' and consequently affected his perspective on *li*, *guishen*, etc. It is also likely that his exposure to Western learning and Christianity in his twenties brought him an intellectual awakening that broadened the range of his thought.[70] That said, Chŏng Yak-yong's early exposure to,

69 Chŏng Ok-cha 鄭玉子, *Chosŏn hugi yŏksa-ŭl ihae* 朝鮮 後期 歷史의 理解 (Seoul: Ilchisa 一志社, 1993), 155.

70 Kŭm Chang-t'ae, *Tasan Chŏng Yak-yong*, 38, 286. Kŭm Changt'ae has said that "by adopting the religious worldview and the natural philosophy of Western learning, ... [Chŏng Yak-yong] established his consistent logic." (ibid., 98).

and indeed immersion in, Christianity does not seem to have greatly affected the overall course of his study and writing, which did not prove very different from those of other Confucian scholars, who followed the standard classical curriculum. We may note in this connection that a great number of Confucian scholars, including even Zhu Xi, had some interest in Buddhism in their youth before becoming ardent Confucians. It should not be so difficult, then, to imagine a similar situation with respect to Chŏng Yak-yong's attitude towards Christianity. It is absolutely conceivable that his earlier faith in Christianity and his later denunciation were both genuine. For, if it is possible for a one-time Buddhist to become a non-Buddhist or an anti-Buddhist, why would it be impossible for a one-time Catholic to become a non-Catholic or an anti-Catholic? Also, if Zhu Xi's synthesis could be seen as an attempt of a Confucian to accommodate the Buddhist challenge, why is it impossible to see Chŏng Yak-yong's effort as a similar attempt of a Confucian to accommodate the challenge of the Catholic ideas that had impressed him deeply in his youth?[71]

5 What Could Chŏng Yak-yong Accept and What Could He Not?

We can ask still another question, which will help define our original question in more concrete terms: How much could Chŏng Yak-yong accept from the ideas and practices lying outside the core of Confucian learning? Based on what he wrote, we can answer this question to some degree.

As we have seen, Chŏng Yak-yong could not accept the five-phase theory and the use of stems and branches (*gan-ji* 干支) in divination. He criticized those who believed that the *Yijing* has 'hidden' (*you* 幽) meanings and is thus 'mysterious' (*shen* 神) and 'cannot be understood' (*bukekui* 不可窺), which, according to him, was not the original "intentions of the sages."[72] In particular, he condemned those who deceive people with such things as "the Yellow River Chart" (*Hetu* 河圖) and "the Luo River Diagram" (*Luoshu* 洛書).[73] We have also seen that he rejected the technique of 'winds and waters' (*fengshui* 風水)

71 Chŏng Yak-yong's attitude toward Buddhism also differs little from what is frequently found among Confucian scholars. He was clearly critical of Buddhism, showing, for example, a great distaste for the tendency to mix in Buddhist ideas when interpreting the *Great Learning* (*Daxue* 大學). See "Tap Chungssi": *Yŏyudang chŏnsŏ* I.20.15b. But he did not always oppose or criticize Buddhism. He showed some respect for '*zuochan* 坐禪' (sitting meditation), which "many teachers of the classics (*jingshi* 經師) practiced in their late years." "Tap Chungssi": *Yŏyudang chŏnsŏ* I.20.28a.

72 "Yŏk ron 1" 易論一: *Yŏyudang chŏnsŏ* I.11.1a-1b.

73 Ibid. I.11.1b-2a.

and various divinatory and fortune-telling techniques. For him, discussing techniques like divination, physiognomy, astrology, which he called 'occult techniques' (*shushu* 術數), was 'deluding [people]' (*huo* 惑), and not 'learning' (*xue* 學).[74] He did not accept the belief that man's 'habits' (*xi* 習) were determined by his 'facial features' (*xiang* 相); he believed, on the contrary, that facial features change over time to reflect one's habits.[75] What was important for him was 'serving heaven' (*shi tian* 事天); he criticized his contemporaries for engaging in divination while not serving heaven.[76] He was convinced that the practices of occult techniques could not be efficacious; he was so convinced that he went as far as to say that on those occasions when they may appear efficacious, in reality they are deluding people through illusive techniques.[77] He even said, "Even if there were *li* [for such occult techniques], a great man would not engage in them. How [much more should they be avoided when in fact] there is absolutely no such *li*?"[78].

Chŏng Yak-yong could not abide the belief in portents (*zaiyi* 災異) either. In particular, he rejected the significance of celestial phenomena – such as solar or lunar eclipses – as portents for rulers.[79] If eclipses were portents, he asked, "How is it that people can predict the time of their occurrence without the slightest error?"[80] Thus he denounced the treatises on astronomy (*tianwenzhi* 天文志) and those on the five phases (*wuxingzhi* 五行志) included in the successive dynastic histories, saying that they are "far-fetched (*fuhui* 傅會) and do not have a single corroboration." For him, "the movements of the stars all have fixed degrees, and cannot be meddled with."[81]

On his list of unacceptable things, Chŏng Yak-yong included disease diagnosis through pulse taking (*maizhen* 脈診). In his essay "On Pulses" (Maek ron 脈論), after talking about the existing theories of pulses – for example that "the *cun* 寸 [position] on the left [wrist] gives a diagnosis of the heart and the *cun* on the right gives a diagnosis of the lungs" – he called them 'absurd' (*wang* 妄).[82]

It must be said, however, that Chŏng Yak-yong did not reject the possibility of portents altogether. He believed in the portentous significance of some

74 "Ohak ron 5" 五學論五: *Yŏyudang chŏnsŏ* I.11.23a-23b.

75 "Sang ron" : *Yŏyudang chŏnsŏ* I.11.13b

76 "Tap Chungssi": *Yŏyudang chŏnsŏ* I.20.15b.

77 "P'ngsu ron 4" 風水論四: *Yŏyudang chŏnsŏ* I.11.31b-32a.

78 "P'ungsu ron 1" 風水論一: *Yŏyudang chŏnsŏ* I.11.30b

79 "Sang Chungssi": *Yŏyudang chŏnsŏ* I.20.24b.

80 *Chungyong kang'ŭi po: Yŏyudang chŏnsŏ* II.4.23a.

81 "Ohak ron 5": *Yŏyudang chŏnsŏ* I.11.23a-23b.

82 "Maek ron 1" 脈論一: *Yŏyudang chŏnsŏ* I.11.12a.

strange occurrences, such as 'grain rain' (*yuli* 雨栗) and a horse with a horn, calling them 'inauspicious things' (*buxiang zhi wu* 不祥之物). He even took them as omens foretelling him that his exile would not end soon.[83] Chŏng Yak-yong did not completely discard the possibility of foreknowledge (*qianzhi* 前知) either. For instance, he stated that "*guishen* can have foreknowledge."[84] Man can also have foreknowledge. He said, "If one is extremely sincere, one can know heaven; if one knows heaven, one can foreknow," and distinguished the foreknowledge of the sages like Zhou Gong 周公 from the foreknowledge of those who used wicked and occult techniques.[85] Still, he did not approve of the widespread tendency of people in his day to always seek out others with foreknowledge.[86]

Regarding *Yijing* divination too, what Chŏng Yak-yong dismissed was not the divination itself but the belief that there is something hidden and mysterious to it.[87] According to him, the *Yijing* was written by the sages in order to "request the command of heaven and to follow its intention." The hexagrams were devised to provide ways for heaven to respond to such request of the sages, and for the sages to interpret the command of heaven.[88] Not only *Yijing* divination, but other methods of divination can also be employed in "requesting the command of heaven." Other divinatory methods could even be more efficient than *Yijing* divination in "receiving the bright [command] of heaven"; however, "in pondering the phrases in order to discern the cause of advancing and retreating, surviving and perishing, and to know how to dispose of oneself, there is only the *Yijing*."[89]

Nor did Chŏng Yak-yong discredit all beliefs associated with occult techniques. For example, he believed in the efficacy of respiratory techniques (*daoyin* 導引).[90] And although he rejected the technique of 'winds and waters,' he accepted such common beliefs about the proper positioning of various parts of a house – windows to the south and walls to the north, the stairs for guests on the west and those for hosts on the east, the inner hall to the north and the outer hall to the south. He referred to such notions as "the spontaneous ten-

83 "Sang Chungssi": *Yŏyudang chŏnsŏ* I.20.25a.

84 *Chungyong kang'ŭi po: Yŏyudang chŏnsŏ* II.4.58b.

85 Ibid. II.4.49a.

86 "Ohak ron 5": *Yŏyudang chŏnsŏ* I.11.23b.

87 "Yŏk ron 1" 易論一: *Yŏyudang chŏnsŏ* I.11.2a.

88 "Yŏk ron 2" 易論二: *Yŏyudang chŏnsŏ* I.11.2a-2b.

89 Ibid. I.11.3a.

90 "Sang Chungssi": *Yŏyudang chŏnsŏ* I.20.24a; "Tap Chungssi": *Yŏyudang chŏnsŏ* I.20.28a.

dency of yin and yang" (*yinyang ziran zhi shi* 陰陽自然之勢) or "the invariable *li* of *qian* and *kun*" (*qiankun buyi zhi li* 乾坤不易之理).[91]

These considerations lead us to ask another related question: Where did Chŏng Yak-yong draw the line between those non-Confucian elements that he found acceptable and those that he did not? In other words, why was he open to some things but not others?

In this connection, it must be pointed out that in terms of what Chŏng Yak-yong accepted and what he dismissed and where he drew that line, he was not so very different from Zhu Xi. Although Chŏng Yak-yong did deviate from Zhu Xi regarding such concepts as heaven, *guishen*, *li*, yin-yang, and the five phases, the relationships between them, and the relationships between these concepts and the objects and phenomena in the world, his ideas about the objects and phenomena themselves were much the same as Zhu Xi's. The scope of these concepts narrowed and, as the result, the nature and content of the things and events he designated with these concepts changed. But his ideas about the things and events themselves stayed more or less the same as Zhu Xi's. For example, as we have seen, he criticized the characterization of *guishen* as 'traces of creative transformations,' but was willing to attribute natural phenomena such as rain, dew, frost, snow, and wind, to these 'traces.' It was simply that in referring to those phenomena he did not use the concept '*guishen*,' the scope of which he had narrowed.[92] The concepts of heaven and *guishen* became more mysterious as the result of this narrowing, because he excluded the 'naturalistic' aspects from the two concepts.[93]

6 Personal Situations and Problems

An obvious direction to take in searching for the answer to our original question 'What did science mean for Chŏng Yak-yong?' is to recognize that he must have been looking for some solution from science to the problems of his day, both of his own and of the country at large. For him personally, the biggest problem was the fall from power that had landed him in a remote place, far

91 "Mun tŏngsŏnambuk" 問東西南北: *Yŏyudang chŏnsŏ* I.9.3a.

92 *Chungyong kang'ŭi po: Yŏyudang chŏnsŏ* II.4.61b-62a.

93 He frequently spoke of "the bright *shen* of heaven" (天之明神) (*Chungyong kang'ŭi po: Yŏyudang chŏnsŏ* II.4.32b, 58b), and at times even said, "Is *guishen* not heaven?" (*Chungyong kang'ŭi po: Yŏyudang chŏnsŏ* II.4.21a). For the traditional naturalistic aspects of the idea of *guishen*, in Zhu Xi for example, see Kim, *Natural Philosophy of Chu Hsi*, Chap. 6.

from the capital; this caused him great hardship, both politically and intellectually. For the country, the problems were the degeneration and decline of the government, and in particular the suffering of the people, to which he came to have ample firsthand witness while in exile. Compared with these problems, moral self-cultivation and reestablishment of the true Way – the key traditional concerns of Confucian scholars – may have appeared less important to him. How, then, did he think that science could help him solve these problems?

It may be helpful to look at others in similar situations. For example, one may look at the seventeenth- and eighteenth-century Chinese scholars of the Jiangnan 江南 region, who, faced with the Jesuit introduction of Western science, turned to evidential research in ancient astronomy and mathematics, considering it to be useful in reasserting the values of their own tradition.[94] This was not the case for Chŏng Yak-yong, however. Indeed, taking into account his harsh criticism of the classical commentary tradition (*xun'gu* 訓詁) for its excessive concern with textual details,[95] it is possible that he would have been equally critical of 'evidential studies' (*kaozheng* 考證) of the Qing.

Chŏng Yak-yong's personal situation – isolated, persecuted, and in exile – must have also had an effect on his intellectual attitude. Generally, scholars in a predicament like his tend to become more critical of central, established values. Chŏng Yak-yong was quite the opposite, clinging even more fervently to Confucian basics and upholding core Confucian values. Even during the period of his exile, he could not bear to think of himself as an outsider to Chosŏn society. Unlike other scholars of the 'Southerner' (*Nam'in* 南人) faction, he could not accept such a status. On the contrary, he seems to have considered himself, perhaps even more so in this period, as a defender of Chosŏn society and culture, from the center of which he had been involuntarily and temporarily displaced. Chŏng Yak-yong was not, even on an intellectual level, a revolutionary or an iconoclast.

People undergoing persecution and isolation sometimes show an inclination towards the mystical. For instance, Zhu Xi showed a great interest in inner alchemy (*neidan* 内丹) and techniques of nourishing life (*yangsheng* 養生) in his later years when he was suffering vicious political persecution.[96] Chŏng

94 Benjamin A. Elman, *From Philosophy to Philology; Intellectual and Social Aspect of Change in Late Imperial China* (Cambridge, Mass.: Harvard University Press, 1984).

95 "Ohak ron 2" 五學論二: *Yŏyudang chŏnsŏ* I.11.20b-21b.

96 Azuma Jūji 吾妻重二, "Shu Ki 'Shueki sandōkei kōi' ni tsuite," 朱熹周易參同契考異について, *Nihon Chūgokugakukaihō* 日本中國學會報, no. 36 (1984): 175-190; Yung Sik Kim, "The *Ts'an-t'ung-ch'i k'ao-i* and the Place of Internal Alchemy (*Nei-tan*) in Chu Hsi's Thought," *Monumenta Serica* 55 (2007): 99-131.

Yak-yong was different in this respect as well: in exile he became a more staunch critic of superstitious and mystical practices. Nor can Chŏng Yak-yong be said to have been like Zhu Xi in his final years, when he developed a deep interest in cosmological and astronomical topics.[97] There is no evidence that Chŏng Yak-yong demonstrated any such attitude, despite the similarity of their situations.

7 Search for Practical Utility

Having explored these related questions, we are now more prepared to ask our original question. What comes up first, however, is an obvious answer and a well-known aspect of Chŏng Yak-yong's attitude, namely his emphasis for practical utility. Science and natural knowledge for him were, above all, something useful. Although science was of only secondary importance, as we have seen, its practical utility could have appeared to him as something that could help solve the problems of his day.

This must have been especially so of the various practical techniques, for whose promotion and importation Chŏng Yak-yong was a tireless advocate. It was his view that whereas such moral virtues as filial piety and brotherly love "have their roots in heaven's nature (*tianxing* 天性) and are evident in the sages' books,'" and thus can be extended and replenished once they have been introduced, techniques change and advance continuously and thus have to be constantly imported anew.[98] For him, China was the place to turn to for advanced foreign techniques, a view in line with the sentiment of 'Northern Learning' (*Pukhak* 北學) that was in fashion among the intellectuals of the capital in the late eighteenth century.[99] He lamented that, among those many people who had visited Beijing, none had returned with "things useful for improving livelihood."[100]

---

97 Kim, *Natural Philosophy of Chu Hsi.*

98 "Kiye ron 3" 技藝論三: *Yŏyudang chŏnsŏ* I.11.11b.

99 On the "Northern Learning," see Yu Pong-hak 劉奉學, *Yŏn'am ilp'a Pukhak sasang yŏn'gu* 燕巖一派 北學思想 研究 (Seoul: Ilchisa 一志社, 1995). His pragmatic attitude was also in line with the slogan, 'Eastern Way and Western Instruments' (*Tongdo sŏki* 東道西器), which was to characterize many Korean intellectuals in the late nineteenth century. For a discussion of the idea of 'Eastern Way and Western Instruments,' see No Tae-hwan 盧大煥, *Tongdo sŏki ron hyŏngsŏng kwajŏng yŏn'gu* 東道西器論 形成過程 研究 (Seoul: Ilchisa 一志社, 2005).

100 "Song Yi ch'amp'an Kiyang sa Yŏn'gyŏng sŏ" 送李参判基讓使燕京序: *Yŏyudang chŏnsŏ* I.13.13a.

This pragmatic attitude led Chŏng Yak-yong to highlight the importance of studying subjects of practical utility. In a letter to his sons written while in exile, he stressed the need to focus their minds on 'the learning of practical utility' (*shiyong zhi xue* 實用之學), in addition to the study of the classics and history. He instructed them: "Always have in mind an intention of benefiting all people and fostering all things. Only then can you be a gentleman (*junzi* 君子) who reads books."[101] His recommendation to the king to study geography and to publish geographical books was also in the same practical vein.[102]

Thus, it is natural that Chŏng Yak-yong was harsh with the Confucian scholars of his time who were bogged down in the abstruse details of abstract metaphysical problems while neglecting such practical subjects as 'names, things, degrees and numbers' (*mingwu dushu* 名物度數), 'singing and dancing,' and the learning of 'punishment, names, usefulness and profit' (*xingming gongli* 刑名功利), and who were reluctant to engage in such practical affairs as "finance, military, litigation, and ceremonies."[103] He added that although those who neglected practical subjects defended themselves by referring to Zhu Xi, they were actually slandering him, because Zhu Xi was not like that. Chŏng Yak-yong praised Zhu Xi's many achievements in practical affairs.[104] Indeed, in his view, "originally, true Confucian learning did not consider it improper to wish to govern the country, to make people comfortable, to expel enemies, to make finances abundant, to be competent both culturally and militarily."[105] He also criticized the tendency to "attend to [problems that are] far away, while neglecting [problems that are] near at hand" (*wuyuan hujin* 務遠忽近), which he thought was especially strong in Korea.[106] He frequently lamented that, unlike in China, the 'learning of names and things' (*mingwu zhi xue* 名物之學), geography for example, was very much neglected in Korea.[107]

Nevertheless, this did not lead Chŏng Yak-yong to think that scholars should study, and become familiar with, the techniques themselves. True, he lamented the fact that astronomical work was left to the functionaries of the 'middle-people' (*chung'in* 中人) class, and proposed, as previously mentioned, that technical specialists, those in the Astronomical Bureau for example, be sent to

101 "Ki ia": *Yŏyudang chŏnsŏ* I.21.4b.

102 "Chiri ch'aek" 地理策: *Yŏyudang chŏnsŏ* I.8.15a-16a.

103 "Ohak ron 1": *Yŏyudang chŏnsŏ* I.19.19b. In the preface to the *Hŭmŭm sinsŏ* 欽欽新書, he complained that the literati bureaucrats of his time were completely ignorant of the concrete problems of legal administration.

104 "Ohak ron 1": *Yŏyudang chŏnsŏ* I.19.20a.

105 "Sok'yu ron" 俗儒論": *Yŏyudang chŏnsŏ* I.12.8a.

106 "Chiri ch'aek": *Yŏyudang chŏnsŏ* I.8.1b.

107 Ibid. I.8.15a-15b.

Beijing to study the developments there, and that the techniques be promoted by treating the technicians themselves well.[108] But what he had in mind was nothing more than the pragmatic concern for recruiting, training, and employing such manpower needed for the government. There also were other proposals of a similarly pragmatic nature: the proposal that medical doctors be paid according to their achievements,[109] the proposal that the decimal system be used in the reform of weights and measures,[110] and the discussion of the smallpox vaccination.[111] Such proposals and discussions have prompted a modern scholar to even suggest that "the techniques became one of the basic objects of scholarly investigation of the literati."[112] Yet no evidence can be found that, in Chŏng Yak-yong's mind, such proposals reflected, or resulted in, an elevation of the intellectual status of technical knowledge and activities.

Even in discussions of moral problems, Chŏng Yak-yong's inclination was towards practicality, evident in his great distaste for metaphysical speculations that had been dominating Confucian discourse during the Chosŏn period. He said, for example, "As for those who take a single phrase for their main tenet, their learning is all 'heterodox' (*yiduan* 異端)."[113] Admittedly, he did write an essay entitled "the Discussion of the Manifestation of *Li* and *Qi*" (Ibal kibal pyŏn 理發氣發辨),[114] but even there he steered clear of speculative discussions; what was important for him was moral practice, rather than debates over details of the relationship between *li* and *qi*. Indeed, he never mentioned the notorious 'Ho-Rak 湖洛' debate that had occupied the mind of Chosŏn Confucian scholars for generations up to his time.[115] It should be noted in this regard that the ethical problem – how to live as a worthy human being – was a practical one for Chŏng Yak-yong.

---

108 "Kun'gi ron 1" 軍器論一: *Yŏyudang chŏnsŏ* I.11.9b; *Chungyong kang'ŭi po: Yŏyudang chŏnsŏ* II.4.40a-40b; *Kyŏngse yup'yo* 經世遺表: *Yŏyudang chŏnsŏ* V.1.6a-6b.

109 *Chungyong kang'ŭi po: Yŏyudang chŏnsŏ* II.4.41b.

110 "Toryanghyŏng ŭi" 度量衡議: *Yŏyudang chŏnsŏ* I.9.26a.

111 "Chongdu sŏl" 種痘說: *Yŏyudang chŏnsŏ* I.10.11b-13a

112 E.g., Kim Yŏng-ho, "Chŏng Tasan-ŭi kwahakkisul sasang," 284.

113 "Ch'i ryangji pyŏn" 致良知辨: *Yŏyudang chŏnsŏ* II.12.18a.

114 *Yŏyudang chŏnsŏ* I.12.17a-18a.

115 On the "Ho-Rak" debate, see, e.g., Han'guk sasangsa yŏn'guhoe 韓國思想史硏究會, *Insŏng mulsŏng ron* 人性物性論 (Seoul: Han'gilsa 한길社, 1994), especially part 3. Of course, as Chŏng Sun-u 丁淳佑 noted, avoiding such metaphysical discussions had the effect of narrowing down the variety of possible questions that Chŏng Yak-yong could treat. See the roundtable discussion: Choadam 座談, "Tasan, Chujahak, kŭrigo Sŏhak" 茶山, 朱子學, 그리고 西學, 219.

Chŏng Yak-yong's practical bent also explains how he could have shown such a great interest in detailed protocols of ceremonies like sacrificial and funeral services. He was also concerned with the matter of the proper way to adopt sons, as can be seen from a discussion in a letter to his brother Chŏng Yak-chŏn.[116] He even wrote an essay called "On Adoption" (Iphu ron 立後論).[117] These, then, were important practical matters for him. In contrast, problems like the structure and the workings of the natural world were not, which may in part explain the relative lack of interest in scientific subjects of a practically minded scholar like Chŏng Yak-yong.

It is in the same practical, pragmatic vein that Chŏng Yak-yong disapproved of vague and mystical ideas and practices that were of no conceivable use. We have seen that he criticized those who interpreted the *Yijing* in such a manner. Some features of his criticism of occult techniques (*shushu* 術數) also show such a pragmatic bent. In particular, the specific points of his argument as to why occult techniques could not actually work indicate that he was a typical pragmatist. For example, he argued that if the masters of 'winds and waters' thought such sites would truly yield great benefits, they would use the auspicious sites for themselves instead of giving them away to others at a much lower profit.[118] He wondered how a corpse inside a coffin could possibly protect descendants when even living parents of great ability and position could not.[119] He mentioned instances in which battles were waged and sacrifices offered successfully without prior recourse to divination, or in which descendants became rich and prosperous although they had not followed the technique of 'winds and waters' in burying their ancestors.[120]

8 Rejection of the Correlative Mode of Thinking

Another constant theme, though not a direct answer to our question, may still be helpful in understanding Chŏng Yak-yong's attitude toward science and the natural world: his rejection of 'correlative thinking,' or classificatory-associative thinking. This mode of thinking flourished among the Chosŏn scholars of

116 "Sang Chungssi": *Yŏyudang chŏnsŏ* I.20.25b-26b.
117 *Yŏyudang chŏnsŏ* I.11.14b-17b.
118 "P'ungsu ron 2" 風水論二: *Yŏyudang chŏnsŏ* I.11.31a.
119 "P'ungsu ron 1": *Yŏyudang chŏnsŏ* I.11.30b.
120 "Kap'ŭl ron 1" 甲乙論一: *Yŏyudang chŏnsŏ* I.11.28a-28b; "P'ungsu ron 1": *Yŏyudang chŏnsŏ* I.11.30b.

the time in the form of the so-called 'studies of images and numbers' (*xiang shuxue* 象數學).[121]

Chŏng Yak-yong's rejection of the five-phase theory was basically a rejection of this correlative mode of thinking. His criticism began by pointing out that "the five phases are nothing but five things among the myriad things, and thus are the same [as other] things," and that it is thus difficult to imagine that all things originated from the five phases. He pointed out, for example, that Metal and Wood are not seen inside the bodies of man and animals that have blood and *qi*.[122] The main point of his attack on the five-phase theory, however, was that associations of various things and ideas with the five phases were frequently far-fetched, and that such an associative scheme thus could not work. He said, for example, that instead of associating Water with 'soaking and descending' (*runxia* 潤下), Wood with 'bending and straightening' (*quzhi* 曲直), and Metal with 'obeying and changing' (*congge* 從革) qualities, Water should have been associated with 'moist and cold' (*shileng* 濕冷), Wood with 'soft and ascending' (*rousheng* 柔升), and Metal with 'hard and falling' (*gangzhui* 剛墜).[123] Of course, the Western four-element theory seems to have played a role in Chŏng Yak-yong's rejection of the five-phase theory. But the effect that the four-element theory really had on him was to open his eyes to the arbitrariness of the number five, namely that the number of the basic, fundamental phases – or elements – underlying all things and events of the world did not have to be five. It could be four, six, or eight.[124] In fact, his criticism could have been directed equally at the four-element theory, which was also of the same associative and classificatory nature.

Chŏng Yak-yong showed a similar attitude toward the associations involving the yin-yang scheme. Again, the basis of his critique was that yin and yang could not be the source of all things. He said, for instance, that yin and yang "originally do not have a body (*ti* 體) or tangible quality (*zhi* 質)," and thus "cannot be considered as 'parents of the myriad things' (*wanwu zhi fumu* 萬物之父母)."[125] Yet, his criticism was directed more sharply at the arbitrariness of

121 Pak Kwon-su, "Sŏ Myŏng-ŭng-ŭi yokhakchŏk ch'ŏnmun'gwan"; Moon Joong-yang, "Traditional Cosmology Associated with the *I-ching*." Chŏng Yak-yong's rejection of correlative thinking was in line with what John Henderson termed the "decline of Chinese Cosmology," a movement of rejecting the correlative cosmology that had been going on in China since the seventeenth century. John B. Henderson, *The Development and Decline of Chinese Cosmology* (New York: Columbia University Press, 1984), especially Chapters 7-8.

122 *Chungyong kang'ŭi po: Yŏyudang chŏnsŏ* II.4.3a.

123 *Sangsŏ kohun* 尙書古訓: *Yŏyudang chŏnsŏ* II.25.32a.

124 Ibid. II.25.31a-31b.

125 *Chungyong kang'ŭi po: Yŏyudang chŏnsŏ* II.4.1b-2a.

the associations, for example such virtues as humaneness (*ren* 仁), righteousness (*yi* 義), propriety (*li* 禮), and knowledge (*zhi* 智) with either yin or yang.[126] After providing an explanation for eyesight defects – nearsightedness and farsightedness – he criticized those who used the yin-yang scheme only to end up making a muddle of something that to him was clear and obvious.[127] His criticism of the 'stem-branch' (*ganzhi* 干支) theory in his "On Stems and Branches" (Kap-ŭl ron 甲乙論), which will be discussed in more detail later, is yet another example of his censure of the associative mode of thinking.

Chŏng Yak-yong's criticism of correlative thinking was not limited to categorical schemes like yin-yang, the five phases, and stems and branches. He was also critical of various other kinds of associations. For example, the reason he dismissed those who tried to base a system of standard weights and measures (*du-liang-heng* 度量衡) on the five musical notes (*wusheng* 五聲) was essentially because he was against arbitrarily associating two kinds of things that are unrelated to each other.[128] In his view, all that was necessary for a system of weights and measures was to select a standard and have it used by everyone.[129] Similarly, the issue he had with pulse diagnosis (*maizhen* 脈診) was with associating a particular visceral organ with a particular pulse point.[130]

Chŏng Yak-yong spelled out the grounds for his rejection of some associations. He noted that the associations which originated in particular places, times or occasions were valid only for those particular situations, and would become problematic when applied to different situations. He mentioned many such cases in his essay "On Stems and Branches." For example, he noted that the use of the stem-branch pairs for counting years began only in the time of Han Wudi 漢武帝 (r. BC 141-87), before which the years had no stem-branch designations and thus could not be associated with good or bad luck.[131] He pointed out that as calendars have changed over the years, different stem-branch pairs, and thus different fortunes, could be assigned to the date, month, and year of a same point in time.[132] The variation of the beginning of the year according to the longitudes can give rise to similar problems, because a same point in time can belong to different calendar years at different locations on earth.[133] He drew attention to the fact that the stem-branch pairs, and divina-

126 Ibid. II.4.2b-3a.

127 "Kŭnsi ron" in Kim Tae-wŏn, "Chŏng Yak-yong-ŭi *Ŭiryŏng*. 1," 348.

128 "Toryanghyŏng ŭi": *Yŏyudang chŏnsŏ* I.9.25b.

129 Ibid. I.9.25b.

130 "Maek ron 3" 脈論三: *Yŏyudang chŏnsŏ* I.11.13a-13b.

131 "Kap'ŭl ron 1": *Yŏyudang chŏnsŏ* I.11.28a.

132 "Kap'ŭl ron 2" 甲乙論二: *Yŏyudang chŏnsŏ* I.11.30a.

133 "Kap'ŭl ron 1": *Yŏyudang chŏnsŏ* I.11.28b-29a.

tions based on them, were formulated in ancient China, and thus should not be used outside China, which made him question why heaven should choose to inform only the people of China about good or bad luck coming their way.[134] He also noted that, unlike the directions of the north and the south, connected with the north and south poles, which are fixed on earth, the directions of the east and the west are relative to each other and cannot be fixed.[135]

Chŏng Yak-yong also objected to associations between two categories consisting of different numbers. He mentioned the difficulty of associating the ten stems with the four compass directions (*sifang* 四方) or the four seasons,[136] and the three virtues of 'knowledge' (*zhi* 知), 'humaneness' (*ren* 仁), and 'courage' (*yong* 勇) with the standard 'five virtues.'[137] He also pointed out that some of the associations of the four compass directions lead to unacceptable consequences. For instance, following on logically from such associations, there would be no compass directions at the north and the south poles, where there were no seasons.[138] On the associations of the four directions with the five constant virtues, he underlined that this would have to mean that in China (associated with the center) there should be only the virtue of 'faith' (*xin* 信) (associated with the center) and not the other four virtues associated with the other four directions.[139] Similarly, he pointed out that China should not be able to be called 'the Red Xian' (Chixian 赤縣) if the associations with the five colors were to be accepted.[140]

Such a critical attitude towards correlative thinking also underlay Chŏng Yak-yong's harsh rejection of what he called 'number studies' (*shuxue* 數學), by which he meant numerological speculations. To him, numerological interpretations of texts were as bad as interpreting the *Great Learning* by means of Buddhist doctrines.[141] While proposing the decimal system be adopted for weights and measures in place of such arcane practices that saw 16 ounces (*liang* 兩) equal one pound (*jin* 斤), he challenged the habit of considering the number 'eight,' which is associated with 'the four images' (*sixiang* 四象), 'the

---

134 Ibid. I.11.29a.

135 Ibid. I.11.28b, 29a; "Mun tŏngsŏnambuk": *Yŏyudang chŏnsŏ* I.9.1a. This understanding of the relative nature of the east and west, of course, was based upon the idea of a spherical earth.

136 Kap'ŭl ron 1": *Yŏyudang chŏnsŏ* I.11.28b.

137 *Chungyong kang'ŭi po*: *Yŏyudang chŏnsŏ* II.4.38a

138 "Mun tŏngsŏnambuk": *Yŏyudang chŏnsŏ* I.9.1b.

139 Ibid., I.9.2a.

140 Ibid.

141 "Tap Chungssi": *Yŏyudang chŏnsŏ* I.20.15a-15b.

eight trigrams' (*bagua* 八卦), etc. as 'the ancestor' (*zong* 宗) of number studies.[142]

The same attitude was also at the basis of Chŏng Yak-yong's dismissal of occult techniques. What he rejected was not superstition or supernatural beliefs in general, but the use of correlative thinking based on associative schemes, such as the stem-branch pairs. He felt much the same about 'occult' uses of the *Yijing*. He discredited attempts to apply the *Yijing* to calendars (*li* 曆), maintaining the opposite – that calendars should be applied to the *Yijing*. He said: "The calendar is the record of the sun, the moon and the five stars. Even if there is the slightest error, the four seasons will be off. How can one leisurely make it imitate the *Yijing*?"[143] He lamented the habit of connecting the "Diagram of the Supreme Ultimate" (Taiji tu 太極圖) to all sorts of things, even in interpreting the *Doctrine of the Mean*.[144]

Chŏng Yak-yong's denunciation of correlative thinking was clearly a deviation from the view of Zhu Xi, who generally accepted the yin-yang and the five-phase systems of association. Underlying this deviation were those key philosophical points where Chŏng Yak-yong disagreed with Zhu Xi. In particular, his separation between the natural world and human values and beliefs may have had the effect of precluding the possibility of connecting the natural world and the human realms through various kinds of associations. It was precisely this variety of associations that was frequently behind the arbitrary and superstitious beliefs which Chŏng Yak-yong criticized. Yet the seeds of that deviation were in Zhu Xi himself, who was also critical of some forms of correlative thinking. Zhu Xi said, for example, that manifestations of numbers in the world were "spontaneously so," and were not "arranged (*anpai* 安排).[145]

9 Concluding Remarks

So far in this essay I have repeatedly asked the question 'What did science mean for Chŏng Yak-yong?' I have tried to frame the question within the actual context of the society and culture in which he lived – his position in it, and the problems he faced. And in doing so, I have raised a number of related questions.

142 "Toryanghyŏng ŭi": *Yŏyudang chŏnsŏ* I.9.26a.

143 *Yŏkhak soŏn* 易學緒言, ch. 2: *Yŏyudang chŏnsŏ* II.46.14a.

144 *Chungyong kang'ŭi po: Yŏyudang chŏnsŏ* II.4.64a.

145 On this aspect of Zhu Xi's thought, see Kim, *Natural Philosophy of Chu Hsi*, 75-76.

The resulting picture is that of a fundamentally conservative scholar who considered himself as a defender of the Chosŏn *yangban* society, from the center of which he was involuntarily and temporarily displaced. His interest in science and the natural world could thus only be secondary to his primary concern – preserving and strengthening the Confucian tradition as the foundation for Chosŏn society. He did not have much interest in scientific subjects, except when their practical utility appeared to him as something that could help solve the problems of the day or better understand the Confucian classics.

PART 2

Social and Cultural Contexts of Science in East Asia

CHAPTER 5

Problems in the Study of the History of Chinese Science: An Early Review of the Field*

One of the questions most frequently asked about Chinese culture in general and about Chinese science in particular is "Why did the Scientific Revolution not take place in China?" or "Why did modern science not develop independently in China?"[1] This question has exerted a great influence on the scholarly study of traditional Chinese science.

For many of the scholars who have asked it, the question has provided a motive for them to study traditional Chinese science and scientific thought. For example, it is doubtless what underlies Joseph Needham's monumental work *Science and Civilisation in China*.[2] Although occasionally too speculative and far-fetched, it is nonetheless rich in information and insight into all the Chinese scientific and technical traditions and the schools of natural philosophy; it is the broadest, most interesting, determined and enthusiastic attempt to ask and answer the question "why not"?

But the 'why not' question has also had a different kind of influence on the field through the detailed works of those scholars who have deliberately put the question to one side or ignored it. Having concluded that asking this 'why not' question and its variations should be postponed until traditional Chinese science itself is much more thoroughly understood, these scholars have

* Originally published as "Natural Knowledge in a Traditional Culture: Problems in the Study of the History of Chinese Science," *Minerva* 20 (1982): 83-104.

1 Explicit formulation of some forms of the 'why not' question and attempts to answer them are found in Joseph Needham's "Mathematics and Science in China and the West," in *Science and Civilisation in China* (Cambridge: Cambridge University Press, 1954-), vol. 3, 150-168; and his "Science and Society in East and West," in *The Grand Titration* (Toronto: University of Toronto Press, 1969), 190-217.

2 For examples of very early works of the same nature, see Fung Yu-lan, "Why China Has No Science – An Interpretation of the History and Consequences of Chinese Philosophy," *The International Journal of Ethics* 32 (1922): 237-263; Homer H. Dubs, "The Failure of the Chinese to Produce Philosophical Systems," *T'oung Pao* Series 2, 36 (1928): 96-109; Derk Bodde, "The Attitude toward Science and Scientific Method in Ancient China," *T'ien Hsia Monthly*, 2 (1936): 139-160. More recent works that can still be considered worthy of attention are Wing-tsit Chan, "Neo-Confucianism and Chinese Scientific Thought," *Philosophy East and West* 6 (1957): 309-332; Hu Shih, "The Scientific Spirit and Method in Chinese Philosophy," in *The Chinese Mind*, ed. Charles A. Moore (Honolulu: University of Hawaii Press, 1967), 104-131.

 | DOI 10.1163/9789004265318-007

engaged in detailed, concrete studies of specific technical topics within Chinese scientific traditions. Nathan Sivin has been the most influential figure among them; he has steadily emphasized this position in his editorship of two series of monographs and a journal devoted to Chinese science, as well as in many of his own works.[3] These studies, together with similar works by Chinese and Japanese scholars[4] standing in the tradition of the 'evidential studies' (*kaozheng* 考證) of the Qing 清 (1644-1911), have greatly added to and deepened our understanding of various Chinese scientific traditions such as those of astronomy, calendrical method, mathematics, alchemy, medicine and harmonics.

During the last decade, a number of studies have even raised the issue of the admissibility of the 'why not' question itself; these studies deal with the historical, philosophical and sociological problems it stirs up. The sharpest and most cogent rejection of the question has come from A. C. Graham.[5] He argues that we cannot ask why an event did not happen unless there is some reason to have expected it, and that even in the conditions of Europe in the sixteenth century nothing justified thinking of the Scientific Revolution as an event which was bound to occur in the near future. Still less can we expect, Graham continues, that the conditions which existed in Western Europe, and which did favor the emergence of modern science there, would also have led to the emergence of modern science in China if they had existed there. Graham notes that explanations of China's failure to develop modern science have usually been

3 Sivin has been editor of the "MIT East Asian Science Series," of which six volumes have been published, of "Science, Medicine, and Technology in East Asia" (Center for Chinese Studies, University of Michigan) and the journal *Chinese Science* since 1973. He has also edited or co-edited two volumes of collections of papers: with Shigeru Nakayama, *Chinese Science: Explorations of an Ancient Tradition* (Cambridge, Mass.: MIT Press, 1973), and *Science and Technology in East Asia* (New York: Science History Publications, 1977). See also Sivin's *Chinese Alchemy: Preliminary Studies* (Cambridge, Mass.: Harvard University Press, 1968); "Cosmos and Computation in Early Chinese Mathematical Astronomy," *T'oung Pao*, Series 2, 6 (1969): 1-73.

4 The most notable examples are the work on the history of Chinese science done by a group at the University of Kyoto led by Yabuuchi Kiyoshi 藪内清 and by the Institute of the History of Natural Science at the Academia Sinica of Beijing. For bibliographical information on the works of the former group, see Shigeru Nakayama, "Kyoto Group of the History of Science," *Japanese Studies in the History of Science* 9 (1970): 1-4; Kiyoshi Yabuuti, "The Study of the History of Chinese Science in Kyoto," *Acta Asiatica* 36 (1979): 1-6. A bibliography of the works on Chinese astronomy by the Beijing group can be found in Xi Zezong, "Chinese Studies in the History of Astronomy, 1949-1979," *Isis* 72 (1981): 456-470.

5 A. C. Graham, "China, Europe, and the Origins of Modern Science," in Shigeru Nakayama and Nathan Sivin, *Chinese Science*, 45-69.

nothing more than proof that China did not follow the same route as Europe. He thus criticizes and rejects the various answers given by many scholars regarding the absence in China of certain conditions such as experimentation, mathematics, language with number and case, "more stringent, logical and intellectual arguments," "linear conception of time," "the concept of a divine legislator," "the rise of a merchant class to power,"[6] which would account for the non-development of modern science. Graham concludes that "the 'why not' question could be fruitfully asked if it should prove possible to detach the factors from their historical situation and show that they are necessary,"[7] and he finishes with a rhetorical question: "Is it necessary to say more than that one set of conditions for the genesis of modern science came together in sixteenth-century Europe, and that since it spread too fast to allow independent occurrence elsewhere this is the only set of conditions of which we can ever know?"[8]

Sivin too has frequently dealt with the same problems.[9] In the most recent expression of his reflections on them, he characterizes the 'why not' question as a 'heuristic' one useful only at the beginning of an enquiry, and he then proceeds to identify the two fallacies that usually accompany it:[10] namely, the arbitrary assumption that a given feature of Western thought at the time of the Scientific Revolution – the 'Baconian' method, for example – amounts to a necessary condition applicable to all cultures, and the confusion that the absence of further scientific development of a set of ideas in China – such as the hexagram system of the *Yijing* – is a result of its role as an 'inhibiting factor.' Sivin also points out how frequently it is assumed that the question can be answered by an examination of intellectual factors alone, or by social and economic factors alone.

Robert S. Cohen has considered the problem of the 'why not' question more abstractly, from a philosophical perspective.[11] He observes that the question predetermines the answers themselves, in other words that a presupposed answer is contained in what is asked and that this restricts the range of what is regarded as an answer to the question. Cohen sees in the 'why not' question a cluster of issues for the philosophy of science. For example, he stresses the

6 Graham, "China, Europe, and the Origins of Modern Science", passim.

7 Ibid., 54.

8 Ibid., 68.

9 Sivin, *Chinese Alchemy*, 1ff.; Nathan Sivin, "Shen Kua," in, *Dictionary of Scientific Biography*, vol. XII, ed. Charles C. Gillispie (1975): 369-393; "Introduction," in Sivin, *Science and Technology*, xi-xxiv; "Why the Scientific Revolution Did Not Take Place in China-or Didn't It?", *Chinese Science*, no. 5 (1982): 45-66.

10 Sivin, "Why the Scientific Revolution Did Not ...", 51ff.

11 Robert S. Cohen, "The Problem of 19(k)", *Journal of Chinese Philosophy* 1 (1973): 103-117.

necessarily comparative nature of the answers to the 'why not' question and the fundamental difficulty of establishing the comparability of cultural variables in different historical settings.[12] According to Cohen, variables such as slavery, feudalism and 'nature-mysticism' elude clear understanding even in one cultural context; some of these variables may themselves be candidates for a 'why not' question, others may be merely "intervening variables" or convenient "conceptual stand-ins" that have to be replaced later. Cohen also points out the difficulty in the "causal conditional arguments" employed in these discussions.[13]

Some sociologists whose primary interest has been the understanding of the nature of social and cultural influences on scientific knowledge have analyzed certain of the social and cultural factors which have been proposed as possible answers to the 'why not' question. Benjamin Nelson, for example, discussed Needham's various answers to the question.[14] He discerned "five interrelated systems or complexes in terms of which the changes in sociocultural patterns can be explained," and arranged under these five categories various factors that could help to resolve, among other things, the 'why not' question.[15]

The problems discussed in these analyses and criticisms of the 'why not' question are, however, not so much about the question itself as about the answers that have been given to it. In particular, they have pointed to a hidden

---

12 Cohen, "The Problem of 19(k)," 107ff.

13 Cohen raises other, more methodological issues, too. For example, he believes that the question must be approached with a "statistical sense," paying due attention to the "logic of small numbers," considering the problems of "norms" and "fluctuations." In his comment on Cohen's article, Arnold Koslow has even provided a logical analysis of typical attempts to answer the 'why not' question. Arnold Koslow, "More on 19(k)," *Journal of Chinese Philosophy* 2 (1975): 181-196.

14 Benjamin Nelson, "Sciences and Civilizations, 'East' and 'West': Joseph Needham and Max Weber," in *Boston Studies in the Philosophy of Science*, XI, ed. R. S. Cohen and M. Wartofsky (Boston: Reidel, 1974), 445-493.

15 Nelson, "Sciences and Civilizations, 'East' and 'West'," 451-452. The five are (1) "geographical-environmental settings"; (2) "socio-economic structures"; (3) "cultural ontologies and philosophical principles"; (4) "symbolic (in the narrow sense) technologies, notably linguistic patterns of the written and spoken languages"; and (5) "community and associational patterns and values, including norms of conduct." Another sociologist, Sal P. Restivo, has made an exhaustive enumeration of all the factors that Needham has seen as "inhibiting" or "conducive to" the emergence of modern science in China and the West. S. Restivo, "Joseph Needham and the Comparative Sociology of Chinese Science and Modern Science," in *Research in Sociology of Knowledge, Science and Art*, II, ed. Robert A. Jones and Henrika Kuklik (Greenwich, Conn.: Aijai, 1979), 25-51.

assumption that underlies many attempts to answer the question, namely the assumption that there is a universal developmental pattern of the growth of scientific knowledge. This assumption can be detected as implicit in much early and current scholarship on traditional Chinese science.

1 Assumption of a Single Developmental Pattern of Universal Science

This assumption forms part of another and still broader assumption concerning the general history of China, namely, the assumption that the difference between Chinese and Western cultures does not reflect different patterns of cultural development, but different stages in a single universal developmental pattern that every culture necessarily follows. This broader assumption has given rise to many debates concerning certain problems in interpreting traditional Chinese history, most notably those concerning its division into periods, and the problems of feudalism, serfdom, capitalism, and cities.[16]

Needham has given the most explicit expression of this assumption with respect to Chinese science. His view is that there is only one "unitary science of Nature" and that all the sciences of the past have progressed towards a common end – modern science – in "a single history and a single ever-growing structure" of this "unitary science."[17] This view appears strikingly in his frequently used image of "the ancient and medieval sciences of all the peoples and cultures as rivers flowing into the ocean of modern science."[18] To be sure, Needham is aware of criticisms of this view and he concedes that the sciences of the past should be studied and understood in their own cultural contexts. But convinced nonetheless that "modern science was their common end," he warns emphatically against the dangers implicit in arguments that deny "the fundamental continuity and universality of all sciences." He calls such arguments "Spenglerian pessimisms" that view sciences in different cultures as

16 John Meskill, ed., *The Pattern of Chinese History: Cycles, Development, or Stagnation?* (Boston: Heath, 1965); James T. C. Liu and P. Golas, eds., *Change in Sung China: Innovation or Renovation?* (Boston: Heath, 1969). See also Mark Elvin, *The Pattern of the Chinese Past: A Social and Economic Interpretation* (Stanford: Stanford University Press, 1973).

17 A most recent expression on this view is in his "Author's Note," in *Science and Civilisation*, vol. 5, part 2. (1974): xxi. Almost identical expressions are repeated in the "Author's Notes" to vol. 5, part 3, (1976) and part 4, (1980).

18 Needham, *Science and Civilisation*, vol. 5, part 2 (1974): xxviii.

developing in their own way and thus as understandable only within their own "frames of reference."[19]

Such a view has in various ways affected the selection and presentation of material in Needham's work. For example, he lays emphasis on materials which suggest Chinese anticipations and precedents of modern scientific theories and concepts; in his translations, he invariably chooses words that carry modern scientific connotations.[20] It is easy to understand how, when he deals with the similarity of ideas or inventions in China and those found in other cultures, Needham, with his imagery of rivers, characteristically infers transmission between cultures rather than independent development.

Needham's view is already contained in the table of contents of *Science and Civilisation in China*, where he uses the names of modern scientific disciplines to classify various areas of Chinese natural knowledge.[21] He begins with mathematics and moves on to astronomy, meteorology, geography, geology, mineralogy, physics and physical technology, chemistry and chemical technology, biology, agriculture and medicine. This classification is consistent with his outlook because, although he fully realizes that such subjects as biology or physics did not exist in traditional China as separate, autonomous areas of knowledge, they are, for him, the ultimate forms into which Chinese science would have had to develop in its march towards a 'universal' science. For this reason, he is consistent when he assigns astrology, alchemy and harmonics, which actually existed as separate specialized traditions in traditional China, to the headings of astronomy, chemistry, and physics (acoustics), while classifying another highly developed tradition, geomancy (*fengshui* 風水), as "pseudo-science." Needham also proceeds in the sequence of mathematics, astronomy, earth sciences, physics, chemistry and biology, which is basically the same as the order followed by modern classifications of the scientific disciplines; he looks upon this as the "logical" sequence.[22]

In contrast, Sivin, moving from the opposite assumption, lists the above-mentioned fields – astrology, alchemy, harmonics and geomancy – along with those of mathematics, astronomy, medicine and what he calls "physical stud-

---

19 Ibid., xxi-xxiii.

20 Needham, *Science and Civilisation*, passim; "Science and China's Influence on the World," in *The Grand Titration*, 55-122. Willard J. Peterson has also pointed out this tendency in Needham's work. W. J. Peterson, "'Chinese Scientific Philosophy' and Some Chinese Attitudes towards Knowledge about the Realm of Heaven-and-Earth," *Past and Present* 87 (1980): 20-30.

21 Needham, *Science and Civilisation*, vol. 1 (1954), xxii-xxxvii.

22 Ibid., vol. 4, part 1 (1962), xxviii.

ies," as the 'disciplines' of traditional Chinese science.[23] When he divides these disciplines into groups, he uses criteria not of modern science but of the sort that were implied by traditional Chinese usage. Thus, Sivin considers as 'fundamental' the distinction between orthodox and non-orthodox sciences which originally appeared in Needham's work.[24] In it, astronomy, astrology and harmonics belong to the former, and alchemy and geomancy to the latter, with medicine lying on the borderline. More recently Sivin has suggested a tripartite division that could employ the traditional Chinese trichotomy of 'heaven-earth-man': the sciences of "the cosmos and its phenomena"; of "the earth and its features, territorial divisions, creatures, and products"; and of "man and his institutions, usages, and accomplishments."[25] Elsewhere, he has also used the distinction between 'quantitative' and 'qualitative' in dividing the sciences with mathematics, astronomy and harmonics falling into the former category and the others into the latter.[26]

2 Traditional Sciences in Their Own Contexts: Traditional China and the Pre-Modern West

Among recent scholars, only Needham has explicitly espoused the notion of a single universal science. But the same point of view is often revealed in various tendencies in the studies of Chinese science and scientific thought. The assumption of a single universal pattern of scientific development accounts perhaps for the frequent comparisons of traditional Chinese science with that of the modern West, which is assumed to be the outcome of a normal development that the former should have followed. A corollary of this assumption is the mistaken belief that the sciences in China and Western Europe were about the same until the Scientific Revolution, when Western science underwent a radical transformation. This belief has led to the emphasis, in attempts to

23 Sivin, "Preface," in Nakayama and Sivin, *Chinese Science*; "Introduction," in *Science and Technology.*

24 Sivin, "Introduction," in *Science and Technology*, xxvi. See also Joseph Needham, "Poverties and Triumphs of Chinese Scientific Tradition," in *The Grand Titration*, 14-54; "Social Position of Scientific Men and Physicians in Medieval China," *Proceedings, XIVth International Congress of the History of Science* (Tokyo and Kyoto: Science Council of Japan, 1974), IV, 19-34.

25 Sivin, "Why the Scientific Revolution Did Not ...", 48.

26 Sivin, "Introduction" in *Science and Technology.*

answer the 'why not' question, on factors external to science[27] – social and economic factors or a new scientific method.

Comparisons between the contents and contexts of scientific knowledge in different traditional cultures, especially in traditional China and the pre-modern West, would be more interesting and fruitful.[28] Such comparisons would remind us that there were differences in those contents and contexts. They would also help us to realize that the Scientific Revolution was a very complex historical phenomenon uniquely rooted in Western culture. Such differences as might be disclosed by these comparisons could be the very ones that might have led to the different developments of scientific knowledge in the two cultures. But we need not claim that much. It would be sufficient to be interested in the differences as means of understanding the various forms of traditional scientific knowledge and their respective cultures.

There were traditions of theoretical knowledge in many scientific subjects in the West before the modern age; they date back to Greek antiquity. These scientific traditions persisted not only in fields like astronomy, mathematics and medicine, in which the possibility of immediate practical application was obvious, but also in subjects like cosmology, the theory of matter and the science of motion. By the High Middle Ages in Europe, though at different times for different subjects, all these subjects became highly technical and often mathematical. As legitimate forms of intellectual endeavor, they even acquired a rudimentary and marginal institutional basis in medieval universities.[29] It was partly the presence of these traditions of theoretical scientific knowledge in Western European culture that enabled its intellectuals to repeatedly ask the kind of questions to which the 'new scientists' of the seventeenth century found new answers.

Of course, there did exist traditions of technical natural knowledge in traditional China. But most often they consisted simply of technical knowledge used for practical purposes, such as calendars, healing, divination, manufacturing and warfare; they were not pursued with a theoretical or 'intellectual'

---

27 Needham characteristically doubts that there could have been many intellectual factors "which could not have been overcome if the social and economic conditions had been favourable." Needham, *Science and Civilisation*, vol. 5, part 2 (1974): xxiii.

28 Needham's *Science and Civilisation in China* is full of just these kinds of comparisons between various traditional sciences – Chinese, European, Islamic and Indian. However, very often, they are there merely to underline similarities between Chinese and pre-modern Western sciences (or more frequently Chinese supremacy) or to suggest the diffusion of scientific ideas and inventions.

29 David C. Lindberg, ed., *Science in the Middle Ages* (Chicago: University of Chicago Press, 1978).

interest. As such, they were rather isolated from the mainstream of the Chinese learned world: their practitioners were distinctly lower in social status than the members of the group called 'literati' (*shi* 士),[30] who were the social, political and intellectual leaders of traditional Chinese society. We do not know sufficiently about the connections between the bearers of those technical traditions and the literati; we do not even know exactly what the literati in the central intellectual tradition knew about the natural world.[31]

It is certain, however, that the position of natural knowledge in traditional Chinese learning was marginal, much more so than in the medieval West. This is clear from a comparison of the syllabuses of the medieval European universities with the tables of contents of the neo-Confucian anthologies, which were the standard textbooks for the education of the literati. The sciences had a place in the medieval universities in the *quadrivium* – arithmetic, music, geometry, astronomy – which every student had to study in addition to the *trivium* – grammar, rhetoric and logic. One of the 'three philosophies' was natural philosophy (the others were moral philosophy and metaphysics), based on the scientific writings of Aristotle. These subjects were always included in various medieval classifications of knowledge.[32] In contrast, knowledge about the natural world occupied much less space in the standard neo-Confucian textbooks such as *Zhuzi yulei* 朱子語類 (Classified Conversations of Master Zhu, compiled in 1270) and *Xingli daquan* 性理大全 (Collected Works on Human Nature and Principle, compiled in 1415). The standard schemes of classification in these books did include a category, '*li-qi* 理氣,' which covered topics in cosmology, astronomy and what could be called 'natural philosophy'; the discussions, however, were almost always weighted heavily toward moral, metaphysical and ontological concerns. Other fields of scientific knowledge, including medicine, were usually placed in the 'miscellaneous category' (*zalei* 雜類).

30 For the lack of a better word, I shall refer to this group consistently as 'literati' throughout this essay. 'scholar-officials' is perhaps a more accurate rendering of the term *shi* in most discussions of Chinese intellectual and social history, but it is too cumbersome.

31 This sentence, and a number of other sentences in this essay originally written more than thirty years ago, does not represent the current situation accurately. For example, a good deal of work has been done on the literati knowledge of, and attitude towards, the natural phenomena and technical traditions in the past three decades since this essay was first published. See the essays in Part I of this book.

32 James A. Weisheipl, "Curriculum of the Faculty of Arts at Oxford in the Early Fourteenth Century," *Mediaeval Studies* 26 (1964): 143-185; "Classification of the Sciences in the Mediaeval Thought," *Mediaeval Studies* 27 (1965): 54-90; "The Nature, Scope and Classification of the Sciences," in Lindberg, *Science in the Middle Ages*, 461-482.

Thus, while the major European thinkers of the seventeenth century, such as Bacon, Hobbes, Descartes, Mersenne and Pascal, when confronted with the intellectual crisis created by the challenge of the skepticism of various non-Aristotelian schools of thought, became deeply concerned with the problems of the natural world,[33] their Chinese counterparts in similar situations – Wang Yangming 王陽明 (1472-1528), and the thinkers of both the radical Taizhou 泰州 school and the orthodox Donglin 東林 group of the late Ming 明 – entered more and more deeply into problems of the human mind and moral cultivation.[34] An investigation into why these divergent paths were pursued in China and in the West may perhaps rest on the same mistaken assumption of a single developmental pattern. But it is certainly legitimate to inquire into the kind of natural knowledge possessed by these Chinese thinkers, with which they apparently were so satisfied that they were not impelled to ask further questions.

The assumption of a universal pattern of scientific development has also affected studies of the contexts of traditional Chinese natural knowledge; it has done so by stressing those factors which were significant in Western scientific development, as if they had to be the same factors which affected the development of Chinese science. This had led scholars working on the history of Chinese science to interest themselves in such social and economic phenomena as capitalism, urbanization, the merchant class and religious beliefs.[35] However, variables chosen for their importance in the Chinese context, rather than for their significance in Western development, might be more rewarding subjects of study.

A most promising subject for this type of study would be the Chinese civil service, especially in the period after the Song 宋 (960-1279), when it offered the possibility of a career as a civil servant, which almost completely overshad-

33 For a discussion of one facet of this intellectual crisis, see Richard H. Popkin, *The History of Scepticism from Erasmus to Spinoza*, rev. ed. (Berkeley: University of California Press, 1979).

34 For Wang Yangming, see Tu Wei-ming, *Neo-Confucian Thought in Action: Wang Yangming's Youth (1471-1509)* (Berkeley: University of California Press, 1976) and Julia Ching, *To Acquire Wisdom: The Way of Wang Yang-ming* (New York: Columbia University Press, 1976). For a more general discussion, covering the entire development of the neo-Confucian tradition, see Thomas A. Metzger, *Escape from Predicament: Neo-Confucianism and China's Evolving Political Culture* (New York: Columbia University Press, 1977), esp. Chap. 3.

35 See, for example, Needham, "Poverties and Triumphs."

owed other career alternatives for Chinese intellectuals.[36] Needham very rightly pays a great deal of attention to this aspect of Chinese culture. His extensive discussions of the relationship between science and the civil service contain information most pertinent to science, although his observations are frequently colored by his excessively negative assessment of the Chinese bureaucracy as an 'inhibiting' factor.[37] Deeper and more balanced knowledge of various aspects of this subject – such as the civil service examination, its role in education for the learned occupations and the place of science in the examination – is essential for a better understanding of the context of traditional Chinese science.

The Chinese civil service did include offices to be filled by experts on the calendar, mathematics and medicine.[38] Furthermore, officials who were not appointed to these specialized offices frequently had, as part of their official duties, tasks connected with technical subjects, for example, in agriculture, military strategy and supplies, surveys and map-making, transportation, public health and finance. At least during the Northern Song (960-1125), civil servants did not totally ignore such technical expertise.[39] Thus, in the very animated cultural life of eleventh-century Song, men like Shen Gua 沈括 (1031-1095) had a genuine interest in technical natural knowledge.[40] Even Zhu Xi 朱熹 (1130-1200), the author of the great neo-Confucian synthesis, showed a very broad concern for many technical subjects.[41]

36 For some general discussions of the Chinese civil service, see E. A. Kracke, Jr., *Civil Service in Early Sung China, 960-1067* (Cambridge, Mass.: Harvard University Press, 1953); Ho Ping-ti, *The Ladder of Success in Imperial China: Aspects of Social Mobility, 1368-1911* (New York: Columbia University Press, 1962); J. M. Menzel, ed., *The Chinese Civil Service: Career Open to Talent?* (Lexington, Mass.: Heath, 1963).

37 In addition to sections scattered throughout *Science and Civilisation*, see his "Thoughts on the Social Relations of Science and Technology in China," "Poverties and Triumphs," and "Science and Society," in *The Grand Titration*. See also "China and the Origins of Qualifying Examinations in Medicine," in Needham, *Clerks and Craftsmen in China and the West* (Cambridge: Cambridge University Press, 1970), 263-287 and "Social Position."

38 Needham, "Qualifying Examinations"; Shigeru Nakayama, *A History of Japanese Astronomy: Chinese Background and Western Impact* (Cambridge, Mass.: Harvard University Press, 1969), 12-23; Ulrich Libbrecht, *Chinese Mathematics in the Thirteenth Century* (Cambridge, Mass.; MIT Press, 1973), 2-21.

39 Sivin, "Shen Kua." On financial expertise in particular, see Robert M. Hartwell, "Financial Expertise, Examinations, and the Formulation of Economic Policy in Northern Sung China," *Journal of Asian Studies* 30 (1971): 281-314.

40 Sivin, "Shen Kua."

41 Yamada Keiji 山田慶兒, *Shushi no shizen-gaku* 朱子の自然學 ("Zhu Xi's Natural Knowledge", Tokyo, 1978).

As the civil service system evolved further, however, and especially following the Yuan 元 (1280-1367), when the government adopted the synthesis formulated by Zhu Xi as the basis for the civil service examination, officials' interest in technical subjects decreased considerably. Men like Shen Gua, and even Zhu Xi, became rare among the literati, whose intellectual interests became more and more literary and philosophical.[42] Although later in the seventeenth century there was an enhanced curiosity in natural knowledge among the literati, this represented in large part responses to the Jesuit publications of Western natural knowledge, and it failed in any case to develop into a sustained interest.[43]

The failure of the Chinese traditions of technical natural knowledge to penetrate the intellectual traditions of the literati needs investigation. There was nothing like the convergence of technical traditions and the main intellectual

---

42 On the process by which Zhu Xi's neo-Confucian synthesis became the state orthodoxy, see James T. C. Liu, "How Did a Neo-Confucian School Become the State Orthodoxy?" *Philosophy East and West* 23 (1973): 483-505; Conrad Schirokauer, "Neo-Confucians under Attack: The Condemnation of *Wei-hsueh*", in *Crisis and Prosperity in Sung China*, ed. J. W. Haeger (Tucson: University of Arizona Press, 1975), 163-198. For a more recent study covering a longer period, see W. Theodore de Bary, *Neo-Confucian Orthodoxy and the Learning of the Mind-and-Heart* (New York: Columbia University Press, 1981). Wing-tsit Chan also discusses an aspect of this 'narrowing down' of the neo-Confucian intellectual outlook in "The Ch'eng-Chu School of Early Ming," in *Self and Society in Ming Thought*, ed. William Theodore de Bary (New York: Columbia University Press, 1970), 29-51. For a discussion of the political and social changes that occurred at about the same time as this intellectual change, see Robert M. Hartwell, "Demographic, Political, and Social Transformations of China, 750-1550," *Harvard Journal of Asiatic Studies* 42 (1982): 365-442.

43 Willard J., Peterson, "Fang I-chih: Western Learning and the 'Investigation of Things'," in *The Unfolding of Neo-Confucianism*, ed. William Theodore de Bary (New York: Columbia University Press, 1975), 369-411; Joseph R. Levenson, *Confucian China and Its Modern Fate* (Berkeley: University of California Press, 1968), 3-14. For a discussion of Western natural knowledge in China at this time, see Willard J. Peterson, "Western Natural Philosophy Published in Late Ming China," *Proceedings of the American Philosophical Society* 117 (1973): 295-322.

Sivin, in discussions more focused on the introduction of Western astronomy, has suggested the presence of a still stronger possibility, referring to it as "scientific revolution of seventeenth-century China": Nathan Sivin, "Copernicus in China," *Studia Copernicana*, 6 (Polska Akademia Nauk, 1973), 63-122; "Wang Hsi-shan," in *Dictionary of Scientific Biography*, vol.14, ed. Charles C. Gillispie, 159-168. Jonathan Porter even sees a "scientific community" of the "specialists in exact sciences" in eighteenth-century China and attempts a prosopographic analysis; J. Porter, "The Scientific Community in Early Modern China," *Isis* 73 (1982): 529-544.

traditions such as took place in early modern Europe.[44] It is reasonable to inquire into the absence of such a convergence in China; this might be another form of asking the question, "Why were the Chinese literati not greatly interested in natural knowledge?" And it is necessary to examine the natural knowledge of the literati, especially men like Zhu Xi, whose teachings subsequently became the orthodox doctrine. Further research into the attitudes of the literati towards the technical traditions and the expert practitioners of those traditions is also desirable.

If the attitudes of Zhu Xi are representative of the literati, they were condescending and disdainful towards technical experts. True to his doctrine of the so-called 'investigation of things' (*gewu* 格物), which greatly emphasized the need to study the actual knowledge of concrete things and events in all the areas of human concerns, Zhu Xi studied and discussed the knowledge contained in such technical traditions as astronomy, calendrical method and harmonics. But at the same time he thought that moral and social problems were more important, and that technical subjects were of only secondary interest. He did not have much respect for the technical experts in those traditions. For example, Zhu Xi thought that the calendrical experts could only observe obvious facts and make practical computations, but could not fathom more profound matters.[45] He said that "contemporary calendar makers do not have any fixed method, but merely add to or subtract from the [observed] movements of the heavens to fit them [with the numbers in the calendar]."[46] Thus, Zhu Xi did not hesitate to criticize them on problems that are clearly related to calendars but that in his view lay outside their narrow technical expertise; sometimes he rejected their views entirely without careful consideration of the relevant technical details.

Scientific methods are given much prominence in modern discussions of traditional Chinese science and scientific thought; this is again a consequence of the assumption of a universal developmental pattern. Since rigorous logical methods, experimentation and mathematization are all considered to have had a role in the emergence of modern science in the West, their presence in traditional Chinese learning has been sought and investigated.[47] The persis-

44 See, for example, Paolo Rossi, *Philosophy, Technology and the Arts in the Early Modern Era* (New York: Harper, 1970).

45 For many examples of the literati's attitudes toward medical practitioners, see also Paul U. Unschuld, *Medical Ethics in Imperial China: A Study in Historical Anthropology* (Berkeley: University of California Press, 1979).

46 *Zhuzi yulei* 朱子語類 (1270 edition, reprinted in 1473), 2.11b.

47 See Fung, "Why China Has No Science"; Dubs, "Failure of the Chinese to Produce Philosophical Systems"; Bodde, "Attitude toward Science and Scientific Method in Ancient China."

tence of this assumption accounts for the scholarly interest in the logical methods of the ancient 'Dialecticians' and the Mohists,[48] despite the fact that the concern of these two groups with rigorous methods was not followed up by later generations of Chinese thinkers. The same persistent assumption underlies the intense focus on certain fundamental ideas that were significant in the development of Western science such as the concepts of the law of nature, time, causality and a creator.[49]

Of course, the modern perspective cannot be eliminated from our attempt to understand traditional Chinese science. A minimal amount of basic modern vocabulary and concepts are needed to expound to others what we have understood. What is to be avoided in the choice of subjects for our investigation into Chinese science, however, is the preponderant, if not exclusive, emphasis, on those concepts and aspects that were significant in the development of modern science in the West.

3 Overemphasis on Detailed Studies of Technical Texts

A different problem in the historiography of Chinese science is related to a reaction against the assumptions of the 'why not' question. The recent emphasis on increasingly more detailed, technical, and even textual, studies of particular texts – often quite marginal ones – can be problematic. The motive of these studies is of course admirable; behind them is a desire to master the content of Chinese science itself before asking more general questions. This

---

48 Among many scholars, including Needham, who have studied the topic since the appearance of the pioneering work by Hu Shih, *The Development of the Logical Method in Ancient China* (Shanghai, 1922), A. C. Graham has been the most persistent and the most reliable: see the synthesis of his research in *Later Mohist Logic, Ethics, and Science* (London: The School of Oriental and African Studies, 1978).

49 Needham has studied all these ideas in various places in his *Science and Civilisation*, but especially in vol. 2, 279-303 and 518-583 and vol. 4, part 1, 3-14. See also his *Time and Eastern Man* (London: Royal Anthropological Institute, 1965). For works by other authors, see Derk Bodde, "Evidence for 'Laws of Nature' in Chinese Thought," *Harvard Journal of Asiatic Studies* 20 (1957): 709-727; "Chinese 'Law of Nature': A Reconsideration," *Harvard Journal of Asiatic Studies* 39 (1979): 139-155; Manfred Porkert, *The Theoretical Foundations of Chinese Medicine: Systems of Correspondence* (Cambridge, Mass.: MIT Press, 1974), 1-8; Donald Leslie, "Les theories de Wang Tch'ong sur la causalité," in *Melanges offertes a Monsieur Paul Demieville* (Paris, 1974), vol. 2, 179ff; N. J. Girardot, "The Problem of Creation Mythology in the Study of Chinese Religion," *History of Religions* 15 (1976): 289-318; John S. Major, "Myth, Cosmology and the Origins of Chinese Science," *Journal of Chinese Philosophy* 5 (1978): 1-20.

perfectly healthy attitude can, however, become excessive.[50] Such studies are prerequisites for the correct understanding of Chinese natural knowledge in its own intellectual terms and its own social context and pattern of development. However, they are not enough; we wish to know more than the scientific content of the technical traditions. After all, we are not interested in re-creating these traditions.[51] Our task is to understand the nature, place, role and impact of these technical branches of knowledge in traditional Chinese culture.

Admittedly, works treating these broader questions have not been lacking. From the writings of Needham on medicine, for example, we have learned much about the variety of the social positions of medical practitioners in traditional China and the role of the state in overseeing and promoting medical knowledge and practice.[52] From the works of Sivin on alchemy, we have learned how the Chinese alchemical practices can be viewed as the simulation, on a scale far reduced both in time and in space, of the hidden operations of nature.[53] Sivin's work on astronomy has shown us the conception of the universe as a congeries of cyclical time relationships that underlies ancient Chinese mathematical astronomy.[54] Shigeru Nakayama's brief essay on astrology has helped us understand the basic assumptions of Chinese astrology and its close connection with court politics and the calendars.[55] There have been a great many speculative works on traditional Chinese medicine from philosophical, sociological, anthropological and even modern scientific perspectives, some of them extremely instructive.[56]

Recently, however, the tendency in the study of Chinese science has on the whole been very markedly towards detailed technical studies. This emphasis

50 See Lawrence Stone's address to the American Historical Association Annual Meeting, Washington, D. C, 27-29 December, 1982.

51 Manfred Porkert, for example, is explicit about the purpose of his studies of traditional Chinese medicine, namely to make it applicable to present-day medical practice: Porkert, *Theoretical Foundations of Chinese Medicine,* 5. This explains, at least in part, the artificial precision in his translation, or over-translation, of many Chinese terms, e.g., "configurational energy" for *qi* 氣; "the active and structive physiological energies" for *xueqi* 血氣.

52 Needham, *Clerks and Craftsmen*, 263-287.

53 Sivin, "Chinese Alchemy and the Manipulation of Time," *Isis*, 67 (1976): 513-526, and its expanded version in *Science and Civilisation*, vol. 4, part 4, 210-323.

54 Sivin, "Cosmos and Computation."

55 Nakayama, "Characteristics of Chinese Astrology," *Isis*, 57 (1966): 442-454.

56 See, for example, Porkert, *Theoretical Foundations of Chinese Medicine*; Nathan Sivin, "Social Relations of Curing in Traditional China," *Nihon ishigaku zasshi* 日本醫史學雜誌 [Japanese Journal of Medical History] 23 (1977): 505-532; Unschuld, *Medical Ethics in Imperial China,* 3-24.

has resulted in works of a very high standard, and has greatly enhanced our understanding of Chinese technical traditions. But it has had negative consequences as well, in part owing to its very merits. For these thoroughly scholarly studies of many Chinese scientific traditions – such as mathematics, astronomy, optics, medicine and various technical arts[57] – tell very little about the extent to which these traditions penetrated the Chinese intellectual world and about the Chinese intellectuals' ideas about the natural world. These works do not make clear that their subject matter actually had a relatively minor position among the intellectual concerns of the literati. At times they even obscure the fact that the technical knowledge described was confined to a quite special group of persons on the fringes of the Chinese intellectual community. Our understanding of medieval Western science was brought into proper perspective by the work of scholars like Anneliese Maier,[58] and more recently of John

---

57 This tendency has been most noticeable in mathematics, in which technical expertise is indispensable. See for example Lay Yong Lam, "The Geometrical Basis of the Ancient Chinese Square-Root Method," in Sivin, *Science and Technology*, 56-66; "The Jih yung suan fa: An Elementary Arithmetic Textbook of the Thirteenth Century," in Sivin, *Science and Technology*, 68-81; Libbrecht, *Chinese Mathematics*; Donald B. Wagner, "Liu Hui and Tsu Keng-chih on the Volume of a Sphere," *Chinese Science*, no. 3 (1978): 1-26; Lay Yong Lam, "Chu Shih-chieh's *Suan-hsueh ch'i-meng* (Introduction to Mathematical Studies)," *Archive for the History of Exact Sciences*, 21 (1979): 1-31; Christopher Cullen, "An Eighth Century Chinese Table of Tangents," *Chinese Science*, no. 5 (1982): 1-33. See also other recent works cited in U. Libbrecht, "Joseph Needham's Work in the Area of Chinese Mathematics," *Past and Present* 87 (1980): 30-39.

For examples of similar works in other areas, see Christopher Cullen, "A Chinese Eratosthenes of the Flat Earth," *Bulletin of the School of Oriental and African Studies* (London) 39 (February 1976): 106-127; Y. Maeyama, "On the Astronomical Data of Ancient China (ca. -100~+200): A Numerical Analysis," *Archives internationales d'histoire des sciences* 25 (1975): 247-276; 26 (1976): 27-58; A. C. Graham and Nathan Sivin, "A Systematic Approach to the Mohist Optics," in Nakayama and Sivin, *Chinese Science*, 105-152; William C. Cooper and Nathan Sivin, "Man as a Medicine: Pharmacological and Ritual Aspects of Traditional Therapy Using Drugs Derived from the Human Body," in Nakayama and Sivin, *Chinese Science*, 203-272; Brian E. McKnight, trans., *The Washing Away of Wrongs: Forensic Medicine in Thirteenth-Century China* (Ann Arbor: University of Michigan Press, 1981); Andre W. Sleeswyk, "Reconstruction of the South-pointing Chariots of the Sung Dynasty: Escapement and Differential Gearing in Eleventh Century China," *Chinese Science*, no. 2 (1977): 4-36; "Celestial River: A Reconstruction," *Technology and Culture* 19 (1978): 423-449.

58 Anneliese Maier, *Studien zur Naturphilosophie der Spdtscholastik*, 5 vols. (Rome: Edizioni di Storia e Letteratura, 1949-58).

E. Murdoch,[59] who have placed medieval scientific texts in a broad and more accurately studied context of philosophical ideas. Only after such a perspective had been attained could the results of more detailed, technical studies of particular scientific texts be properly assessed.[60] Such an understanding has scarcely been achieved in scholarship dealing with traditional Chinese science. The concentration on detailed technical studies of particular texts has even had the effect of discouraging the formation of such a perspective – on the bigger picture.

The difficulty lies in part in the overemphasis on technical details. This restriction of attention to those limited areas of knowledge in traditional China which attained a degree of technical refinement comparable to that of modern scientific disciplines has led to the neglect of many other fields of knowledge about the natural world that did not reach such a level of sophistication. This can be seen as another consequence of an ahistorical imposition of the standards of modern science. For, though not restricting its focus on the subjects which belong to modern scientific disciplines, this tendency makes the scholars consider rather only those subjects that reached the technical level of modern science.[61] If one of our objectives in studying the scientific knowledge achieved by a traditional culture is to understand the style, method and degree of technical complexity of its knowledge about the natural world,

59 John E. Murdoch, "Philosophy and the Enterprise of Science in the Later Middle Ages," in *The Interaction between Science and Philosophy*, ed. Yehuda Elkana, (Atlantic Highlands, N.J.: Humanities Press, 1974), 51-74, 91-113. See also the articles in John E. Murdoch and Edith D. Sylla, eds., *The Cultural Context of Medieval Learning* (Boston: Reidel, 1975).

60 See, for example, the following works in the series "University of Wisconsin Publications in Medieval Science" (Madison: University of Wisconsin Press, 1952-): Ernest A. Moody and Marshall Clagett, *The Medieval Science of Weights* (1952); Curtis Wilson, *William Heytesbury: Medieval Logic and the Rise of Mathematical Physics* (1960); Marshall Clagett, *The Science of Mechanics in the Middle Ages* (1959); *Nicole Oresme and the Medieval Geometry of Qualities and Motions* (1968); David C. Lindberg, *John Pecham and the Science of Optics* (1970); Edward Grant, *Nicole Oresme and the Kinematics of Circular Motion* (1971).

61 In this respect, it is interesting to note that Needham, who covers traditional Chinese knowledge in all the areas dealt with by modern scientific disciplines – irrespective of their technical level – does not omit any subject of traditional Chinese natural knowledge, or of its philosophical context. In contrast, in Sivin's list of Chinese scientific 'disciplines,' the vaguely defined field of 'physical studies' ("the area in which fundamental concepts were adapted and applied to explain particular physical phenomena, as well as chemical, biological, and psychological phenomena, that were not distinguished or were thought to be closely related") has to cover all the remaining subjects of natural knowledge that do not belong to the other 'disciplines': Sivin, "Preface," in Nakayama and Sivin, *Chinese Science*, xxiii.

the emphasis of recent scholarship precludes that understanding from the beginning by its decision to study only those subjects characterized by a certain style, method and level of technical complexity. This emphasis has certainly resulted in detailed knowledge of the substance of these technical traditions, which in the case of traditional China also included bureaucratic, financial and military administration and ceremonial protocols, as well as various areas of the arts and technology.[62] These are extremely interesting subjects that are essential to the scholarly studies of traditional Chinese science. However, knowing about them is not enough.

The issue under discussion here is not simply the difference between the 'internalist' and the 'externalist' approaches to the history of science,[63] though not altogether unrelated to it. What is at issue is not just the exclusive interest in the contents of the texts; rather it is the exclusive interest in certain kinds of texts, the importance of which is not measured primarily by the contemporary Chinese intellectual context, nor by the influence which they exerted on the subsequent intellectual development of Chinese natural knowledge, but rather by the achievement, represented in these texts, of a level of technical complexity comparable to that possessed by modern scientific disciplines.

Now that some foundations have been laid in concrete studies, scholarly research in Chinese science should broaden its scope and perspective. Above all, it should be integrated into Chinese intellectual history, of which it is a legitimate part. A particularly pertinent subject for understanding the intellectual context of traditional Chinese science is neo-Confucianism, which was the dominant mode in the thought and values of the literati. Thus far, however, the study of Chinese science has been conducted largely in isolation from the increasingly more learned and sophisticated studies that have been appearing on neo-Confucianism.[64] Nor have scholars working on neo-Confucianism paid

---

62 This can be seen from the contents of the most famous book in this genre, *Mengxi bitan* ("Brush Talks from Dream Brook") by Shen Gua, analyzed by Needham in *Science and Civilisation*, vol. 1, 136, and by Sivin in "Why the Scientific Revolution Did Not ...", 66.

63 For a brief discussion of this distinction, see Thomas S. Kuhn, "History of Science," in *The International Encyclopedia of the Social Sciences*, vol. 14 (New York: Macmillan, 1968), 74-83.

64 De Bary, *Self and Society*; de Bary, *The Unfolding of Neo-Confucianism*; William Theodore de Bary and Irene Bloom, eds., *Principle and Practicality: Essays in Neo-Confucianism and Practical Learning*, (New York: Columbia University Press, 1979); Hok-lam Chan and William Theodore de Bary, eds., *Yüan Thought: Chinese Thought and Religion under Mongols* (New York: Columbia University Press, 1982). The themes of two of the most recent international conferences on neo-Confucianism, one on Korean neo-Confucianism (held in Bellagio, Italy, 3-7 July, 1981) and the other on Zhu Xi (in Honolulu, 6-15 July,

much attention to the recent studies on Chinese science.[65] That cannot be just because the level of technical complexity of Chinese science has been beyond the reach of those who study neo-Confucianism. It is conceivable that they have found little relevance in the technical studies, which do not even tell them what kind of knowledge the literati possessed about the physical world in which they lived.

There is at least one field in the study of Chinese intellectual history, namely the study of Daoism, which has established a connection with the studies of traditional Chinese science.[66] The work of Maspero, Needham, Welch and others has shown that many Chinese scientific traditions, including those of alchemy, medicine and astrology, originated in part from, and maintained connections with, the various beliefs and practices called 'Daoism.'[67] Yet this connection between science and Daoism may have been overemphasized. Many of the scientific traditions, such as those of mathematics, astronomy and harmonics, which Needham calls 'orthodox,' had little place in Daoist concerns. And even when we are dealing with those other sciences apparently connected with Daoism, only some association with the traditions of the literati could

---

1982), testify to the growing breadth and depth of recent scholarship on neo-Confucianism. However, 'science' or 'natural knowledge' were characteristically absent from the list of topics discussed there. See William Theodore de Bary and Ja Hyun Kim Haboush, eds., *The Rise of Neo-Confucianism in Korea* (New York: Columbia University Press, 1985); Wing-tsit Chan, ed., *Chu Hsi and Neo-Confucianism* (Honolulu: University of Hawaii Press, 1986).

65 Needham's *Science and Civilisation*, vol. 2 still seems to be the single work most frequently read by Chinese intellectual historians.

66 See, for example, the special issue ("Current Perspectives in the Study of Chinese Religions") of *History of Religions*, vol. 17 (February and May 1978), especially Nathan Sivin, "On the Word 'Taoist' as A Source of Perplexity. With Special Reference to the Relations of Science and Religion in Traditional China," 303-330, K. Schipper, "The Taoist Body," 355-386 and M. Saso, "What is the Ho-t'u?" 399-416; Holmes Welch and Anna Seidel, eds., *Facets of Taoism: Essays in Chinese Religion* (New Haven: Yale University Press, 1979), esp. M Strickmann, "On the Alchemy of T'ao Hung-ching," 123-192 and Hou Ching-lang "The Chinese Belief in Baleful Stars"; Sakai Tadao 酒井忠夫, ed., *Dōkyō no sōgōdeki skenkyū* 道教の綜合的研究 ("Synthetic Studies on Daoism", Tokyo, 1977), which contains articles (in Japanese) by Sivin (on "ritual medicine") and by Needham (on social manifestations of Daoist alchemy).

67 Henri Maspero, *Taoism and Chinese Religion* (Amherst, Mass.: University of Massachusetts Press, 1981); Needham, *Science and Civilisation*, vol. 2, 33-164; vol. 5, part 2, 62-127; vol. 5, part 3, 1-117; Holmes Welch, *Taoism: The Parting of the Way*, rev. ed. (Boston: Beacon, 1965), part 3; M. Kaltenmark, *Lao Tzu and Taoism* (Stanford: Stanford University Press, 1969), Chap. 5.

have permitted a sustained intellectual development. It is time to redress the balance and to look more closely into the relationship between science and Confucian literati. In this respect, it would be far more instructive to examine how the abstract philosophical tenets of Daoism exerted a formative influence on neo-Confucian natural philosophy.[68]

Thus it is that we return to the need to examine the traditions of the literati. We need to study not just the natural knowledge of the literati but also their thought in general as the context of that natural knowledge. In particular, we should look at the basic beliefs and the intellectual assumptions of the literati and consider the roles these beliefs and assumptions played in their natural knowledge. To be sure, the 'literati' were never a clearly defined group. They included a variety of persons ranging over a wide array of values and beliefs.[69] Their natural knowledge and the basic ideas and the implicit assumptions underlying it – their 'worldview' – were neither identical nor unchanging. Many of the literati said little about the natural world. But there were others from whose works we can reconstruct particular examples of the presumably widely shared, though changing, 'worldview' of the literati. Men like Zhang Zai 張載 (1020-1077), Zhu Xi, Wang Fuzhi 王夫之 (1619-1692) and Dai Zhen 戴震 (1723-1777) are good examples, although each of them was also very unique. Each of them left excellent texts that can be fruitfully studied. Other excellent sources for such studies are anthologies like the *Xingli daquan*, which, based essentially on the orthodox teachings of Zhu Xi, presented the basic ideas of neo-Confucian philosophy topic by topic in a textbook-like manner. The existence of such anthologies both testified to and maintained the presence in traditional China of the widely shared and fairly stable 'neo-Confucian worldview.'

These subjects, however, have largely been ignored by specialists on traditional Chinese science.[70] There are only some fragmentary treatments of the topics scattered in the writings of a handful of scholars of Chinese intellectual and cultural history. The works of Fung Yulan, Wing-tsit Chan and of A. C. Graham discuss concepts like *li* 理, *qi*, the 'numinous (*shen* 神),' 'heaven (*tian*

68 This subject has been neglected by students of neo-Confucianism as well as of Daoism. For a very brief treatment of the topic, see Wing-tsit Chan, "Chu Hsi's Completion of Neo-Confucianism," in *Sung Studies, In Memoriam Etienne Balazs*, second series, vol. 1 (Paris: Mouton, 1973), 59-90.

69 A. F. Wright, "Values, Roles, and Personalities," in *Confucian Personalities*, ed. A. F. Wright and D. Twitchett, (Stanford: Stanford University Press, 1962), 3-23.

70 Needham and Yamada Keiji, who have frequently dealt with the natural knowledge of the literati and its general philosophical context, are almost the only exceptions. See, for example, Needham, *Science and Civilisation*, vol. 2, and Yamada, *Shushi no shizengaku*.

天),' and the 'investigation of things,' which are concepts in the basic vocabulary of the natural philosophy of the literati.[71] The writings of Derk Bodde contain information on the Chinese conceptions of the objects and phenomena at the boundaries of what we see as 'natural,' such as myths and the supernatural.[72] D. C. Lau's work provides an example of how Chinese thinkers, in the absence of a formal logical method, went about arguing their cases.[73] Only very recently, Willard J. Peterson has supplied us with an intelligible account of the intentions of the "Great Treatise" (Xici zhuan 繫辭傳) of the *Yijing* 易經, which is one of the basic sources of Chinese natural philosophy.[74]

4 The Fundamental Intellectual Outlook of the Neo-Confucian Literati: Zhu Xi as an Example

Comprehensive studies of the basic concepts and assumptions in the thought of the literati can throw light on the particular course of the development of natural knowledge in China. The thought of Zhu Xi provides illustrations of what such studies can reveal.

The concept of *li* occupied such an important place in neo-Confucian philosophy that it is sometimes referred to as 'the learning of *li*' (*lixue* 理學). Frequently translated as 'principle,' *li* has many layers of meaning.[75] In the thought of many neo-Confucian literati like Zhu Xi, however, the *li* of an object or phenomenon does not explain the object or phenomenon. *Li* is merely that

71 Fung, Yu-lan, *A History of Chinese Philosophy*, 2 vols., (Princeton: Princeton University Press, 1937, 1952); Wing-tsit Chan, *A Source Book in Chinese Philosophy* (Princeton: Princeton University Press, 1963); "The Evolution of the Neo-Confucian Concept *Li* as Principle," *Tsing Hua Journal of Chinese Studies*, new series, 4 (1964): 123-149; A. C. Graham, *Two Chinese Philosophers: Ch'eng Ming-tao and Ch'eng Yi-ch'uan* (London: Lund Humphries, 1958).

72 D. Bodde, "The Chinese View of Immortality" (originally published in 1942), "The Chinese Cosmic Magic Known as 'Watching for the Ethers'" (1959), "Myths of Ancient China" (1961), and other articles reprinted in Bodde, *Essays in Chinese Civilization* (Princeton: Princeton University Press, 1981).

73 Lau, "On Mencius' Use of the Method of Analogy in Argument", in *Mencius*, trans. D. C. Lau (Harmondsworth: Penguin, 1970), 235-263; "A Note on *Ko-wu*," *Bulletin of the School of Oriental and African Studies* (London) 30 (1967): 353-357.

74 Willard J. Peterson, "Making Connections: The 'Commentary on the Attached Verbalizations' in the *Book of Changes*," *Harvard Journal of Asiatic Studies* 42 (1982): 67-116.

75 In a recent paper Chung-ying Cheng has provided an analytical list of the various meanings of *li*: "Chu Hsi's Methodology and Theory of Understanding", in Chan, *Chu Hsi and Neo-Confucianism*, 169-196.

because of which the object exists or because of which the phenomenon occurs as it actually does. In other words, *li* represents a kind of necessary condition for the existence of the object or the occurrence of the phenomenon. In Zhu Xi's own words, "When there is *li* for this, only then is there this thing or event."[76] For Zhu Xi, therefore, *li* is not a simple, basic 'principle' that can be used in an analysis of a complex object or phenomenon. It refers to a given object or phenomenon as a whole. The *li* of an object or phenomenon has little additional content beyond the object or phenomenon itself. Although Zhu Xi spoke of the *li* of movement and rest, it is merely something that determines that there should be movement and rest, and not something that permits an explanation of the nature and details of movement and rest. Similarly, although Zhu Xi said, "[the fact that] complex interconnection of the yin and yang and the five phases (*wuxing* 五行) does not lose order is just [because there] is *li*,"[77] *li* does not tell us anything about this order except that it exists.

It is therefore not surprising that Zhu Xi did not analyze or study *li*, but merely invoked it. What he asked were questions like "Is there such *li*?", implying of course that, if there is, the corresponding object or phenomenon would also exist. But he did not discuss the content of that *li*. This interest in questions about the existence, and not about the content, of *li*, in turn, led Zhu Xi and his followers to ask questions like "Where does *li* exist?", "Where can I find *li*?" and even "How can I attain *li*?" And it is for this reason that the central neo-Confucian endeavor of 'investigation of things,' though it aimed at the *li* of 'things,' always had a moral – as well as an intellectual – intent, involving not only exertion in study, but also immersion in quietude (*jing* 靜) and composure (*jing* 敬).

Another fundamental concept in the thought of the neo-Confucians is *qi*, which constitutes every object and underlies every phenomenon in the world. What is interesting for us about this is while *qi* undergoes various forms of action and movement such as aggregation and dispersion, ascent and descent, contraction and expansion, these are its innate qualities that do not require the action of external causes.

This conception has an important implication regarding the way in which the literati understood natural phenomena. For when certain phenomena had been attributed to certain actions and movements of *qi*, they were considered to have been sufficiently accounted for. These actions and movements were inherent to *qi* and there was no need to look for their external causes. For example, while explaining the beginning of heaven and earth, Zhu Xi attrib-

76 *Zhuzi yulei*, 13.12a.

77 Ibid., 1.2b.

uted the formation of the earth to a rapid turning of *qi*: "When the turning became rapid, a large quantity of the 'sediments' of *qi* was compressed, and as there was no outlet [for the sediments] they consolidated to form the earth at the center."[78] But he paid no attention to the cause of the turning, as if turning were the thing for *qi* to do. Similarly, Zhu Xi conceived the production of life as occurring through the spontaneous congelation of the essences of yin *qi* and yang *qi*, but he did not inquire into the cause of this congelation.

Moreover, the absence of the need to look for external causes for the actions and movements of *qi* had the further effect of precluding detailed analyses of such actions and movements, which would have raised questions as to why certain actions and movements should produce a certain particular phenomenon. Even when the *li* of these actions and movements was mentioned, it was merely invoked and was never used to seek a deeper understanding or explanation of them. Thus, Zhu Xi not only did not ask what caused the essences of the *qi* to be congealed, but he also did not ask why the congelation of those essences of *qi* could produce life.

My next example is not about a single concept but about an assumption underlying much of neo-Confucian thought, namely the dichotomy of what is 'above physical form' (*xing er shang* 形而上) and what is 'below physical form' (*xing er xia* 形而下). *Li, dao* 道 (the Way), and benevolence (*ren* 仁) – abstract and sublime concepts without manifest physical forms (*xing* 形) – belong to the former, whereas actual concrete things with tangible physical forms belong to the latter. It was commonly assumed that what has physical form and is visible is easy to understand and even trivial, whereas what is without physical form is difficult to understand and hence worthy of further investigation. Zhu Xi had this idea in mind when he said, "The error in the weights and lengths of 'things' is easy to see; the error in the weights and lengths of the mind is difficult to see. The error in things does no harm; the error in the mind does harm."[79]

This idea also seems to have influenced Zhu Xi's attitude towards natural phenomena, especially his relative lack of interest in them. Most natural phenomena are directly perceived by man. They are accompanied by tangible qualities and physical effects, and belong to what is 'above physical form.' Natural phenomena were thus thought to be obvious and were taken for granted; they were in most cases simply accepted in the way human beings observed them. What was more important and worthy of consideration for Zhu Xi were the more difficult problems in the moral and social realms. When Zhu Xi discussed natural phenomena, frequently it was not for their own

78 Ibid., 1.4b.

79 *Zhuzi yulei*, 51.5a.

intrinsic interest but to explain moral or social problems. He drew on knowledge of common natural phenomena as analogies for more challenging moral and social issues.[80] For example, Zhu Xi used the fact that once a cart has started to move, no great exertion of force is needed to keep it moving, to support the view that in study as well, great exertion is needed only at the outset, after which it becomes easy. Similarly, in order to demonstrate how a little evil mixed with goodness spoils the goodness entirely, he argued analogically that when gold is mixed with a small quantity of silver all of the gold loses its value as gold. Obviously, we can learn about Zhu Xi's ideas on some natural phenomena from these examples; we can know what he thought, for example, about the movements of objects and about the properties of mixtures of metals. But in neither of these instances were the natural phenomena themselves of intrinsic interest to Zhu Xi. To him, they were obvious facts that needed no further consideration. What he was really interested in was to argue for two propositions – that the beginning is the most important phase of study, and that even the slightest corruption of evil should not be tolerated.

As Zhu Xi simply accepted most common natural phenomena as obvious in this manner, so he did not attempt further theoretical or abstract examination of them beyond their phenomenal surface. Therefore, there was little prospect of considering such complex, yet fundamental, abstract problems as those of the void, the infinite, the mover versus motion, the indivisibility of atoms, matter versus space, and the way elements exist in mixtures. The consideration of such problems could have generated controversies as they did in the medieval West, and thus led to a closer examination.[81]

Another neo-Confucian attitude seems to have played a role in this respect: the emphasis on the reality of the external world. Characteristic of Confucian thought from the beginning of its development, this emphasis was sustained in its neo-Confucian phase, despite the powerful infusions of influence from the Daoist and the Buddhist philosophies, both of which denied the reality of the phenomenal world. Neo-Confucians like Zhu Xi considered acceptance of the reality of the world to be what distinguished them from the Daoists and

80 John E. Murdoch mentions instances in medieval Europe in which certain theological points seem to have been used as excuses to discuss such natural phenomena as light, motion and heavenly spheres, cases which appear to be exactly opposite to what Zhu Xi does: Murdoch, "From Social into Intellectual Factors: An Aspect of the Unitary Character of Late Medieval Learning," in Murdoch and Sylla, *The Cultural Context of Medieval Learning*, 271-339, esp. 278-279.

81 William A. Wallace, "The Philosophical Setting of Medieval Science," in Lindberg, *Science in the Middle Ages*, 91-119; Edward Grant, "Cosmology", in Lindberg, *Science in the Middle Ages*, 265-302.

Buddhists with their otherworldly tenets. Zhu Xi thus rejected concepts like 'emptiness,' 'nothingness' and 'infinity' and avoided discussions of them because of their association with the Daoists and Buddhists, who tended to lead men to concentrate on introspection and ignore the real world. To Zhu Xi, such concepts as 'void,' 'infinite,' 'matter' and 'space' were imaginary. They were of no help in reckoning with the reality of the actual world, nor were they useful in dealing with the moral and social problems of that world.

Speculation about these abstract, theoretical concepts frequently contributes to a better understanding of natural phenomena. In the West at least, controversies over the interpretation of precisely such concepts eventually led to their resolution during the Scientific Revolution.[82] Ironically, the emphasis of the neo-Confucians on the reality of the actual world made it difficult for them to consider in detail those very concepts that could have proved productive in understanding that world.

5 Concluding Remarks

I return now to the 'why not' question. Although negative opinions have increasingly been expressed about its validity, the 'why not' question remains an interesting one. That many scholars have asked the question and many will continue to do so, even if only to reject it, shows that the question indeed triggers genuine interest. Sivin, who was the first to raise an explicit doubt about the significance of the question, is also the author of the most recent essay dealing with it.[83]

There is nothing inherently wrong with the 'why not' question itself. The real issue lies in the suggested answers to the question; these have tended to single out the factors that were significant in Western scientific development while neglecting the context of science in China. This tendency has also led discussions of the 'why not' question to stress social and economic factors at the cost of intellectual ones. In the selection of subjects, the same tendency has led to an emphasis on the level of technical complexity rather than on the content of the natural knowledge itself. We should seek to avoid these pitfalls.

82 It is interesting to note that much of the medieval European discussion of concepts like 'void' and 'infinite' was undertaken in a theological context. See, for example, Murdoch, "From Social into Intellectual Factors," 274ff. A similar theological context did exist in Chinese Buddhism and in Daoism but they were both opposed by the dominant Confucian literati.

83 Sivin, *Chinese Alchemy*; "Why the Scientific Revolution Did Not ...", 1

However, restricting ourselves to detailed studies of technical scientific texts, although they are of the utmost importance, cannot be a satisfactory alternative to posing the question.

It might perhaps be possible to reduce the difficulties connected with the 'why not' question by reformulating it as "Why did the Chinese develop their knowledge of the natural world in the way they did and not in the way the Europeans developed modern science?" Framing it in this manner, we may be able to divide the range of issues raised by the 'why not' question into more manageable parts. We may, for example, inquire, as Arnold Koslow has suggested, into the reason for the 'discontinuance' of technical traditions such as optics, acoustics and even mathematics,[84] without necessarily assuming that these traditions, even if they had continued, would have become something like modern scientific disciplines. Consideration of the connection of the technical traditions with, and especially the degree of their integration into, the intellectual traditions of the literati might well prove useful in this regard. Or perhaps, following the sociological reformulation of Sal P. Restivo, we may ask "Why did scientific activity become functionally differentiated and institutionalized in Western Europe and not in China or elsewhere?"[85] Commenting briefly on this aspect, Sivin has recently pointed to the absence in traditional China of groups in other fields that could serve as models for scientific studies, as the medical profession did in the West.[86] The fact that the civil service was regarded as the only possible career by Chinese intellectuals may also throw some light on the matter. The questions, even after reformulation, continue to be negative, looking as they do for the reasons for the *absence* of 'interest,' 'continuance' and 'convergence.' Nonetheless, the inquiries that would result from this approach would be far better focused, with more clearly delineated areas of study. They would not lead us to a blind search for answers from every aspect of Chinese science and civilization.

Ultimately, what is necessary is the realization that the 'why not' question, expressed in whatever form, is a complex question, to which there is no simple answer. It requires us first to understand Chinese natural knowledge in its own intellectual and social context without any inappropriate assumptions, before embarking on comparisons with the conditions under which modern science emerged in the West. With such a qualification we may, and even should, keep asking the question. For many, that problem will continue to provide the single most important motivation for their studies of traditional Chinese science.

84 Koslow, "More on 19(k)", 185.

85 Restivo, "Joseph Needham and the Comparative Sociology", 29.

86 Sivin, "Why the Scientific Revolution Did Not ...", 61.

Whatever its limitations, the 'why not' question helps us to keep a broad perspective in our investigation of Chinese natural knowledge.

CHAPTER 6

Confucian Scholars and Specialized Scientific and Technical Knowledge in Traditional China, 1000-1700*

In modern society we are faced with a clear, often severe, separation between the scientific and technical knowledge of specialists and the broader interests and reach of general intellectuals.[1] This demarcation is so clearly and widely found today that people tend to assume that it is the natural state that had existed all along. Yet it is a rather recent phenomenon, a result of the specialization of science over the last few centuries. And it was a uniquely Western phenomenon; it was not an event bound to happen in every culture in the course of its historical development. For example, such a development did not take place in China – at least not to the degree it did in the West. In traditional China, and in the West before this process of separation took place, there was no reason for general intellectuals – scholars, writers, or 'thinkers' – to exclude science, or knowledge about the natural world, from their range of interests.[2] In this essay, I will examine the position of specialized knowledge, and scholars' attitudes towards it, in traditional China.

All that can be done in this brief essay, however, is a sketch. This is unavoidable because the situation surrounding those scholars and specialized knowledge in traditional China was very complex. In fact, even all the words constituting the topic – 'scholars,' 'specialized knowledge,' and 'traditional China' – are very complex in themselves. None of them denotes a single homogeneous group, knowledge, or period. The expression 'specialized knowledge,' to begin with, refers to various different subjects which do not form a single

* Originally published as "Confucian Scholars and Specialized Scientific and Technical Knowledge in Traditional China, 1000–1700: A Preliminary Overview," *East Asian Science, Technology and Society: an International Journal* 4 (2010): 207-228.

1 C. P. Snow, for example, characterized this separation as a 'two-culture' problem in his famous book: Charles P. Snow, *The Two Cultures and the Scientific Revolution* (Cambridge: Cambridge University Press, 1959).

2 There existed a separation in the Islamic world as well, but that was between the Arabic religious sciences and the ancient philosophical sciences. See A. I. Sabra, "Situating Arabic Science: Locality versus Essence," *Isis* 87 (1996): 654-670; J. L. Berggren, "Islamic Acquisition of the Foreign Sciences: A Cultural Perspective," in *Tradition, Transmission, Transformation*, ed. F. Jamil Ragep and Sally P. Ragep (Leiden: E.J. Brill, 1996), 270-271.

 | DOI 10.1163/9789004265318-008

whole. Subjects like calendrical astronomy (*li* 曆), harmonics (*lü* 律), mathematics (*suan* 算) and medicine (*yi* 醫) were often mentioned alongside other technical subjects like techniques (*gong* 工), agriculture (*nong* 農) and horticulture (*pu* 圃). Such a list routinely included subjects that cannot be called 'scientific' today: divination (*zhanbu* 占卜), alchemy (*liandan* 煉丹), and geomancy (*fengshui* 風水), and such non-scientific subjects as criminal justice (行刑), financial administration (*caizheng* 財政), military strategy (*bingfa* 兵法) and taxation (*shuiwu* 稅務).[3] When we speak of the 'specialists' themselves, the situation is even more complicated, as we are faced with all sorts of people possessing and applying various kinds of specialized knowledge: artisans, craftsmen and peasants, functionary bureaucrats, various medical practitioners, and even scholars with specialized interest and knowledge in the above subjects. The term 'scholars' also covers various kinds of people: some were genuine 'scholars' devoted to study and learning; some were career bureaucrats; some were landowners; some were engaged in mercantile, industrial and other practical activities; some were teachers; some were recluses. And many of them were more than one of these at the same time. These 'scholars' had vastly different predilections and inclinations, and naturally had different attitudes to different subjects. Finally, the expression 'traditional China' also covers a long time span over which great differences existed in the situation under examination.

To reduce the complexity of such a situation, I will confine my attention in this essay to specialized scientific and technical subjects, to scholars devoted to Confucian doctrines and learning, and to the period after about the year 1000, when the distinct intellectual pattern of such 'Confucian scholars' took shape. Nonetheless, this does not solve the problem, as there were various elements in the historical background of traditional China that had different effects on the scholars' attitudes toward specialized knowledge. I will begin the essay by analyzing such elements: the classical ideas and phrases that were influential in forming scholars' attitudes toward specialized subjects; and the practical reasons that scholars could not shy away from them. Then I will look at the actual examples showing the varying relationships between scholars and specialized subjects, and try to provide an overview of the relationship between Confucian scholars and specialized scientific and technical knowledge in China roughly between 1000 and 1700. I will end with a comparative look at the situation in the West, where there were tendencies towards both

3 On the situation of such specialized subjects in traditional China, see, e.g., Yung Sik Kim, *The Natural Philosophy of Chu Hsi, 1130-1200* (Philadelphia: American Philosophical Society, 2000), Sec. 12.1.

separation and convergence between general scholars and specialized scientific and technical knowledge.

On the whole, I will try to provide a global picture with a minimum of detail. My sketchy discussion in this essay cannot properly depict such a complex, multifarious situation, as I will be forced to simplify – frequently oversimplify – details that are extremely rich and intricate, varying greatly from scholar to scholar, from subject to subject, and from period to period. I believe, however, that in spite of such an inevitable deficiency there is still some value in attempting to see the bigger picture.

1 Classical Sanction to Become Generalists

There was a common tendency of Confucian scholars in traditional China to become generalists, which in turn had the effect of discouraging scholars' interest in specialized scientific and technical subjects. Some key ideas in the Confucian classics played important roles in this.

1.1 *Dichotomy of "Above Physical Form" and "Below Physical Form"*

Firstly, there was the dichotomy between what was considered 'above physical form' (*xing er shang* 形而上) versus what was 'below physical form' (*xing er xia* 形而下). This dichotomy can be traced back to the famous passage in a key Confucian classic, the *Yijing* 易經: "What is 'above physical form' is called '*dao*' 道 (way); what is 'below physical form' is called 'tools' (*qi* 器)." (*Xing er shang zhe wei zhi dao; xing er xia zhe wei zhi qi* 形而上者謂之道, 形而下者謂之器: "Xicizhuan" 繫辭傳, A12) There were other similar dichotomies of *li* 理 (principle, pattern, etc.) versus *qi* 氣, *li* versus 'numbers' (*shu* 數), *li* versus 'tools' (*qi* 器), and so on.[4] These dichotomies inevitably influenced Confucian scholars' attitudes towards natural phenomena and scientific knowledge.

According to such a dichotomy, *dao*, *li*, human nature (*xing* 性), the mind (*xin* 心), and humaneness (*ren* 仁) – abstract and sublime concepts without manifest 'physical forms' (*xing* 形) – belong to what is 'above physical form,'

4 On these various terms, see Ho Peng Yoke, *Li, Qi and Shu: An Introduction to Science and Civilization in China* (Hong Kong: Hong Kong University Press, 1985), Chap. 1; Kim, *The Natural Philosophy of Chu Hsi*, Chaps. 2-5. In *Mencius* (*Mengzi* 孟子) there even was an assertion that could be seen as a dichotomy between those who work with their mind versus those who work with their hands: "Some work with the mind; some work with [physical] force. Those who work with the mind rule other people; those who work with force are ruled by others." (3A4: "或勞心, 或勞力. 勞心者治人, 勞力者治於人.")

whereas concrete things with tangible physical forms – mundane, particular, practical matters – belong to what is 'below physical form.' It was commonly thought that what has physical form and is visible is easy to understand whereas what is without physical form is difficult to grasp. Naturally, what is difficult to understand was thought to be important, and thus worthy of further consideration, whereas what is easy to understand was considered obvious and even trivial. Such an attitude can be seen in the following remark by Zhu Xi 朱熹 (1130-1200), who, as the author of the so-called 'neo-Confucian synthesis,' exerted an overwhelming influence on the Chinese and East Asian intellectual world for many centuries:

> Things are easy to see; the mind has no physical form or measure. The weight and length of things are easy to measure; the weight and length of the mind are difficult to measure. When a thing is in error, only [that] one thing is in error. When the mind is in error, myriad things are in error.[5]

Since most common natural phenomena are accompanied by tangible qualities and physical effects and thus belong to what is 'below physical form,' it is understandable that they were thought to be obvious and were simply accepted in the way they were perceived; no further investigation was attempted beyond the surface of the phenomenal realities of empirical data. Thus, Confucian scholars were easily led to think lightly of natural phenomena and scientific knowledge about them. There developed a hierarchy between metaphysical and moral philosophies that were 'above physical form' on the one hand and scientific knowledge and techniques that were 'below physical form' on the other, which in turn gave rise to a hierarchical character of the distinction between generalist scholars and specialists seeking knowledge of what was 'below physical form.'[6] This hierarchical sentiment was instituted into the bureaucratic structure: generalist scholar-officials ranked above specialist functionaries, the former supervising the work of the latter.[7] In this frame of

5 *Zhuzi yulei* 朱子語類 (Taipei: Zhengzhong shuju 正中書局, 1962), 51.5a. Earlier in the same passage, Zhu Xi expressed the same idea differently: "The error in things has no harm; the error in the mind has harm."

6 Although the existence of 'literati physicians' (*ruyi* 儒醫) does show the acceptability of medicine as a subject for Confucian scholars, it was nevertheless unmistakable that their status was lower than that of general Confucian scholars. See Zhu Pingyi 祝平一 (Chu Pingyi), "Song Ming zhi ji de yishi yu 'ruyi'" 宋明之際的醫史與儒醫, *Zhongyang yanjiuyuan Lishiyuyan yanjiusuo jikan* 中央研究院歷史語言研究所集刊 77 (2006): 401-449.

7 Thatcher Deane has even noted a conscious government policy of the Han 漢 period (206, BC–AD 220) to lessen the status of calendrical specialists vis-à-vis generalist officials. See

mind, there was ample justification to separate subjects that were 'above physical form' from those 'below physical form.'[8]

1.2 *"Gentlemen Are Not Mere Tools."*

A phrase from the *Analects* (*Lunyu* 論語), another influential Confucian classic – "Gentlemen are not [mere] 'tools' (*Junzi buqi* 君子不器, 2.12) – was combined with the above dichotomy, or hierarchy, of *dao* versus tools, and formed the basis of Confucian scholars' attitudes towards the specialized subjects. The phrase, attributed to Confucius himself, was interpreted as meaning that to become a true 'gentleman' (*junzi* 君子) one should pursue such high ideals like *dao*, *li*, and human nature (*xing*), and should not concern oneself with mere 'tools' – concrete, practical, technical problems and knowledge. In other words, a gentleman should not be confined to a narrow scope, performing instrumental functions as a mere tool, but rather should aim at a broad learning and cultivation. This naturally led to Confucian scholars' general neglect and even contempt for what they considered 'mere tools,' which included specialized scientific and technical subjects.

1.3 *The 'Small Dao'*

Another *Analects* passage (19.4) exerted a considerable influence on Confucian scholars' attitudes towards specialized subjects: "Even the 'small *dao*' (*xiaodao* 小道) must have something worth 'looking at' (*guan* 觀). But if pursued too far, one may be bogged down. For this reason, gentlemen do not 'do' (*wei* 爲) them." ("雖小道, 必有可觀者焉. 致遠恐泥, 是以君子不爲也.") By calling certain subjects 'small *dao*,' and declaring that "gentlemen do not 'do' them," the phrase had the effect of discouraging Confucian scholars from engaging in activities that could be considered 'small *dao*,' which usually included specialized scientific and technical subjects.

2 Elements Conducive to Scholars' Interest in Specialized Subjects

While the ideas and passages discussed in the last section had the effect of discouraging scholars' interest in specialized scientific and technical knowl-

Thatcher E. Deane, "The Chinese Imperial Astronomical Bureau: Form and Function of the Ming Dynasty *Qintianjian* from 1365 to 1627" (Ph.D. diss., University of Washington, 1989), 68.

8 In editing the *Tianxue chuhan* 天學初函, the influential late Ming compilation of books on Western learning, Li Zhizao 李之藻 (1565-1630) divided them into two parts – the part on *li* (*li bian* 理編) and the part on 'tools' (*qi bian* 器編), although in this case the distinction was between Western religion and scientific knowledge.

edge, there were some elements in traditional Chinese history that facilitated scholars' interest in specialized subjects.

2.1 *Ambivalence of the Classical Sanctions*

To begin with, the actual situation surrounding the influence of those Confucian ideas and classical passages on scholars' attitudes was not straightforward. For example, alongside the tendency to separate *dao* and 'tools' and to value *dao* over 'tools,' there was also a tendency to take the two to be inseparable ("*daoqi buli*" 道器不離) and to consider them both important. Those who were inclined toward this trend maintained that *dao* exists in the tools, and that without the tools *dao* cannot exist. The same went for the other pairs like *li* and *qi* 氣, *li* and 'numbers,' and *li* and 'tools' – that *li* cannot exist in isolation from *qi*, or numbers, or tools. Thus, to many Confucian scholars, not only *dao* and what was 'above physical form,' but also tools and what was 'below physical form' – including specialized scientific and technical subjects – were important and worth pursuing. Indeed, the unity of *li* and 'numbers,' which was present in neo-Confucian thought ever since Shao Yong (邵雍, 1011-1077) and Zhu Xi, provided a powerful rationale for the inclusion of mathematics and calendrical astronomy into Confucian learning. For, in this view, the study of numbers (*shu*) could itself become the search for *li* 理, the goal of Confucian learning.

The effect of the slogan "Gentlemen are not [mere] tools" was not so straightforward or one-sided, either. While it emphasized that a gentleman should aim at broad learning and cultivation, this also included the study of the natural world, and the knowledge – if not the actual practices – of the techniques as well. Knowledge and speculation about the natural world, and about the phenomena and objects in it – 'heaven and earth' (*tiandi* 天地), the 'ten thousand things' (*wanwu* 萬物), and man (*ren* 人) – were not things to be brushed off as mere 'tools,' but rather were a legitimate part of a gentleman's study and cultivation. What the slogan did was to warn Confucian scholars against becoming mere specialists confined to a narrow subject – be it scientific, technical, or any other specialized subject. But it did not exclude scientific and technical subjects from their wide-ranging interests and learning.

When it comes to the 'small *dao*,' the ambivalence is even more apparent. The original *Analects* passage expressed both a positive attitude, namely that they "must have something worth 'looking at,'" as well as a negative one, that "gentlemen do not 'do' them." Indeed, the ambivalence exists in the word 'small *dao*' itself. For, while the character '*xiao*' 小 indicates that we are dealing with something minor, secondary, or less important, it is nonetheless a '*dao*,' and thus could potentially be a part of Confucian learning. This ambivalence –

seeing these things as something not to 'do,' but still worth 'looking at' – gave rise to many different opinions and debates among Confucian scholars concerning the proper status of specialized subjects. For example, although Zhu Xi highlighted that one should study those specialized branches that he referred to as 'small *dao*,' his actual words clearly implied that they were not quite the real '*dao*':

> The small *dao* are not heterodox; they are also '*dao*.' It is simply that they are small. [Subjects] like agriculture, horticulture, medicine, divination and 'the hundred techniques' also have *dao* and *li* in them. If one seeks *dao* and *li* only in the loftier realms, they will not be comprehended.[9]

Also, there were different opinions among scholars as to which subjects belonged to the 'small *dao*.' Calendrical astronomy in general was not included but it was considered by some to be one of them.[10] Medicine, agriculture, and techniques, on the other hand, were usually included in the list.[11]

2.2 *Key Philosophical Concepts*

Many of the specialized subjects were associated with key concepts of Confucian philosophy, and thus were important for Confucian scholars of traditional China. The significance of the concept of 'heaven' (*tian* 天), for example, made calendrical astronomy, the subject that deals with the physical heavens, important for classically-trained scholars. Geography and geomancy, dealing as they do with the other half of the term 'heaven and earth,' were also valued. The position of music as part of Confucian rituals also lent significance to the related subject of harmonics. Similarly, the authority of the classic *Yijing* and the ideas and diagrams it contained could carry over into the subjects of 'images and numbers' (*xiangshu* 象數), as well as divination and alchemy. Alchemy, especially in the form of 'inner alchemy' (*neidan* 內丹), was related to the concept of '*dao*' because it was among the techniques practiced by those who seek *dao*, the so-called 'masters of *dao*' (*daoshi* 道士). Confucian scholars must have felt that the study of these subjects would help them understand the above-mentioned ideas and concepts. And indeed, they were interested in such subjects as calendrical astronomy, harmonics, geography, alchemy, divination, etc., which became important parts of their intellectual concern.

9 *Zhuzi yulei*, 49.2a.

10 This was noted by an eighteenth-century Korean thinker Hong Tae-yong 洪大容 (1731-1783): *Tamhŏnsŏ* 湛軒書, naejip 內集, 3.22b.

11 Kim, *The Natural Philosophy of Chu Hsi*, 246, 276.

2.3 *The Commentary Tradition and Standard Texts*

Knowledge of various specialized subjects was present in the standard texts widely studied by Confucian scholars. In particular, scholars frequently commented upon natural objects and phenomena referred to in the ancient classics, which were supposedly written by – or at least contained the intentions of – the sages. Sometimes they made use of specialist knowledge in their comments on such passages. The most salient examples are the two passages in the *Book of Documents* (*Shujing* 書經). One is from the "Shundian" 舜典 (Cannon of Shun) chapter: "Set in order the 'Seven Governors' (*qizheng* 七政: the sun, the moon, and the five planets) by the *xuanji yuheng*" (*xuanji yuheng yi qi qizheng* 璿璣玉衡以齊七政). The other is from the "Yaodian" 堯典 (Cannon of Yao) chapter: "One year [has] 366 days" (*Ji sanbai you liuxun you liuri* 朞三百有六旬有六日). Many Confucian scholars presented detailed discussions of astronomical instruments and methods of placing intercalary months (*zhirunfa* 置閏法) in their comments on these two passages. The expression "*xuanji yuheng*," usually interpreted as astronomical instruments, became such a common part of Confucian scholars' vocabulary that in the *Boxue hongru* 博學鴻儒 (Broad Learning and Great Scholars) examination during the Kangxi 康熙 (r. 1661-1722) period, the examinees were asked to write a rhapsody (*fu* 賦) on it.[12] Commentaries on the "Yueling" 月令 (Monthly Ordinances) and the "Yueji" 樂記 (Record of Music) chapters of the *Record of Rites* (*Liji* 禮記) also contained discussions of specialized knowledge in calendrical astronomy and harmonics. Important examples of passages from other classics that provided occasions for discussing specialized knowledge were: the poem titled "the Crossing of the Tenth Month" (*shiyue zhi jiao* 十月之交) in the *Book of Poetry* (*Shijing* 詩經); the phrase, "Multitude of stars pay homage to it [i.e., the North Star]" (*zhongxing kong zhi* 衆星共之), in the *Analects* (2.1); and the *Mencius* passage (4B26) about "heaven being high and the stars being far" (*Tian zhi gaoye; Xingchen zhi yuanye* 天之高也. 星辰之遠也). Confucian scholars also discussed animals and plants whose names appear in the classics, the *Book of Poetry* in particular. There even emerged as a legitimate branch of classical scholarship a tradition of identifying and describing the plants and animals whose names appear in the *Book of Poetry*.[13]

12 Ping-yi Chu, "Technical Knowledge, Cultural Practices and Social Boundaries: Wan-nan Scholars and the Recasting of Jesuit Astronomy, 1600-1800" (Ph.D. diss., University of California, Los Angeles, 1994), 150-157.

13 Joseph Needham et al., *Science and Civilisation in China*, vol. 6, part 1 (Cambridge: Cambridge University Press, 1986), 463ff.

Standard commentaries of the above-mentioned classics frequently contained quite detailed discussions of specialized scientific knowledge, which became standard sources for such knowledge. Wang Yinglin's 王應麟 (1223-1296) compilation of astronomical passages from the classics and commentaries to them, titled *Liujing tianwen bian* 六經天文編 (Astronomical Compilations from the Six Classics), is a good example. In his introduction to the *Liujing tianwen bian* for the *Siku quanshu zongmu tiyao* 四庫全書總目提要 (Complete Books of the Four Treasuries, Annotated Catalogues), Dai Zhen 戴震 (1724-1777) said that the key topics of calendrical astronomy can be found in the Six Classics (*Liujing* 六經).[14] Scholars also frequently discussed knowledge about various techniques in connection with the "Kaogongji" 考工記 (Record of the Investigation of Techniques) chapter of the *Rites of Zhou* (*Zhouli* 周禮).[15]

There were other standard texts that contained discussions of specialized knowledge. The official dynastic histories (*zhengshi* 正史), for example, almost always included treatises on astronomy, calendars, harmonics, geography, as well as on rituals and music. These treatises dealt with such subjects in enough detail to become the standard sources of such knowledge. There were also treatises by scholars themselves written about various topics relating to specialized subjects. The most famous examples were the *Tongdian* 通典 (Comprehensive Canon) by Du You 杜佑 (735-812) and the *Mengxi bitan* 夢溪筆談 (Brush Talks from Dream Brook) by Shen Gua 沈括 (1031-1095). Numerous later Confucian scholars studied and cited these treatises.

Confucian scholars routinely studied the relevant portions of the above-mentioned commentaries and treatises, at times reaching a considerable level of understanding. Zhu Xi, for example, felt confident that he could make his own judgment as to which of the commentaries and treatises were best suited to studying a particular subject or problem.[16] For instance, he praised the subcommentary (*shu* 疏) to the "Yueling" chapter for its clear statement that heaven makes one full rotation plus one degree a day, and the subcommentary on the "Shundian" chapter for its account of the structure of the heavens in general. He also mentioned the subcommentary on the "Yaodian" chapter in connection with the problem of the intercalary month (*runyue* 閏月). Another old source he valued highly was the astronomical treatise of the *Jin History*

14 *Siku quanshu zongmu tiyao* 四庫全書總目提要 (Beijing: Zhonghua shuju 中華書局, 1965), 892.

15 For example, Wang Anshi 王安石 (1021-1086) and Lin Xiyi 林希逸 (*jinshi* 1235) wrote a treatise titled "Kaogongji jie" 考工記解, and Dai Zhen wrote the *Kaogongji tu* 考工記圖.

16 Kim, *The Natural Philosophy of Chu Hsi*, 251-252, 263.

(*Jinshu* 晉書). Zhu Xi also pointed out some of their weak points. For example, he pronounced that "the calendrical treatise of the Former Han [i.e., *Hanshu* 漢書] is not as good as that of the *Later Han History* (*Hou Hanshu* 後漢書)." Concerning the knowledge of harmonics and music, he frequently praised the "*Lüshu*" 律數 (Pitchpipe Numbers) chapter (Chapter 25) of the *Record of the Historian* (*Shiji* 史記), which, in his view, covered all the essential points.[17]

2.4 Gewu

The idea of *gewu* 格物 (investigation of things), the basis of the Confucian intellectual and moral endeavor from Song 宋 (960-1279) times on, also provided an important motivation for scholars to study specialized subjects. Since Confucian scholars took the term *gewu* to mean investigating 'the *li* 理 of things' (*wu zhi li* 物之理), and since every 'thing' (*wu* 物) in the world has its *li* – which is a manifestation of the single universal *li*, i.e. the *li* of heaven (*tianli* 天理) – everything was worth investigating for them.[18] Zhu Xi, for example, said repeatedly that every thing or event in the world has its *li* and should be studied and understood.[19]

Other expressions from the classics were interpreted in this spirit of *gewu*. For one, the expression 'broad study' (*boxue* 博學), again from the *Analects*,[20] became the goal pursued by many Confucian scholars. Zhu Xi, for example, used the emphasis on broad study to back his assertion that everything should be studied and understood.

---

17 Indeed, Zhu Xi must have felt that this was what he had to do, for, having asserted the importance of the specialized subjects and the necessity to study them, he had to decide which texts were the best or proper ones to be studied, just as he had done for the moral and social philosophies. For this aspect of Zhu Xi's 'program of learning' in general, see, e.g., William Theodore de Bary, "Chu Hsi's Aims as an Educator," in *Neo-Confucian Education: The Formative Stage*, ed. William Theodore de Bary and John W. Chaffee (Berkeley: University of California Press, 1989), 186-218; Daniel K. Gardner, *Chu Hsi: Learning to Be a Sage* (Berkeley: University of California Press, 1990), 35ff.

18 To be sure, they glossed the word '*wu*' as 'events' (*shi* 事), and their discussions of *gewu* were weighted heavily toward moral and social concerns. But 'the things' – concrete objects and phenomena – were not excluded from their conception of '*wu*,' and thus, knowledge and understanding of natural phenomena and technical artifacts have a place in their *gewu* endeavor. For more on *gewu* and the concept of *li*, see Kim, *The Natural Philosophy of Chu Hsi*, Chap. 2.

19 *Zhuzi yulei* 15.4b, 18.22b, 34.33b, 116.13b, 117.12b.

20 The expression '*boxue*' appears, e.g., in the *Analects*, 12.5, 19.6.

> The way of the *Great Learning* (*Daxue* 大學) must begin with 'investigation of things and extension of knowledge' (*gewu zhizhi* 格物致知) and then [move] to the *li* 理 of all under heaven. Of the books under heaven, there is none not to be 'broadly studied' (*boxue*).[21]

Another phrase from the *Analects* (14.37) had a similar influence: 'study down below and attain up above' (*xiaxue shangda* 下學上達), meaning that one should start with concrete things that are clearly manifest and easy to understand.

This aspect of *gewu*, with its stress on the study of concrete things, persisted in the period after Zhu Xi, despite the occasional reaction, like that of Wang Yangming 王陽明 (1472-1529). Scholars like Yang Shen 楊慎 (1488-1559), Li Shizhen 李時珍 (1518-1593), and Song Yingxing 宋應星 (1587-1666), who authored important books containing an extremely broad range of specialized scientific and technical knowledge, all resorted to the *gewu* doctrine when presenting specialized knowledge in their texts and underlining the importance of studying such topics.[22] Indeed, many seventeenth-century Confucian scholars used the *gewu* doctrine as a rationale for embracing Western scientific knowledge into Confucian learning.[23]

2.5 *Practical Utility*

In addition to the conceptual, textual, or ideological factors we have dwelled upon so far, the practical utility of specialized subjects also played a role in fostering scholars' interest in them.

Confucian scholars could not steer clear of specialized subjects because as potential officials ('scholar-officials') they had to deal with – or at least manage

21 *Huian xiansheng Zhuwengong wenji* 晦庵先生朱文公文集 (Sibu beiyao 四部備要 edition), 60, 16b.

22 Willard J. Peterson, "Confucian Learning in Late Ming Thought," in *The Cambridge History of China*, vol. 8, ed. Denis Twitchett and Frederick W. Mote (Cambridge: Cambridge University Press, 1998), 783-786.

23 Willard J. Peterson, "Fang I-chih: Western Learning and the 'Investigation of Things,'" in *The Unfolding of Neo-Confucianism*, ed. William Theodore de Bary (New York: Columbia University Press, 1975), 369-411; Zhang Yongtang 張永堂, *Mingmo Fangshi xuepai yanjiu chubian: Mingmo lixue yu kexue guanxi shilun* 明末方氏學派研究初編: 明末理學與科學關係試論 (Taipei: Wenjing 文鏡, 1987); Zhang Yongtang 張永堂, *Mingmo yu Qingchu lixue yu kexue guanxi zailun* 明末與清初理學與科學關係再論 (Taipei: Xuesheng shuju 學生書局, 1994). Zhang Yongtang even says that Fang Yizhi 方以智 (1611-1671) actually performed some 'physical research' (*wuli yanjiu* 物理研究) as part of his *gewu* endeavor: Zhang, *Shilun*, 84ff.

and supervise – many tasks involving scientific and technical matters: calendars, agriculture, the military, surveying, building, water control, medicine, etc. Thus, specialized subjects like calendrical astronomy, mathematics, agriculture and medicine frequently commanded scholars' interest. Scholar-officials often produced handbooks on administration containing specialized knowledge of various subjects. It was because of the practical utility of the specialized knowledge that, even during the Burning of Books at the time of the First Emperor of the Qin (Shi Huangdi 始皇帝, 259-210 BC), an exception was made for books on medicine, pharmacy, divination, agriculture and horticulture, and that, during the sack of Kaifeng by the Mongols, the calendrical specialists at the Astronomical Bureau were spared.[24] Many late Ming 明 (1368-1643) scholars were drawn to various subjects of practical knowledge as potential means to ameliorate the dismal conditions of the country.[25] That was why Xu Guangqi and Wang Zheng 王徵 (1571-1644), for example, were interested in technical things and problems even before they were exposed to Western scientific and technical knowledge through the Jesuits.[26]

Among the various specialized subjects, it was calendrical astronomy that attracted the greatest attention from Confucian scholars. Knowledge of the subject was needed for the essential government task of establishing accurate calendars, and thereby 'granting [the correct] time' (*shoushi* 授時), an accurate calendar being a necessary sign of assurance that the ruler was in accord with the 'mandate of heaven' (*tianming* 天命). This had clear political significance, for both the rulers and the officials. Astrology and divination were also closely connected to calendrical astronomy, as they frequently borrowed knowledge and data from it; also, astrological and divinatory tasks were important duties performed by calendrical specialists working in the imperial astro-

24 R. Kent Guy, *The Emperor's Four Treasuries: Scholars and the State in the Late Ch'ien-lung Era* (Cambridge, Mass: Harvard University Press, 1987), 11; Yamada Keiji 山田慶兒, "Jujireki no michi. Genchō chika no tenmondai to tenmongakusha" 授時暦への道 – 元朝治下の天文臺と天文學者, in *Chūgoku no kagaku to kagakusha* 中國の科學と科學者 (Kyoto: Kyōto daigaku jinbunkagaku genkyūsho 京都大學人文科學研究所, 1978), 20. It is interesting to note that in the destruction of the books on Western learning in the Chŏngjo 正祖 period (1776-1800) of Chosŏn Korea also, scientific books were excluded.

25 Francesca Bray and Georges Métailié, "Who Was the Author of the *Nongzheng quanshu*?" in *Statecraft and Intellectual Renewal in Late Ming China: The Cross-Cultural Synthesis of Xu Guangqi (1562-1633)*, ed. Catherine Jami, Peter Engelfriet, and Gregory Blue (Leiden: Brill, 2001), 323.

26 Jacques Gernet, "A Note on the Context of Xu Guangqi's Conversion," in Jami, Engelfriet, and Blue, *Statecraft and Intellectual Renewal in Late Ming China*, 186-190.

nomical bureau. It was because of this astrological and divinatory significance that successive dynasties tried to monopolize the knowledge of calendar making, and even imposed bans on the private study of calendrical astronomy.[27]

Other specialized subjects also attracted Confucian scholars' interest owing to their practical value. A good example was medicine: maintaining health and curing disease undeniably met the practical needs of the scholars themselves, their families and their country. Thus, there were many Confucian scholars who developed an interest in the knowledge, and sometimes practice, of various aspects of medicine. The same sense of practicality can be seen from the widespread interest among Confucian scholars in the techniques of 'nourishing life' (*yangsheng* 養生) in order to maintain health and achieve longevity.[28] Agriculture was another subject of continual interest to Confucian scholars, who generally considered it a basis for social order and the welfare of the people and the country.[29] There were thus not just pragmatic, but also cosmological and moral reasons for emperors and officials to promote agriculture.[30] In fact, 'encouraging agriculture' (*quannong* 勸農) was an important task for scholar-officials, who frequently issued 'proclamations for encouraging agriculture' (*quannongwen* 勸農文).[31] Moreover, most Confucian scholars were landowners, some of them active in managing farms, especially when off from official service. Indeed, many scholars authored agricultural treatises and handbooks, which for them became an "important minor genre" of writing.[32]

---

27 Sin Minchŏl 申敏澈, "Myŏngdae ch'ŏnmun 'sasŭp'-ŭi kŭmjiryŏng-kwa ch'ŏnmun sŏjŏk-ŭi ch'ulp'an: kŭ yinyŏm-kwa silchae" 明代 天文 '私習'의 禁止令과 天文書籍의 出版: 그 理念과 實在 [The Ban on the Private Study of Astrology and Publication of Books on Astrology in the Ming Dynasty: Ideas and Reality], *Han'guk Kwahaksa Hakhoeji* 韓國科學史學會誌 29 (2007): 231-260.

28 Yung Sik Kim, "The *Ts'an-t'ung-ch'i k'ao-i* and the Place of Internal Alchemy (*Nei-tan*) in Chu Hsi's Thought," *Monumenta Serica* 55 (2007): 99-131.

29 Indeed, agriculture was frequently referred to as the '*ben*' 本 (root, basis) or the '*benye*' 本業 (basic task). See, e.g., Francesca Bray "Science, Technique, Technology: Passages between Matter and Knowledge in Imperial Chinese Agriculture," *British Journal for the History of Science* 41 (2008): 321, 328.

30 Bray "Science, Technique, Technology," 330.

31 On the 'proclamations for encouraging agriculture,' see Kim, *The Natural Philosophy of Chu Hsi*, 280; Sudō Yoshiyuki 周藤吉之, *Sōdai keizaishi kenkyū* 宋代經濟史研究 (Tokyo: Tokyo University Press, 1962), 45-48. Even the agricultural work of Xu Guangqi 徐光啓 (1562-1633) was derived mainly from his commitment to the official duty of 'encouraging agriculture.' See Bray and Métailié, "Who Was the Author of the *Nongzheng quanshu*?" 322-359.

32 Timothy Brook, "Xu Guangqi in His Context: The World of the Shanghai Gentry," in Jami, Engelfriet, and Blue, *Statecraft and Intellectual Renewal in Late Ming China*, 92-93.

Related to agriculture, hydraulic engineering was another important subject in which many scholars showed interest.[33] Also of practical value for officials, and scholars as potential officials, were various techniques. Architectural and military techniques were among the obvious examples, but techniques involving key products under state control – salt, silk and porcelain for example – also attracted their attention.

Naturally, there were differences among the Confucian scholars in their attitudes towards specialized subjects of practical utility, reflecting their individual intellectual predilections. A key factor underlying such differences was their official position. Officials in charge of practical tasks usually showed a more active interest in the relevant specialized subjects, often as much as the specialist practitioners themselves. Scholar-officials' knowledge could therefore be broken down into two types: practical knowledge for statecraft tasks versus scholarly knowledge for the practice of self-cultivation.[34]

3 Scholars Interested in Specialized Knowledge

Given the various elements discussed in the last two sections, the attitude of scholars toward specialized knowledge in traditional China could be nothing but complex. Over the long period of traditional Chinese history, we see various aspects of both 'separation' and 'convergence' in the relationship between Confucian scholars and specialized scientific and technical knowledge.

There indeed were many Confucian scholars with interest in such subjects. Some became as knowledgeable in the subjects as contemporary specialists; some even wrote specialized treatises on them. Such scholars were not rare in the early imperial period. Ever since Han times, there were many scholar-officials actively involved in the debates over, and actual work of, frequent calendar reforms, and those who had mastered the specialized knowledge of such subjects as agriculture, medicine, and techniques.[35] In the Song as well, there

33 See Choying Li, "Contending Strategies, Collaboration among Local Specialists and Officials, and Hydrological Reform in the Late-fifteenth-Century Lower Yangzi Delta," *East Asian Science, Technology and Society: An International Journal* 4 (2010): 229-253.

34 Francesca Bray, "Introduction: The Powers of *Tu*," in *Graphics and Text in the Production of Technical Knowledge in China: The Warp and the Weft*, ed. Francesca Bray, Vera Dorofeeva-Lichtmann, and Georges Métailié (Leiden: Brill, 2007), 1-78, esp. 46.

35 Jin Qiupeng 金秋鵬, *Zhongguo kexue jishushi: Renwu juan* 中國科學技術史: 人物卷 (Beijing: Kexue chubanshe 科學出版社, 1998).

were many scholars with such interest.[36] Indeed, some of the most important scholars of the Northern Song period – Fan Zhongyan 范仲淹 (989-1052), Hu Yuan 胡瑗 (993-1059), Ouyang Xiu 歐陽脩 (1007-1072), Wang Anshi 王安石 (1021-1086), Sima Guang 司馬光 (1019-1086), and Su Shi 蘇軾 (1037-1101), who were officials most of the time – showed interest in various specialized subjects.[37] Not only the scholar-officials but emperors too displayed a curiosity for specialized knowledge, particularly in medicine. Many Northern Song emperors had a deep interest in medical therapies and books.[38] Huizong 徽宗 (r. 1100-1125), for example, tried to attract Confucian scholars to study, and even to practice, medicine.[39]

A number of Confucian scholars of the Song period mastered a broad range of specialized subjects, and wrote treatises on them. The above-mentioned Shen Gua and Zheng Qiao 鄭樵 (1104-1162) are two major examples. Another notable example was Cai Yuanding 蔡元定 (1135-1198), who wrote a treatise devoted to harmonics, titled *Lülü xinshu* 律呂新書 (New Book on the Pitches), which was circulated widely among scholars as it was included in the *Xingli daquan* 性理大全 (Great Compendium on Human Nature and *Li*). Many devoted themselves to the study of mathematics, and attained a high level of mathematical knowledge; some wrote the best mathematical treatises of the period. For instance, both Qin Jiushao 秦九韶 (1202-1261), author of the *Shushu jiuzhang* 數書九章 (Mathematical Book in Nine Chapters), and Li Ye 李冶 (1192-1279), author of the *Ceyuan haijing* 測圓海鏡 (Sea Mirror of Circle Measurements), were scholars who had received a classical Confucian education. Qin Jiushao, an official in the late Southern Song, stated in his preface to the above-mentioned book that "numbers and *dao* are not of two [different] roots." (*Shu yu dao fei erben* 數與道非二本). Li Ye dedicated himself so intensely to the study of mathematics that he feared the possibility that people would criticize him for "losing the [correct Confucian] will" (*sang zhi* 喪志).

As for Zhu Xi, he did not achieve the level of learning in specialized scientific and technical subjects as the above-mentioned scholars. But he did have a wide range of interests and a considerable degree of knowledge in many specialized subjects, and he urged other scholars to study them. This attitude of

36 Le Aiguo 樂愛國, *Songdai de ruxue yu kexue* 宋代的儒學與科學 (Beijing: Zhongguo kexuejishu chubanshe 中國科學技術出版社, 2007).

37 Ibid., 10-37.

38 Asaf Goldschmidt, "The Song Discontinuity: Rapid Innovation in Northern Song Dynasty Medicine," *Asian Medicine* 1 (2004): 53-90.

39 Charlotte Furth, "The Physician as Philosopher of the Way: Zhu Zhenheng (1282-1358)," *Harvard Journal of Asiatic Studies* 66 (2006): 435.

Zhu Xi's had a great influence upon scholars that came after him: because of his endorsement, the specialized subjects were considered legitimate for later Confucian scholars. Admittedly, after him Confucian scholars' scope of interests narrowed, no longer mirroring Zhu Xi's wide-ranging brand of learning.[40] But this did not cancel out the significant contribution Zhu Xi made in finding a place for specialized scientific and technical knowledge in his system of learning. It encouraged later Confucian scholars, most of whom were his followers, to develop an interest in, and study, specialized subjects. Consequently, there continued to be some degree of convergence between Confucian scholars and specialized scientific and technical knowledge.

Such a convergence was especially noticeable in the field of calendrical astronomy. During the Ming, many Confucian scholars were drawn towards the study of calendrical astronomy, long before the Jesuits' introduction of Western astronomy aroused interest in the subject. In spite of the difficulty for scholars to gain access to astronomical books due to the official ban on private study of the subject, interest in astronomical knowledge endured.[41] The *Ming History* (*Mingshi* 明史), for example, mentioned scholars like Zhu Zaiyu 朱載堉 (1536-1611) and Tang Shunzhi 唐順之 (1507-1560) as examples of those who "were knowledgable in calendrical astronomy although they were not calendrical officials" (*fei liguan er zhilizhe* 非曆官而知曆者).[42] Xing Yunlu 邢雲路 (1549-?) even said that calendrical astronomy – literally 'the learning of calendars, images, and granting time' *lixiang shoushi zhi xue* 曆象授時之學 – was "our basic task, as Confucian scholars" (*wuru zhi benye* 吾儒之本業).[43] This

40 Part of the reason may have been the breadth of the Zhu Xi corpus itself. It was almost as if, once Zhu Xi had produced a corpus of knowledge that covered everything, those who came after him did not feel the need to worry about problems other than their principal concern – morality and self-cultivation; everything else, including the specialized scientific and technical knowledge, was already there in Zhu Xi after all. This tendency is especially pronounced among the Korean followers of Zhu Xi in the Chosŏn 朝鮮 dynasty (1392-1910). Many of them devoted themselves to establishing the so-called 'established theory of Master Zhu' (*Zhuzhi dinglun* 朱子定論), while showing little interest in the scientific knowledge coming from the West. See Ku Man-ok 具萬玉 (Koo Mhan-ock), *Chosŏn hugi kwahak sasangsa yŏn'gu 1. Chujahakchŏk ujuron-ŭi pyŏndong* 朝鮮後期科學思想史硏究 I. 朱子學的 宇宙論의 變動 (Seoul: Hyean 慧眼, 2004), Chap. 5.

41 Sin Minchŏl 申敏澈, "Myŏngdae ch'ŏnmun 'sasŭp'-ŭi kŭmji," 25-26. From the mid-sixteenth century onward, thriving publishers began to print astronomical books to meet scholars' growing demand: ibid., 27.

42 *Mingshi* 明史 (Zhonghua shuju 中華書局 edition) 31.544.

43 Xing Yunlu 邢雲路, *Gujin lülei kao* 古今律曆考 (Wenyuange siku quanshu 文淵閣四庫全書 vol. 787), *juan* 65, "Bian Datongli zhi shi" 辯大統曆之失.

situation was noted by Matteo Ricci (利瑪竇, 1552-1610), who observed that "no people esteem mathematics as highly as the Chinese."[44]

4 Limited Convergence between Scholars and Specialized Knowledge

On the whole, however, the convergence between scholars and specialized scientific and technical knowledge in traditional China was not complete. Typically, Confucian scholars were not fully committed to the study of such subjects, which they often did not consider intellectually challenging. Nor did they consider them to be that difficult that they could not have mastered them if they had wanted to.

Such an attitude was apparent in Zhu Xi. While emphasizing the need to study and understand specialized subjects, Zhu Xi did not hide his feeling that those subjects were secondary to the more important ones – moral and social problems. That was what he meant when he said repeatedly that one has to understand 'the basis' (*ben* 本), or what is 'great,' before moving on to 'small' matters. Zhu Xi also admitted that it was not necessary to try to achieve a complete understanding of all the details of specialized subjects. He felt that he could have fully comprehended them if only he tried, however difficult these subjects may have appeared. It was simply that these subjects did not command his full interest that he did not apply himself fully to studying them.[45] Zhu Xi did not have very high opinions of the specialists in these fields, either. For him, they were mere functionary experts in subjects that he had no interest in mastering himself. He even blamed the calendrical specialists for the deviations of calendars from the actual motions of the heavens; in his view, they were merely busying themselves observing, recording, computing, and predicting the motions of the heavens, seeking only to get numbers using expedient methods, rather than trying to understand the actual workings of the heavens, which proper calendrical method should have aimed at.[46]

This attitude of Zhu Xi's more or less became the basic attitude of Confucian scholars in the subsequent period. On the whole, Confucian scholars remained

44 Louis Gallagher, tr., *China in the Sixteenth Century: the Journals of Matteo Ricci, 1583-1610* (New York: Random House, 1953), 476.

45 Kim, *The Natural Philosophy of Chu Hsi*, Chap. 12, esp. 245-250. For a subject that he truly considered difficult and beyond his understanding, namely inner alchemy (*neidan* 內丹), especially that of the *Cantongqi* 參同契, Zhu Xi indeed felt the intellectual challenge to master it and exerted great efforts to do so. See Kim, "The *Ts'an-t'ung-ch'i k'ao-i*," 99-131.

46 Kim, *The Natural Philosophy of Chu Hsi*, 252-253.

'scholars,' with some intellectual curiosity about, but never fully committed to, specialized knowledge. It could hardly have been otherwise, because their main object was to grasp the one universal *li*, 'the *li* of heaven.' Because this universal *li* is manifest in every individual thing and event of the world as the *li* of that thing or event, Confucian scholars had the obligation to study all objects and phenomena, including specialized scientific and technical subjects. Yet what they were truly concerned with was reaching the universal *li*, rather than engaging themselves in the particular, individual *li*. The abovementioned concern of Li Ye's about the possibility of people blaming him for focusing too much on mathematics very well illustrates the common attitude of Confucian scholars towards specialized fields. Devoted study of a subject like mathematics was likely to appear to them as 'losing the [correct] will.'

This mind-set was reflected in the status of specialized knowledge in the civil service examination, which was of undeniable importance for the intellectual standing of any scholarly subject in traditional China.[47] To be sure, topics relating to specialized subjects had been included in the civil service examination from early on, and especially during the Ming.[48] Zhu Zaiyu 朱載堉 (1536-1611), for example, recorded that in 1497 and 1522 examinations were administered to recruit those who were well versed in calendrical astronomy among officials and their sons, as well as among students, soldiers, and even recluse scholars.[49] On the whole, however, specialized scientific and technical subjects never became a full-fledged part of the civil examination curriculum;

---

47 Benjamin A. Elman, *A Cultural History of Civil Examinations in Late Imperial China* (Berkeley: University of California Press, 2000). Flemish Jesuit Ferdinand Verbiest (南懷仁 1623-1688) seems to have fully understood the importance of the civil examination curriculum, as he tried to secure imperial permission to include scholastic philosophy in the civil examination along with the Confucian classics. See Nicolas Standaert, ed., *Handbook of Christianity in China, Volume One: 635-1800* (Leiden: Brill, 2001), 608.

48 Elman, *A Cultural History of Civil Examinations*, 461-481.

49 Zhu Zaiyu 朱載堉, *Shengshou wannianli* 聖壽萬年曆 (Wenyuange siku quanshu 文淵閣四庫全書, vol. 786), appendix. There was also an instance in 1521 when officials were promoted several ranks solely because of their mastery of calendrical astronomy. (*Wuzong shilu* 武宗實錄, 1521, 7th month, 6th day.) Although the Qing 清 dynasty (1644-1911) discontinued the practice, it was not because the subjects were considered unimportant, but because of the fear that the controversies involving calendars would create political unrest. See Benjamin A. Elman, *On Their Own Terms: Science in China, 1550-1900* (Cambridge, Mass.: Harvard University Press, 2005), 133. Topics of calendrical astronomy are found in the civil examination questions of early sixteenth-century Korea as well: Ku, *Chosŏn hugi kwahak sasangsa yŏn'gu*, 74.

they remained secondary subjects. And as the result, there were no satisfactory career prospects in the government for scientific and technical specialists.[50]

What convergence there was between scholars and specialists was in one direction only: many scholars did have an interest in specialized technical knowledge, but the specialist practitioners themselves – artisans, craftsmen, functionaries, peasants, etc. – did not write about their trades. Nor was there, on the part of the specialist practitioners, a broadening of interest or a deepening of intellectual curiosity, of the kind that can be seen among the technical men of early modern Europe.[51] Admittedly, there was a broadening of readership of books on specialized subjects in the late imperial period. For example, there was a great increase in the late Ming of officially sponsored compendia (*leishu* 類書) on classics, civil service, and statecraft, and of private practical encyclopedias on miscellaneous subjects including agriculture, health, and mathematics, which were read by literati, artisans, merchants and clerks.[52] The topics discussed in these books were extremely broad, covering astrology, topography, medicine, mathematics, 'nourishing life' techniques, sericulture, plants, as well as dream interpretation, writing skills, and legal matters.[53] But there is no evidence that specialist practitioners wrote about their fields of expertise; they were busy doing the work required for their trades.

The inclination of these practitioners to be secretive about the craft knowledge of their trades may have had an effect on this situation. Physicians, usually belonging to hereditary physician households with craft secrets, tended to keep their medical recipes to themselves; this was true until the Song-dynasty examination culture and the spread of printing slowly changed the situation, which culminated in the rise of medical case histories (*yi'an* 醫案) in the late

50 Again, Verbiest keenly noted this in explaining the lack of able calendar astronomers in the Qing court. See Willy Wande Walle, "Ferdinand Verbiest and the Chinese Bureaucracy," in John W. Witek, ed., *Ferdinand Verbiest (1623-1688): Jesuit Missionary, Scientist, Engineer and Diplomat* (Nettetal: Steyler Verlag, 1994), 511-512. It is also interesting to point out in this connection that some of the specialized subjects, notably medicine, frequently attracted scholars who had failed, or had been blocked from, the civil examination.

51 Paolo Rossi, *Philosophy, Technology and the Arts in the Early Modern Era* (New York: Harper, 1970); Pamela O. Long, *Openness, Secrecy, Authorship: Technical Arts and the Culture of Knowledge from Antiquity to the Renaissance* (Baltimore: Johns Hopkins University Press, 2001), Chaps. 6-7.

52 Benjamin A. Elman, "Collecting and Classifying: Ming Dynasty Compendia and Encyclopedias (*Leishu*)," *Extrême-Orient, Extrême-Occident* hors série (2007): 131-157.

53 Elman, "Collecting and Classifying," 136.

Ming.[54] There must have been the same need for secrecy in subjects like alchemy, geomancy, and divination. Calendrical astronomers were also reticent to share their specialized skills. Even government was not free from a certain secretiveness concerning the specialized craft knowledge that they had monopolized, as can be seen from the ban on the private study of calendrical astronomy during the Ming (although it was not strictly enforced).[55]

In these circumstances, most of the writing on specialized crafts was done by, and frequently for, the scholars themselves. To be sure, some peasants' knowledge made its way into scholarly writing on agriculture. The authors of the major agricultural treatises – Jia Sixie 賈思勰, author of the sixth-century *Qimin yaoshu* 齊民要術 (Essential Techniques for Common People), Wang Zhen 王楨, who wrote the fourteenth-century *Nongshu* 農書 (Agricultural Book), and Xu Guangqi 徐光啓 (1562-1633), author of the seventeenth-century *Nongzheng quanshu* 農政全書 (Complete Book of Agricultural Agenda) – all consulted farmers and craftsmen in composing their books.[56] But this was a far cry from specialist practitioners writing about their own trades. It is significant that the *Bencao gangmu* 本草綱目 (Outlines of Pharmacopoeia, 1578) by Li Shizhen, the *Tiangong kaiwu* 天工開物 (Heaven's Work in Opening Things, 1634) by Song Yingxing, and the *Nongzheng quanshu* (1639) by Xu Guangqi, the three most important works of specialized knowledge in the late Ming, were all written by scholars. Naturally, intellectual interest was pronounced in these cases, in which the scholar-authors included techniques and products for their own and their readers' – often fellow scholars' – intellectual benefit. In addition, the medical treatises written by Confucian scholars, especially by 'literati physicians' (*ruyi* 儒醫), were not entirely for practical purposes; there was intellectual interest there as well.[57]

There were genres other than books and treatises in which scholars entered the realms of specialized subjects. Yet the principal aims of these genres did not lie in the technical details, though they did contain technical information. Such disregard for technical details can be seen, for example, in the 'proclama-

---

54 Charlotte Furth, "Introduction: Thinking with Cases," in Charlotte Furth, Judith T. Zeitlin, Ping-chen Hsiung, eds., *Thinking with Cases: Specialist Knowledge in Chinese Cultural History* (Honolulu: University of Hawaii Press, 2007), 13-14.

55 Sin Minchŏl, "Myŏngdae ch'ŏnmun 'sasŭp'-ŭi kŭmji."

56 Francesca Bray, "Agricultural Illustrations: Blueprint or Icon?" in Bray, Dorofeeva-Lichtmann, and Métailié, *Graphics and Text*, 521-567, esp. 541-542.

57 Angela Ki Che Leung, "Medical Learning from the Song to the Ming," in Paul J. Smith and Richard von Glahn, eds., *The Song-Yuan-Ming Transition in Chinese History* (Cambridge, Mass.: Harvard University Press, 2003), 374-398; Furth, "The Physician as Philosopher of the Way."

tions for encouraging agriculture' (*quannongwen*).[58] Sometimes scholar-officials wrote poems to disseminate useful information – farming techniques, medical prescriptions, etc. Some scholar-artists drew illustrations of technical artifacts and processes. However, no significant technical improvements were described in these illustrations or poems. An independent genre, the so-called 'pictures of farming and weaving' (*gengzhi tu* 耕織圖), emerged from such illustrations and poems. Usually traced back to Lou Chou 樓璹 (1090-1162), who first put together poems and pictures depicting the main processes of rice farming and sericulture, the genre flourished in the subsequent periods.[59] These pictures and poems, however, aimed more at the rulers and the elite than the farmers themselves; their emphasis was on showing the skills, and the 'toil and effort' of farming and weaving, rather than technical details, thereby conveying a moral message to the members of the ruling class.[60]

The level of Confucian scholars' proficiency in specialized subjects was generally not very high. Seldom was it the case that they became authorities in a scientific or technical field. Again Zhu Xi is a case in point; although he studied many specialized subjects and attained considerable familiarity with them, his understanding was never up to that of the specialists of his time. In fact, the level of knowledge he demonstrated in his commentaries often lagged behind that of his contemporary specialists by several centuries. For example, for the number of days in a year he continued to use 365.25, the number of the ancient *Sifenli* 四分曆 (Quarter Remainder Calendar) – the calendrical system used during the Later Han when the commentary tradition first began – instead of 365.2425 already used by his contemporary calendrical specialists. Confucian scholars after him, mostly his followers, duly followed him in this respect as well, for the *Sifenli* calendrical numbers went on to become canonical in subsequent Confucian discussions. Indeed, most Confucian scholars had only a superficial interest in, and partial grasp of, specialized scientific and technical subjects. A disciple's response to Zhu Xi's proposal to include subjects like calendrical astronomy, geography, music and harmonics in the civil examination – that "there may be no [qualified] examining officer after all" – clearly shows that this was indeed the case.[61]

58 Bray, "Agricultural Illustrations," 549.

59 Bray, "Agricultural Illustrations," 524-535; Zhongguo nongye bowuhan 中國農業博物館 ed., *Zhongguo gudai gengzhi tu* 中國古代耕織圖 (Beijing: Xinhua shudian 新華書店, 1995).

60 Bray, "Agricultural Illustrations," 533-534.

61 *Zhuzi yulei*, 109.8b. In fact, Shen Gua had made a similar observation: during the Huangyou 皇祐 period (1049-1054) the examiners gave a top grade to all the answers to a question on

Our survey so far has shown a lack of full convergence between Confucian scholars and specialized scientific and technical knowledge. This endured even in the later period when Western scientific and technical knowledge was introduced to China. Qing writings on the paper industry, for example, were mostly focused on hard work and the moral values of such toil, as Jacob Eyferth has pointed out. Some officials did write in more detail about paper-making techniques; but these were not read by practitioners. Many paper-making artisans neglected such written forms of knowledge, as they had developed other ways to reliably reproduce knowledge.[62]

5 The Situation in the West: A Comparative Look

Separation of specialized scientific and technical knowledge from the general concern of intellectuals began in Europe some time after the Scientific Revolution,[63] and continued in a long, complicated process in which science became technically utilizable, big and expensive, intellectually challenging, and above all specialized. This process of separation proceeded along with the following two developments:

i) the emergence of the idea of 'science' as a single area covering all knowledge about the natural world.
ii) the emergence of 'scientists' as a separate, self-respecting, group.

These two events had been preceded by an earlier development:

iii) the convergence of, or breaking down of the boundaries between, the traditions of philosopher-scholars and the practitioners ('specialists') of various specialized subjects.

It is worthwhile to note that this convergence was in the direction opposite to events i) and ii), which instead signaled a separation of specialized knowledge

---

astronomical instruments of which they themselves did not have adequate knowledge. *Mengxi bitan* 夢溪筆談 (Beijing: Wenwu chubanshe 文物出版社), 7.9a-9b.

62 Jacob Eyferth, "Craft Knowledge at the Interface of Written and Oral Culture," *East Asian Science, Technology and Society: An International Journal* 4 (2010): 185-205.

63 For the circumstances of the Scientific Revolution, see, e.g., A. Rupert Hall, "The Scholar and the Craftsman in the Scientific Revolution," in *Critical Problems in the History of Science*, ed. Marshall Clagett (Madison: University of Wisconsin Press, 1959), 3-23; Rossi, *Philosophy, Technology and the Arts*.

from scholarly interest. But this apparent contradiction suggests that the separation was possible because of – and not in spite of – the convergence, of type iii), which, in turn, explains the fact that in the West specialization occurred not simply as a 'narrowing down' of the interest of the specialists – 'scientists' – but as some kind of 'broadening up,' in which the specialists were taking over something from the mainstream learned culture of the scholars and incorporating them in their own domains.[64]

The specialists' self-respect for their own activity – event ii) – explains then how the specialization of natural knowledge took place in the West with an eventual increase of prestige of both the knowledge and its practitioners, the result of which is that in the modern world science has in many ways a more prestigious status than non-science – humanities, arts, etc. This sort of development did not take place in China. If any degree of specialization of knowledge took place in traditional China, it did so with a decrease rather than an increase of prestige.[65] One difference in the hierarchy of learning in the West and China appears to be pertinent in this respect. While in the European medieval university structure, the faculties of professions – law and medicine – stood as 'higher' faculties above the faculty of philosophy,[66] it was more or less the opposite in China: generalist philosophy always occupied an uncontested higher intellectual position above specialized knowledge.

This different situation may shed light on the lack of development of specialized professions in traditional China. Yet it was not due to the absence of a model that such professionalization did not occur in traditional China. As a matter of fact, the literati (*shi* 士) were a powerful group with unmistakably 'professional' features that would have provided a good model for other profes-

64 Rossi, *Philosophy, Technology and the Arts*; Long, *Openness, Secrecy, Authorship*. Also, this may be the reason why in the modern science of the West we can see both universalizing and specializing tendencies. Throughout the process of specialization in the eighteenth and the nineteenth centuries, individual scientific subjects become linked to one another, the result of which was the emergence of the concept of 'science' covering all the scientific subjects – that is, event i).

65 Cf. Elman, *On Their Own Terms*, xxix, where Elman suggests that in seventeenth- and eighteenth-century China, mathematical sciences contributed to the rise of the social status of the Confucian scholars who mastered them. However, this was largely because the practical utility of Western science had already been accepted by many Chinese scholars.

66 On the situation surrounding medieval universities, see, e.g., Pearl Kibre and Nancy G. Siraisi "The Institutional Setting: The Universities," in *Science in the Middle Ages*, ed. David C. Lindberg (Chicago: University of Chicago Press, 1978), 120-144.

sionalizing groups to emulate.[67] Indeed, physicians, especially those called 'literati physicians' (*ruyi*), modeled themselves on Confucian literati in their learning, teaching, and transmission of the tradition.[68] We may consider, in this connection, the so-called 'middle people' (*chung'in* 中人), a social class unique to Korea, which in the late Chosŏn period came to monopolize specialized scientific and technical activities. These 'middle people' formed a closed and basically hereditary class of specialists, who, although looked down upon by the ruling *yangban* 兩班 class and given no roles in the major affairs of society, nevertheless enjoyed great class stability and lucrative privileges associated with their specialized activities.[69]

Another feature where the West and China differed yet again was the supposed distinction between science and technology, which may be key to understanding the different ways in which the relationship between scholars and specialized knowledge played out in the two cultures. For the bulk of Western history from ancient times till just a few centuries ago, science and technology were separate, and followed separate paths. The connection between the two in the West began to take shape only around the time of the Scientific Revolution, and has undergone a very complex process to get to where we are today, where they are closely tied to each other.[70] As we start to consider the Chinese

67 For the 'professional' characteristics of the Confucian literati, see, e.g., John W. Dardess, *Confucianism and Autocracy: Professional Elites in the Founding of the Ming Dynasty* (Berkeley: University of California Press, 1983), Chap. 1. Benjamin A. Elman actually uses the word 'professionalization' in referring to a group of Qing scholars: Benjamin A. Elman, *From Philosophy to Philology: Intellectual and Social Aspects of Change in Late Imperial China* (Cambridge, Mass.: Harvard University Press, 1984), Chap. 3

68 Yüan-ling Chao, "The Ideal Physician in Late Imperial China: The Question of *Sanshi*," *East Asian Science, Technology, and Medicine* 17 (2000): 66-93, esp. 82-86; Leung, "Medical Learning," 387-396; Furth, "The Physician as Philosopher of the Way."

69 Many features of these 'middle people' make it difficult to think of them as 'professionals.' For one, they were not specialists of a single specialized subject: they crossed over many branches – calendrical astronomy, harmonics, mathematics, divination, medicine, law, långuage skills, etc. Their specialty was not a subject, but an occupation with a well-defined niche in the social and bureaucratic hierarchy of Chosŏn society. For brief comments and references on the 'middle people,' see Yung Sik Kim, "Problems and Possibilities in the Study of the History of Korean Science," *Osiris* 13 (1998): 48-79, esp. 76. For more detailed studies, see Yŏnse taehakkyo Kukhak yŏn'guwon 延世大學校 國學研究院 ed., *Han'guk kŭndae yihaenggi chung'in yŏn'gu* 韓國近代移行期中人研究 (Seoul: Sinsŏwon 新書苑, 1999).

70 At present the two have even combined into one kind of enterprise, frequently called 'science-technology', or even 'techno-science'. In modern Chinese they are even abbreviated as one word, *'keji'* 科技. For the complex development of the relationship between

situation, however, we are faced with an immediate challenge. For it is not clear whether one can distinguish 'science' and 'technology' in traditional China in the first place, or, what is more important, whether the traditional Chinese themselves distinguished the two.[71] Although there were many specialized subjects that we would now call 'scientific' or 'technical,' these existed individually, as separate, independent subjects, and there was no sharp distinction between which were 'scientific' and which were 'technical.' One may even wonder whether there was anything that could be called 'science' or 'technology' at all in traditional China.

The difficulty of making a distinction between science and technology in the Chinese case becomes clearer when we look at the following three dichotomies that are routinely used in attempting to distinguish the two:

1) nature versus man (natural versus artificial)
2) scholars versus artisans
3) intellectual curiosity versus practical utility

None of these dichotomies works in dealing with the situation in traditional China.

Firstly, while the Aristotelian dichotomy of nature versus artificial – and its translation into the dichotomy of science versus technical arts – dominated the Western mind till the time of the Scientific Revolution,[72] there was no clear distinction between nature and man in traditional China. To be sure, there was a sort of nature versus man distinction, characterized by such pairs of Chinese characters as heaven (*tian* 天) versus man (*ren* 人), and the 'divine' (*shen* 神) versus the 'sagely' (*sheng* 聖). Natural phenomena were usually attributed to heaven. And when the marvelous (or mysterious, or superior) aspect of natural phenomena was highlighted, it was referred to as the 'divine.' Technologies and cultural institutions, on the other hand, were said to have been established or inaugurated by the sages (*shengren* 聖人), the superior men who were nevertheless still men. For example, the famous chapter (B2) in the "*Xicizhuan*"

science and technology in Western history, see, e.g., John M. Staudenmaier, *Technology's Storytellers: Reweaving the Human Fabric* (Cambridge, Mass: MIT Press, 1985), Chap. 3.

71 Of course, a simple clear-cut distinction between science and technology is not possible even in the West, if one examines the situation more closely. But still such a distinction can be made at least in principle.

72 Lorraine Daston, "The Nature of Nature in Early Modern Europe," *Configurations: A Journal of Literature, Science, and Technology* 6 (1998): 149-172. After the Scientific Revolution, one sees both a sharpening of the distinction of nature versus man (from Descartes on) and a merging of the two (the study of man as a natural object).

繫辭傳 (Commentary on the Appended Words) of the *Yijing* attributed the basic techniques used in agriculture and fishing and those used in making clothes, boats, arrows, houses and so on to various ancient legendary sages. The techniques were also considered part of the 'rites' (*li* 禮), and were thus discussed in treatises like the "Kaogongji" 考工記 chapter of the *Rites of Zhou*. This led to something like traditional China's own version of the 'natural versus artificial' distinction: natural phenomena attributed to heaven and the divine (*shen*) versus human acts attributed to man and the sages (*sheng*). Yet the distinction was not that rigorous. The attributes 'heaven' and the 'divine' (*shen*) were not confined to the natural realm, and techniques could also at times be associated with heaven and the divine. Expressions like 'divine work' (*shengong* 神工) and 'divine technique' (*shenji* 神技) were frequently used to characterize supreme human skills. We even find the expression 'heaven's work' (*tiangong* 天工) in the *Tiangong kaiwu*.[73]

Nor do the other two dichotomies work in characterizing the science-technology distinction in traditional China. We cannot simply associate science with scholars and intellectual curiosity on the one hand, and techniques with artisans and practical utility on the other. As we have seen, practical utility came hand in hand with intellectual curiosity in the form of scholarly interest in specialized subjects. Most books on specialized scientific and technical subjects were written not by, or even for, specialist practitioners, but by scholars out of intellectual interest, and for scholars. Also, Confucian scholars, with their primary focus on human problems, frequently leaned towards the 'artificial' side, while artisans, generally associated with Daoism, leaned towards the 'natural.'[74]

Numerous other differences in the circumstances surrounding the scholars and the specialized scientific and technical subjects – both their contents and the ways they were practiced – in the West and traditional China affected the

73 Yung Sik Kim, "'Natural Theology of Industry' in Seventeenth Century China?: Ideas About the Role of Heaven in Production Techniques in Song Yingxing's *Heaven's Work in Opening Things* (*Tiangong kaiwu*)," in *A Master of Science History: Essays in Honor of Charles Coulston Gillispie*, ed. Jed Z. Buchwald (New York: Springer, 2012), 197-214 (Chapter 3 of this book).

74 On the affinity of Daoism with the natural world, and the predilections of Daoists in scientific activities, see Joseph Needham, *Science and Civilisation in China*, vol. 2 (Cambridge: Cambridge University Press, 1956), 33-164. Modern science as well is marked by both 'natural philosophical' and 'instrumentalist' features. See, e.g., Peter Dear, *The Intelligibility of Nature: How Science Makes Sense of the World* (Chicago: University of Chicago Press, 2006), 1-14.

relationship between generalist scholars and specialized knowledge in the two cultures.

6 Concluding Remarks

In the preceding sections, we have seen various aspects of Confucian scholars' ambivalent attitudes towards specialized knowledge in traditional China. On the one hand, Confucian scholars did develop an interest in specialized scientific and technical subjects, with some of them attaining relatively in-depth knowledge. However, their interest in the subjects was only secondary, which put limits on the intellectual and social status of learning in those fields. The specialized subjects were not fully accepted in traditional Chinese learning; nor was there a full convergence between scholars and specialized knowledge.

It is significant, however, that Confucian scholars did not find such a state of affairs acceptable. Many of them insisted that it was not right for specialized subjects to be neglected or ignored. Zhu Xi both exemplified and influenced such an attitude. He believed that in ancient times there had existed a perfect knowledge of specialized subjects, and suggested that ancient scholars had studied, and achieved a high level of knowledge in, subjects like calendrical astronomy, mathematics and music.[75] Based on such a belief, Zhu Xi urged scholars to follow suit, and many Confucian scholars of later times did in fact nurture an interest in, and study, specialized subjects. At the very least, they had no excuse for not studying and understanding these subjects when their official duties, or social responsibilities, called for some degree of specialized knowledge; they could not, and did not, shun them.

75 Kim, *The Natural Philosophy of Chu Hsi*, 254-255.

CHAPTER 7

Science and Religion in Traditional China*

1 Introduction

In the history of Western society and culture, both science and religion have been important and powerful elements. Christianity has been a force to reckon with ever since it took root in ancient Roman times: it has been the single most dominant factor in Western culture for the greater part of the past two millennia. Neither has science played a trivial role in Western history: since having earned some degree of importance during the late Renaissance period, its significance in society has been ever growing. It is thus to be expected that the two key elements in a society would have deeply influenced each other. This is all the more so because the objects of science and religion – Nature and God – were directly linked in the Western conception in that God created the natural world. Thus, throughout Western history there has always been an obvious potential for a strong connection and interaction between science and religion. Thirteenth-century Christian scholasticism, the Galileo affair, science in Puritan and Restoration England, eighteenth-century natural religion, and the Darwinian controversies are only the more conspicuous cases; there are numerous other examples that can be found in almost all other times.[1] In fact, when we look at the relationship between science and religion in the West, we are really looking at the relationship between the two pillars of Western civilization – Athens and Jerusalem.

In traditional China, however, the situation was drastically different. Science and religion were not as important in China as they were in the West; nor was there an obvious link between the objects of the two. Of course, there were exceptional periods in which religious elements, sometimes even scientific interests, carried more weight. For example, several centuries following the fall of the Han 漢 dynasty, around AD 200, Daoist and Buddhist beliefs and prac-

---

* Originally published as "Some Reflections on Science and Religion in Traditional China," *Han'guk Kwahaksa Hakhoeji* 韓國科學史學會誌 [Journal of the Korean History of Science Society], 7 (1985): 40-49.

1 For general discussions of the relationship between science and religion in Western history, see, e.g., C. A. Russell, ed., *Science and Religious Belief: A Selection of Recent Historical Studies* (London: Oxford University Press, 1973); I. A. Barbour, *Issues in Science and Religion* (Englewood-Cliffs, NJ: Prentice-Hall, 1966).

 DOI 10.1163/9789004265318-009

tices became prominent. That same period corresponded as well to a heightened interest in various 'proto-scientific' activities – alchemy, astrology, hygiene and other forms of physical and physiological therapies.[2] Yet there never was a period in which a single religion was dominant in China like Christianity was in the West or Islam was in Arabic societies. Even at its very height, Buddhism represented just one among many religious currents of the time. Nor were those scientific – and proto-scientific – activities ever integrated fully into the scholarly pursuits of the dominant intellectual class. They were the activities mainly of marginal – both socially and intellectually – groups of society; even when in some periods interest in them was strong among the ruling class, it was always secondary to their main intellectual concern – moral and social problems. As a matter of fact, as the traditional pattern of Chinese society and culture evolved in the Song 宋 dynasty (960-1270), both religion and science's hold on society became even weaker still.

Given this situation, it is not surprising that the relationship between science and religion in traditional China was quite different from that in the West, as many previous studies have pointed out. Yet too much attention to this point of difference has sometimes given rise to problems that hinder a proper understanding of the unique nature of the science-religion relationship in traditional China. In the present paper, I shall discuss these problems and suggest some ways to look at the relationship more fruitfully.

2 Hidden Assumptions in the Previous Studies

Discussing the relationship between science and religion in traditional Chinese society amounts to discussing the relationship between two things of relatively little importance in that society. This could easily lead, if one is not very scrupulous, to the danger of assuming that religion must have played a significant role in the development of science in China, as it did in the more religiously-oriented Western societies. Indeed, most of the previous work on the science-religion relationship in China reflects the influence of such an assumption, which consists of the following two pitfalls: 1) emphasis and exaggeration of

2 For brief overviews of the relevant aspects of this period, see Max Kaltenmark, *Lao tseu et le taoisme* (Paris: Editions du Seuil, 1965), Chap. 3; Holmes Welch, *Taoism: The Parting of the Way*, revised edition (Boston: Beacon, 1965), part 3. For more detailed studies, see Henri Maspero, *Le taoïsme et les religions chinoises* (Paris: Gallimard, 1971), especially books 5-7; Holmes Welch and Anna Seidel, eds., *Facets of Taoism: Essays in Chinese Religion* (New Haven: Yale University Press, 1979).

some tenuous links between science and religion; and 2) the search for – and often speculation about – the possible effects that the lack of religious elements may have had upon scientific development.

A prominent example of the first pitfall is the frequently observed link between science and Daoism. Needham, Maspero, and Welch – and others who follow their pioneering work – have noticed the importance of the scientific and proto-scientific activities practiced by various groups associated with Daoism.[3] Needham, in particular, has gone a step farther and stressed the connection, or the congruence, between Daoist philosophy and the study of nature.[4] Perhaps there is no need to present yet another critique of Needham regarding this latter aspect.[5] In any case, there clearly are problems with the many very detailed studies that are supposed to demonstrate links between scientific and proto-scientific activities and Daoism. Although these interesting and informative studies have taught us immensely about some important alchemical, astrological and medical texts and practices, they raise more questions when they are viewed in connection with the issue of science-religion relationship. For, apart from the fact that they were practiced by those groups for some religious purposes, what was the significance – or relevance – of such practices for the relationship between science and religion in China? What exactly do these activities have to do with their practitioners' religious beliefs and practices? Are we sure that their proto-scientific activities were influenced by, or at least linked with, their religion? Are we not looking at a mere coexistence of, and not an essential or even causal connection between, these proto-scientific activities and certain Daoist religious currents?[6] On this note, we may need to be reminded of our experience with the long controversy over a

3 In addition to the references cited in note 2, see Joseph Needham, *Science and Civilisation in China*, 7 vols. (London: Cambridge University Press, 1954-), vol. 5, part 2, 62-127; part 3, 1-117; the special issue ("Current Perspectives in the Study of Chinese Religions") of *History of Religion* 17 (1978); Sakai Tadao 酒井忠夫, ed., *Dōkyō no sōgōteki kenkyū* 道教の總合的┌究 (Tokyo: Kokusho Kankōkai 國書刊行會, 1977).

4 E.g., Needham, *Science and Civilisation*, vol. 2, 33-164.

5 For works containing such critiques, see, e.g., Wing-tsit Chan, "Neo-Confucianism and Chinese Scientific Thought," *Philosophy East and West*, 6 (1957), 309-332; Shigeru Nakayama, "Joseph Needham, Organic Philosopher," in *Chinese Science: Explorations of an Ancient Tradition*, ed. Shigeru Nakayama and Nathan Sivin (Cambridge, MA: MIT Press, 1973), 23-43; Willard J. Peterson, "Chinese Scientific Philosophy and Some Chinese Attitudes towards Knowledge about the Realm of Heaven-and-Earth," *Past and Present*, no. 87 (1980): 20-30.

6 Nathan Sivin even questions whether this group can be properly called "Daoist": "On the Word 'Taoist' as a Source of Perplexity. With Special Reference to the Relations of Science and Religion in Traditional China," *History of Religions* 17 (1978): 303-330.

certain group of seventeenth-century Englishmen among whom propensities for science and religious fervor coexisted.[7] Furthermore, if connections were to be established, are we then to identify the 'origins' of science in China in those activities, despite the fact that they did not develop further into full-fledged scientific activities? How can we call them 'origins' if little development originated from them? Would we then have to go on looking for causes for the absence of such development?

The second kind of misguided focus mentioned above has centered on the concept of God. It has been noted repeatedly that the Chinese lacked the concept of 'creator-god,' 'divine legislator,' or a 'personal,' 'purposive' and 'active' god, who created the world and rules over that world constantly.[8] It has been shown how the lack of such a concept of god influenced the way the Chinese understood the natural world. Most obviously, because of the lack of the concept of a creator-god, phenomena and objects of the natural world were quite irrelevant for religious beliefs in God, who did not create that world. And, as a consequence, the connection between science and religion could only be sociological. Also, because the concept of a divine legislator – a god who ordained a divine law to be obeyed by everything in the world – was lacking, there was among the Chinese much less interest in studying the natural world in order to find, and admire, a law underlying it.[9]

To be sure, not all agree on these points, and indeed such matters as creation mythologies and laws of nature have been found in China.[10] However, whichever position one takes, the discussions themselves are flawed. For they seem to be based upon a hidden assumption that the way science and religion interacted in the West is the norm, the standard, which should have been followed by other cultures. When one tries to see the effect of the absence in

---

7 The controversy began with the publication of Robert K. Merton's famous work, "Science, Technology and Society in Seventeenth Century England," *Osiris* 4 (1938): 360-632. For a recent critical review of the controversy surrounding the so-called "Merton Thesis," see Gary A. Abraham, "Misunderstanding the Merton Thesis," *Isis* 74 (1983): 368-387.

8 Typical, and influential, examples can be found in various works by Joseph Needham: e.g., *Science and Civilisation*, vol. 2, 58ff. For more recent comments on the subject, see Shu-hsien Liu, "Theism from a Chinese Perspective," *Philosophy East and West* 28 (1978): 413-417; John S. Major, "Myth, Cosmology and the Origins of Chinese Science," *Journal of Chinese Philosophy* 5 (1978): 1-20.

9 E.g., Needham, *Science and Civilisation*, vol. 2, 518-583.

10 E.g., N. J. Girardot, "The Problem of Creation Mythology in the Study of Chinese Religion," *History of Religions*, 15 (1976): 289-318; Derk Bodde, "Evidence for 'Laws of Nature' in Chinese Thought," *Harvard Journal of Asiatic Studies* 20 (1957): 709-727; "Chinese 'Law of Nature': A Reconsideration," *Harvard Journal of Asiatic Studies* 39 (1979): 139-155.

China of a certain religious element which existed in the West and played a certain significant role in Western scientific development, one is likely to come to a conclusion that the absence of that particular element in China caused the lack of such an influence in China. Of course, such findings are not altogether trivial and can be interesting in their own ways. They have led to many interesting explanations of various ways that Chinese natural knowledge developed – or failed to develop. Nonetheless, looking for the effects of the lack of a given historical factor is at least as hazardous as looking for the causes of the lack of a certain historical event.[11]

3 Culture-Dependence of the Science-Religion Relationship

The relationship between science and religion is profoundly culture-dependent. Understanding this relationship in China cannot be achieved by simply checking for the presence or absence of those religious concepts, practices, groups, and connections that were relevant in Western scientific development (as if the relationship that developed in the West could be the only one that religion can have with science). The study of this relationship in China should examine different kinds of connections that were meaningful in the specific context of Chinese science, religion and society. We should be looking at different elements and aspects of science and religion than those we would normally analyze when studying the science-religion relationship in the West.

The science-religion relationship is culture-dependent not only in the sense that it varies from culture to culture. It is culture-dependent also in that the meaning, or importance, of the relationship is different in different cultures. As we have seen, science and religion were two of the most important elements in the West, with an inherent link between them. Thus, the relationship between the two was strong and crucial to understanding scientific and religious developments in the West. One cannot expect, however, to find this relationship in a different culture to be as strong and crucial as one finds it in the West. One could even go so far as to say that the science-religion relationship may be meaningless in the case of traditional China, and that one should stop looking for it altogether. Of course, we do not have to go that far. But we do

11 For a discussion of the problems involved in the latter type – search for the causes of the lack of an event – , see my "Natural Knowledge in a Traditional Culture: Problems in the Study of the History of Chinese Science," *Minerva* 20 (1982): 83-104 (Chapter 5 of this book). For a more abstract philosophical discussion of the problems, see Robert S. Cohen, "The Problem of 19(k)," *Journal of Chinese Philosophy* 1 (1973): 103-117.

have to be clear that we are dealing with a different kind of science, a different kind of religion, in a different kind of social and cultural context. Thus, we are looking at a different kind of relationship between science and religion, different kinds of links, interactions and dependences that may be incommensurable with those we look at when studying the science-religion relationship in the West.

Perhaps at the heart of this 'culture dependence' is the concept of 'religion' itself. For in China we are dealing with very different kinds of 'religions' from such Western ones as Christianity and Islam. For example, Chinese religions lacked an established priesthood that was tied with state power yet possessing an independent religious power. Nor was there a strong otherworldly zeal in Chinese religions. We may even have to interpret the term 'religion' differently from the way we do in the Western context so that the notion may fit with Chinese religious practices and beliefs. And certainly, we have to guard ourselves against the tendency to look only at those that could exert a strong influence upon science, as seen in the West. Perhaps what we are looking at may be 'belief systems' rather than the Western type of exertive, institutionalized religion. In any case, only when we fully grasp such differences will we be able to come up with answers – relationships, influences, and interactions between science and religion – that are more meaningful, more interesting, more so than a mere 'presence' or 'absence.'

4 Confucianism and Science

One subject that cannot be ignored in this context is Confucianism, which I offer as a query and not as any kind of solution. Should we not perhaps include Confucian beliefs and practices when we look at the relationship between science and religion in traditional China? For if we were to do so, would we not come up with more interesting comparisons with the West? After all, Confucianism was the only belief system that dominated post-Tang 唐 (618-906) traditional Chinese society and culture to a degree comparable to Christianity in the medieval West. Most of the Chinese intellectuals in that period were at least potentially bureaucrats dedicated to Confucian values and beliefs, just as intellectuals in the medieval West were potential clerics devoted to Christian values and ideology. Thus, regardless of one's view as to whether Confucianism can be referred to as 'religion' or not (and have we not already talked about Daoism as a 'religion' in spite of how problematic it is to do so?),[12] there is no

12 In a much broader context, Max Weber refers to Confucianism and Daoism as 'religions': *The Religion of China, Confucianism and Taoism* (New York: Free Press, 1951).

denying that looking at the ways in which the major belief system of a society influenced its natural knowledge will throw light upon science, religion and the relationship between them in that society.

Again, when we examine Confucianism, what we should look at is not the presence or absence of certain religious concepts and practices that were important in the West, but rather those Confucian concepts and practices that were important in the Chinese context. A foremost example will be the concept of heaven (*tian* 天). However different it may be from the Western concept of god, it cannot be denied that for Confucians heaven is responsible for the coming into being and workings of everything in the world, including man. The neo-Confucian anthologies like *Xingli daquan* 性理大全 ("Collected Works on Human Nature and Principle," compiled in 1415) and *Zhuzi yulei* 朱子語類 ("Classified Conversations of Master Zhu Xi," compiled in 1270) contain many utterances explicitly saying that heaven 'created,' or 'generated' (*sheng* 生), man and the ten thousand things, that heaven 'conferred' (*fenfu* 分付) something that gives life to man and things, that heaven is what is 'in charge of' (*zhuzai* 主宰) everything, etc.[13] The more we examine such aspects of the concept of heaven, the more we come to realize that this concept has some connotations of 'creator-god' and 'divine legislator.' At the least, the argument that the lack of such a concept in China inhibited, or failed to encourage, the study of nature loses much persuasive power. Another term that sometimes has a religious connotation and is related to 'heaven' is 'the sages' (*shengren* 聖人). In most Confucian, especially neo-Confucian, sources it is generally held that, whereas heaven first brought man and things into being, it was through the sages (who, of course, were themselves brought into being by heaven) that heaven created all the various human artifacts – tools, cultural objects, social institutions, including even the classics, the words of the sages manifesting the Way.[14] Bearing this in mind, and understanding the term better, will help us greatly as we try to make sense of the many assertions in and about the classics that are related with things and events of the natural world and with knowledge about them. The Confucian practice of offering sacrificial service (*jisi* 祭祀) to dead ancestors, the central activity of Confucian ancestor worship, has a clear religious implication, and provides still another example. Study of various explanations of the practice, especially through the concept of '*guishen*' 鬼神, which itself has different levels of meaning, has much to tell us about the ideas of human life, death and about the way the practice

---

13 Such utterances can be found in various places in the anthologies: e.g., *Zhuzi yulei* 朱子語類 ("Classified Conversations of Master Chu Hsi", 1270 edition), 4.4b, 16.2b, 17.14b, 18.17b.

14 E.g., *Zhuzi yulei*, 4.20b, 76.4a, 87.23a.

influenced such ideas.[15] Lastly, when dealing with the religious aspects of Confucianism, we can also include the clearly religious character of the intense mental and intellectual endeavors of the so-called 'investigation of things' (*gewu* 格物) of serious neo-Confucian thinkers like Wang Yangming 王陽明 (1472-1528) and those of the Donglin 東林 group.[16] What effect, one may ask, did this religious character have upon their knowledge about the natural world?

5 Buddhism and Science

When it comes to Buddhism, no one questions whether it can be called a 'religion,' though one quite different from Western religions. And as a religion it was certainly very powerful during the middle centuries of the first millennium in China; it never lost its grip completely on the minds of Chinese intellectuals in the later centuries.[17] For example, even as late as the twelfth century, Zhu Xi 朱熹 (1130-1200) felt a need to vehemently criticize Buddhism,[18] a testimony to its strength at the time. Nevertheless, Buddhism has rarely been studied in connection with science, almost the only exception being Needham's section on Buddhist thought.[19]

Needham's verdict on the effect of Buddhist thought upon Chinese scientific development is essentially negative. Basically, in his view, Buddhism

15 For a detailed discussion of the concept of '*guishen*', see my article, "*Kuei-Shen* in Terms of *Ch'i*: Chu Hsi's Discussion of *Kuei-Shen*," *Tsing Hua Journal of Chinese Studies*, new series 17 (1985): 149-163.

16 Such religious features can be discerned in many modern discussions of these thinkers: e.g., Wei-ming Tu, *Neo-Confucian Thought in Action: Wang Yang-ming's Youth (1471-1509)* (Berkeley, CA: University of California Press, 1976); Julia Ching, *To Acquire Wisdom: The Way of Wang Yang-ming* (New York: Columbia University Press, 1976); Heinrich Busch, "The *Tung-Lin* Academy and Its Political and Philosophical Significance," *Monumenta Serica* 14 (1949-1955): 1-163.

17 For historical surveys of Buddhism in China, see, e.g., Arthur F. Wright, *Buddhism in Chinese History* (Stanford: Stanford University Press, 1959); Kenneth K. S. Ch'en, *Buddhism in China* (Princeton: Princeton University Press, 1964); Erik Zürcher, *The Buddhist Conquest of China: the Spread and Adaptation of Buddhism in Early Medieval China* (Leiden: Brill, 1959).

18 Zhu Xi's criticism of Buddhism is found in many places of the Zhu Xi corpus. But important passages are gathered in *Zhuzi yulei*, *juan* 126, titled "Shishi" 釋氏. For some discussions of these materials, see Qian Mu 錢穆, *Zhuzi xin xuean* 朱子新學案, 5 vols. (Taipei: Sanmin shuju, 1971), vol. 3, 489-579.

19 Needham, *Science and Civilisation*, vol. 2, 396-431.

inhibited, or at least did not foster, the growth of science in China. This verdict might very well turn out on balance to be correct, but if so, it will not be exactly for the reasons that Needham gives. He sees in Chinese Buddhism all kinds of 'inimical factors' for science, pointing for example to such tendencies as the denial of reality, agnostic and mystic attitudes, and even hostility toward rational inquiry into nature. These factors, of course, are incompatible with an inquiring attitude into the natural world. However, they could not by themselves have inhibited the development of scientific knowledge in China. In the West of the early Middle Ages, when Christian theology allied with Platonism was dominant, we see all of these same attitudes – the Platonic denial of empirical reality, early Christian agnosticism and mysticism, and hostility of Christian fathers towards science and rationalism.[20] Although these attitudes must have negatively affected scientific development, they were not able to inhibit a reemergence of scientific inquiries in the high Scholasticism of the later Middle Ages.

Moreover, as Needham himself admits, Buddhist doctrines were relatively rich in fundamental abstract concepts that could have generated theoretical discussions about matters of potential scientific significance. For example, the Buddhist emphasis on emptiness (*kong* 空 or *xu* 虛) and nothingness (*wu* 無) could have led to discussions of the concept of the void. Similarly, doctrines of rebirth, hell and paradise could have raised questions about the uniqueness, eternity and infinity of the universe and questions about life, the soul and the body. Although these Buddhist doctrines never produced the kind of intense discussions as similar concepts did in medieval Western Christianity,[21] which eventually led to very fruitful results during the Scientific Revolution, this does not allow us to conclude that Buddhist beliefs and doctrines by themselves were inhibitory to scientific development.

Certain Buddhist doctrines did generate debates of some scientific interest, sometimes even among Confucians. A foremost example is the doctrine of 'the wheel of transmigration of the soul' (*lunhui* 輪廻, *samsara*), the concept that at death the soul passes into another body, either human or animal. Not only did this doctrine generate an interest among Buddhists about the origins of life

20 For brief discussions of the situation concerning religion, science and learning in this period, see Edward Grant, *Physical Science in the Middle Ages* (New York: Wiley, 1971), Chap. 1; Brian Stock, "Science, Technology, and Economic Progress in the Early Middle Ages," in *Science in the Middle Ages*, ed. D. C. Lindberg (Chicago: University of Chicago Press, 1978): 1-51.

21 E.g., John E. Murdoch, "From Social into Intellectual Factors: An Aspect of the Unitary Character of Late Medieval Learning," in *The Cultural Context of Medieval Learning*, ed. John E. Murdoch and Edith D. Sylla (Boston: Reidel, 1975), 271-339.

and the soul,[22] but it also became a target of criticism by neo-Confucians. In order to refute the theory of transmigration, Confucians referred to the general belief that man's *qi* 氣 is dispersed after death.[23] But since that belief was in conflict with the practice of offering sacrificial services to dead ancestors whose *qi* is already dispersed, they had to provide quite an elaborate set of explanations for the practice.[24] Naturally, such explanations did not reach the level of those extremely technical and theoretical discussions on the principle of life, structure of matter, change, etc. that the Christian Eucharist doctrine, for instance, created in the medieval West.[25] But it was not because Confucian or Buddhist beliefs and doctrines precluded or inhibited interest in such matters.

Consequently, it is not differences in religious ideas and doctrines alone that we should look at when we study and compare the science-religion relationship in the West and in China. As we have seen, there were religious ideas and doctrines in China – both in Buddhism and in Confucianism – that could have led to scientifically fruitful discussions. The truly important differences were in the attitudes towards such discussions – differences in the intensity of interests. To illustrate, both Confucians and Buddhists thought such theoretical questions as the eternity or infinity of the universe, the void, etc. were useless, whereas medieval Christian scholastics were deeply involved in debates about such issues. Some Christian natural philosophers even used certain theological points as excuses for discussing such problems as motion, heavenly bodies and light.[26]

6 Concluding Remarks

My queries and reflections in the preceding pages do not yield a set of neat conclusions. Nonetheless, it is important to realize that the relationship between science and religion is not so straightforward in China or in any other traditional culture, and that to understand it we need to look at many other

22 Joseph Needham, for example, sees an influence of the doctrine in Buddhist interest in embryology: *Science and Civilisation*, vol. II, 402.

23 A typical example of neo-Confucian criticism of this doctrine can be found in Chen Chun 陳淳 (1153-1217), *Beixi ziyi* 北溪字義 (c. 1200), 2, 39.

24 For a discussion of such explanations, see my article "*Kuei-Shen* in Terms of *Ch'i*."

25 E.g., John E. Murdoch and Edith D. Sylla, "The Science of Motion: The Role and Scope of Motion in Medieval Natural Philosophy," in *Science in the Middle Ages*, ed. David C. Lindberg, 219-221.

26 Murdoch, "From Social to Intellectual Factors," 278-279.

aspects of the society and culture, not just its science and religion. The result will enrich our understanding of the entire society and culture, of which science and religion are parts. When dealing with a society in which science and religion are indeed important elements, we tend to note connections, interactions, and sometimes even causal influences between them; seventeenth-century England is a classic example. In traditional China, however, such examples are hard to come by.

CHAPTER 8

Science and Bureaucracy in Traditional China*

Ever since the beginning of the imperial history in the late third century BC, bureaucracy was an important element in Chinese society. By the Song period (960-1279), the Chinese bureaucracy had acquired a set of distinct characteristics, and the bureaucrats – the *shi* 士, 'scholar-officials' or 'mandarins' as they were referred to – came to dominate all aspects of Chinese society, which characteristically lacked alternative occupants of power, such as a clerical, noble, military or merchant elite. As Etienne Balazs wrote several decades ago:

> The class of *scholar-officials* (or mandarins), numerically infinitesimal but omnipotent by reason of their strength, influence, position, and prestige, held all the power and owned the largest amount of land. This class possessed every privilege, above all the privilege of reproducing itself, because of its monopoly of education. ... [It] drew its strength from the function it performed – the socially necessary, indeed indispensable, function of coordinating and supervising the productive labour of others so as to make the whole social organism work. All mediating and administrative functions were carried out by the scholar-officials. They prepared the calendar, they organised transport and exchange, they supervised the construction of roads, canals, dikes, and dams; they were in charge of all public works, especially those aimed at forestalling droughts and floods; they built up reserves against famine, and encouraged every kind of irrigation project. Their social role was at one and the same time that of architect, engineer, teacher, administrator, and ruler.[1]

Bureaucracy thus had a profound effect not only upon government and politics, but also upon nearly all aspects of society and culture in traditional China. Clearly, science was no exception. Thus, the bureaucracy should naturally rank near the top on a list of subjects to be studied if one is attempting to understand the social and intellectual contexts of Chinese science. It has a bearing on almost all the other subjects in that list. The Chinese bureaucracy had an

* Originally published as "Science and Bureaucracy in Traditional China," *Medieval History Journal* 3 (2000): 363-379.

1 Etienne Balazs, "China as a Permanently Bureaucratic Society," in *Chinese Civilization and Bureaucracy*, tr. H.M. Wright (New Haven: Yale University Press, 1964), 16.

 | DOI 10.1163/9789004265318-010

overall influence not only over the more obvious subjects like education and examination, recruitment and promotion, and income and social status of scientific practitioners, but also over the bigger themes – science, religion, and technology.

Yet the relationship between science and bureaucracy in China has not been adequately studied.[2] Although the subject has been frequently discussed, or often simply mentioned, what have been abundant are impressionistic statements concerning the 'stimulating' or 'inhibiting' (in actuality, mostly 'inhibiting') effects of the bureaucracy on the development of science in China. This 'either/or' kind of approach reflects an overly simplistic mental outlook, ever intent on classifying a given cultural element as having either a stimulating or an inhibiting effect. Clearly, a far more nuanced approach will be necessary to gain a better understanding of this very complex and multifaceted relationship.

In this essay, I shall explore the relationship between science and bureaucracy in traditional China, and will try to identify what I consider to be the problems and possibilities within the study of this relationship. I shall first look at previous discussions on the matter, note some common problems raised in these discussions, and speculate about the causes behind the problems that have come to characterize the study of the science-bureaucracy relationship. In examining the nature of this relationship more generally, and raising some issues of intercultural comparisons, I shall draw upon my views about the relationship between science and another cultural element in traditional China, that is, religion. I will suggest some new perspectives that can be brought into the study of the relationship between science and bureaucracy in China, and will conclude by mentioning some topics, old and new, which can be studied fruitfully in a new light.

Of course, I do not claim that my own discussion will rise above the level of impressionistic statements about certain aspects of the relationship between science and bureaucracy in China, and it will inevitably contain a fair number

2 Among those who have commented upon the relationship between science and bureaucracy in China, Joseph Needham has been the most influential: see his essays gathered in *The Grand Titration: Science and Society in East and West* (London: Allen & Unwin, 1969), especially Chap. 6, titled, "Science and Society in East and West." Most of the recent comments on the subject have been derived from Needham's view: see Derk Bodde, *Chinese Thought, Society, and Science: The Intellectual and Social Background of Science and Technology in Pre-modern China* (Honolulu: University of Hawaii Press, 1991), Chap. 5; Toby E. Huff, *The Rise of Early Modern Science: Islam, China, and the West* (Cambridge: Cambridge University Press, 1993), Chaps. 7 and 8; H. Floris Cohen, *The Scientific Revolution: A Historical Inquiry* (Chicago: University of Chicago Press, 1994), Chap. 6.

of unguarded speculations touching upon the significance of such a relationship on scientific development in China. In addition, my treatment of many of the concrete facts and events will be very general, tentative and exploratory, thus often simplistic and superficial. I present my remarks fully aware of these risks, yet driven to raise these vital questions and lay out some problems that should be studied and examined further. I hope that these will provide new insight into the nature of the relationship between science and bureaucracy in traditional China and point in some new directions.

1 Did Bureaucracy 'Inhibit' Scientific Development in Traditional China?

As previously mentioned, most of the previous discussion on the interaction between science and bureaucracy in China, or the role of the bureaucracy in the development of science in China, has concluded with judgments focusing simply on whether bureaucracy fostered scientific development or whether it blocked it. It has been inevitable that such simplistic interpretations, though at times leading to extremely insightful observations, have been fraught with problems.

The Chinese bureaucracy has frequently been considered an institution that stifled the development of various elements that would have been necessary for the development of science – a merchant class, democracy, cities, universities, and the like. It is obvious, however, that these views assume that the above elements, had they developed in China, would have had the same stimulating effects on scientific development there as they did in certain times and on certain aspects of scientific development in the West. Joseph Needham, for example, notes that "the failure of the rise of the merchant class to power in the State lies at the basis of the inhibition of the rise of modern science in Chinese society."[3] Yet this is an assumption never proven, even in the case of Western scientific development. The explanation that Needham proposes linking the growth of a merchant class with scientific development – namely that merchants "were interested in the properties of things because on such properties alone could mercantile and quantitative economics develop" – is far from convincing.[4] Furthermore, no one has ever shown exactly how the bureaucracy might have stifled the development of features like the merchant class, democracy, cities and universities in China. It is not even clear whether

3 Needham, *The Grand Titration*, 186.

4 Ibid., 136.

the bureaucracy did indeed stifle them. Besides, it is difficult to completely deny that cities and universities developed in China, although in forms quite different from those in the West.[5]

Another frequently encountered view is that once science was incorporated into the bureaucracy in China, scientific activities were confined to narrow parameters and stagnated. Proponents of this view point to the situation of stagnation and decline of Chinese science in the Ming period (1368-1643) and afterwards.[6] One scholar has even seen the bureaucracy's limiting effect on the astronomical activities in the Qing dynasty (1644-1911), when imported Western astronomy was "confined to a bureaucratically prescribed form" in the Imperial Astronomical Bureau (*Qintian jian* 欽天監).[7] Yet there have been studies questioning whether scientific activities indeed stagnated or declined in the Ming and afterwards.[8] In the case of mathematics at least, what we know about Ming society and economy – the continuing and flourishing activities of commerce, surveying, and the like, as well as the publication of mathematical books – makes it difficult to believe that mathematical activities could have declined in such an atmosphere; in fact, there was a great popular interest in mathematics in the late Ming period. But even if it can be granted that science did stagnate and decline in the Ming, the question of the bureaucracy's connection with such stagnation and decline still remains. How did bureaucracy 'confine' or 'stifle' scientific activities that came under its purview to make such activities 'stagnate' and 'decline'? Was it a conscious intention or an unconscious motive, or both, on the part of bureaucrats that had such an effect? Or is there something inherent in the bureaucracy in general, or in the Chinese bureaucracy in particular, that necessarily 'confines' or 'stifles' scientific activi-

5 E.A. Kracke, Jr., "Sung K'ai-feng: Pragmatic Metropolis and Formalistic Capital," in *Crisis and Prosperity in Sung China*, ed. John W. Haeger (Tucson: University of Arizona Press, 1975), 49-77; Jacques Gernet, *Daily Life in China on the Eve of the Mongol Invasion 1250-1276* (London: Allen & Unwin, 1962), Chap. 1; Thomas H.C. Lee, *Government Education and Examinations in Sung China* (Hong Kong: Chinese University Press, 1985), Chap. 4.

6 On the decline of science in the Ming, see Joseph Needham, *Science and Civilisation in China*, vol. 3 (Cambridge: Cambridge University Press, 1959), 50-52; Li Yan and Du Shiran, *Chinese Mathematics: A Concise History* (Oxford: Clarendon Press, 1987), 175; Liu Dun, "400 Years of the History of Mathematics in China: An Introduction to the Major Historians of Mathematics since 1592," *Historia Scientiarum* 4 (1994): 103-104.

7 Jonathan Porter, "Bureaucracy and Science in Early Modern China: The Imperial Astronomical Bureau in the Ch'ing Period," *Journal of Oriental Studies* 18 (1980): 72.

8 Willard J. Peterson, "Calendar Reform Prior to the Arrival of Missionaries at the Ming Court," *Ming Studies* 21 (1986): 45-61. Roger Hart, *Imagined Civilizations: China, the West, and Their First Encounter* (Baltimore: Johns Hopkins University Press, 2013). Chap. 4.

ties? If the answer to the last question is 'yes,' what then can we make of the dynamic scientific activities and the flourishing scientific developments in the Song, which happened to be the formative period of Chinese bureaucracy? And what about the fact that many of the major 'scientists' of the day – Shen Gua 沈括 (1031-95) and Su Song 蘇頌 (1020-1101), to name only the most famous ones – were bureaucrats and that various scientific and technical activities were bureaucratized in the Song?

Questions can be raised, in this connection, as to whether the bureaucracy's role should be viewed solely in terms of 'control' of scientific and technical activities by the state, resulting in a 'stifling' effect. Could the bureaucracy/state not have 'sponsored' these activities and those engaged in them?[9] Indeed, Needham finds aspects of the Chinese bureaucracy that can be counted as having a stimulating effect on scientific and technological development: for example, the government's need to predict important natural events; the bureaucratic habit of classification; the establishment of bureaucratic routines, and so on. This led the Chinese state to take an interest and engage in such scientific and technological activities as astronomical observation and recordkeeping, making instruments and calendars, publication of encyclopedias and specialized treatises, scientific expeditions, and large-scale construction works. Needham even sees a connection between the 'non-interventionist' attitude of the Chinese bureaucracy, which valued 'no action' (*wuwei* 無爲), and the preference for 'action at a distance,' which he believes to have characterized Chinese physics.[10] For him, such 'stimulating' effects of the bureaucracy were a reason for the fact that China was ahead of Europe in science and technology in the early imperial period. In the end, however, Needham's verdict on the Chinese bureaucracy was negative. Not only did the bureaucracy 'strangle' the merchant class, which to him was so essential for scientific development, but it also created a gulf between the scholar and the artisan, thereby preventing fruitful interactions between the two groups. It is natural, then, that the bureaucracy provides Needham with a key for answering his grand question about "the failure of China to give rise to distinctively modern science while

9 Even before beginning to pursue these questions further, formidable supporting evidence presents itself, namely the fact that science and technology are thriving in modern society, in which they are thoroughly bureaucratized. I shall comment on this later in the essay.

10 For a summary of Needham's views on science and bureaucracy, see *Grand Titration*, 211-212; Bodde, *Chinese Thought*, 190-191; Cohen, *Scientific Revolution*, 451-453. Cf. Harold Dorn's idea of the 'hydraulic' society in which a centralized bureaucracy and utilitarian science coexist: *The Geography of Science* (Baltimore: Johns Hopkins University Press, 1991).

having been in many ways ahead of Europe for some fourteen previous centuries":[11] The Chinese bureaucracy "at first favored the growth of natural knowledge and its application to technology for human benefit, while later on it inhibited the rise of modern capitalism and of modern science."[12]

It would be interesting to compare the situation in China – especially in the post-Ming period when the bureaucracy is supposed to have 'inhibited' scientific development – with that of early modern Europe, where there were many forms of royal patronage for scientists, who could sometimes be aptly characterized as 'courtiers.'[13] One may wish to study, for example, the situation of the astronomers working in the Chinese Imperial Astronomical Bureau and compare it with the situation of those working in the Royal Academy of Sciences in Paris, which, though quite bureaucratized, nevertheless seems to have facilitated the scientific activities of astronomers. Can it still be maintained, after such studies, that the bureaucratization of astronomy in China only had negative effects? Can one say anything more than that (if even that) the overall effect of the exclusive governmental monopoly of astronomical activities *may* have been harmful to astronomy in China as a whole? And finally, even if one is still willing to grant that bureaucracy was responsible for the 'decline' and 'stagnation' of science in post-Ming China, should it assume the whole, or the principal, responsibility?

There is one aspect of the Chinese bureaucracy cited frequently as an inhibiting factor that – on the surface, at least – is more persuasive, namely that the Chinese bureaucracy, through the civil service examination system with its exclusive emphasis on literary, classical and ethical studies, appropriated most young talents who could have gone into other fields, including science.[14] However, the literary and moral emphasis of the civil service examination does not seem to have been as exclusive as has been supposed, and certainly not sufficient to block other interests completely. In the Tang (618-906), for example, there were examinations for specialists in mathematics, astronomy and medicine, even though most candidates took the literary *jinshi* 進士 examinations for generalists. These specialist examinations, it must be noted, did not

11 Needham's 'grand question' is stated in this form in the "Author's note" to *Science and Civilisation in China*, vol. 5, part 2, xxiii.

12 Needham, *Grand Titration*, 197.

13 For the situation in the West during the Scientific Revolution, see Richard S. Westfall, "Science and Patronage: Galileo and the Telescope," *Isis* 76 (1985): 11-30; Mario Biagioli, *Galileo, Courtier: The Practice of Science in the Culture of Absolutism* (Chicago: Chicago University Press, 1993).

14 Needham, *Grand Titration*, 39.

continue into the Song.[15] We learn from Benjamin Elman's study that even in the Ming, the period in which the bureaucracy supposedly 'stifled' scientific developments, civil service examinations did in fact include questions on astronomy and other scientific subjects, which many examination candidates had to study.[16] Moreover, even if one admits that the civil service examinations had an overwhelming literary and moral emphasis, the fact that most talents chose to enter the generalist bureaucracy by taking such examinations rather than to seek other career alternatives cannot constitute a real explanation for the lack of scientific development in traditional China, for it still needs to be explained why most Chinese talents were attracted to the generalist-dominated bureaucracy in the first place and why successive Chinese dynasties chose to maintain such an overwhelming imbalance in their allocation of rewards for their young talents.[17]

All previous discussions mentioned so far share one thing in common: they all conclude with a negative verdict about the effect of the Chinese bureaucracy on scientific development. And it is not difficult to see why the judgment has been so unfavorable. The analysis of the role of bureaucracy in Chinese scientific development is a rare example in the comparative study of Chinese science, for in it one deals with the effect on Chinese science of a cultural element which was important in China while not so in the West. In almost all other cases, the focus has been on the roles of certain elements that were important in Western scientific development but were not important, or were supposedly absent, in China: for example, religious notions such as God, creation and the divine law of nature; methodological tools such as proof, formal logic, geometry and symbolic algebra; and such socio-economic elements as the merchant class, the city and slavery. In these latter cases, it has usually been deduced that it was the *absence* in China of the elements important in Western scientific development that inhibited the development of Chinese science. It is not surprising then that those with this mental outlook would conclude, in the case of the bureaucracy (which was not important in the West but dominant in

15 Bodde, *Chinese Thought*, 213-214.

16 Benjamin A. Elman, *A Cultural History of Civil Examinations in Late Imperial China* (Berkeley: University of California Press, 2000), Chap. 9.

17 It should be noted, on the other hand, that in European countries too, when modern bureaucratic systems were established, regular training and examination of bureaucrats did not cover subjects such as science, mathematics, engineering, agriculture or medicine, and that experts in those areas were specially recruited and were confined to lower positions. See Sir Laurence Helsby, "Recruitment to the Civil Service," in William A. Robson (ed.), *The Civil Service in Britain and France* (London: Hogarth Press, 1956), 44.

China), that it was the *presence* in China of such a factor that inhibited Chinese scientific development.

2 Problems of Intercultural Comparisons

Many problems facing studies of the relationship between science and bureaucracy in China are really problems of intercultural comparisons. This is so because while discussing the relationship one is frequently led to make comparisons of various elements and aspects of China with those of the West. One compares not only scientific development or bureaucracies in the two cultures, but also the relation between science and bureaucracy in both cultures.

Yet in making these comparisons, one should keep in mind the fact that the elements being compared are culture-dependent and thus vary from culture to culture. Science and bureaucracies in the two cultures were clearly different: although we use the same words, 'science' and 'bureaucracy,' in referring to them, they had a different construct in China as compared to the West. Perhaps no one needs be reminded that the 'sciences' of the two cultures were different, but the bureaucracies in China and the West were equally different. Not only did the West lack a fully developed bureaucracy until quite late, but also when the bureaucracy finally emerged there, it did so in a vastly different form: in the modern West the bureaucracy was essentially made up of specialists largely separated from political power, while in China it was dominated by generalists employed by the state.[18] Clearly, one should study how these different forms of bureaucratic developments influenced scientific and technological development in the two cultures.

It is not just scientific development and bureaucracies that differed across the two cultures; the relationship between the two – and the meaning and importance of that relationship – was also very different. When one compares the relationship between science and bureaucracy in different cultures, one is actually dealing with different kinds of science and different kinds of bureaucracy in different kinds of social and cultural contexts. Thus, one is looking at different kinds of links, interactions, dependencies and influences that are likely to be incommensurable with each other. It is clear, then, that there cannot be a single kind of relationship that is meaningful for all the different con-

18 Needham notes the ways in which the Chinese bureaucracy differs from that of the modern West and characterizes it as 'feudal.' He even suggests the possibility of using the Chinese term '*guanliao* 官僚' to designate the characteristic features of Chinese bureaucracy: *Grand Titration,* 206.

texts in which science and bureaucracy exist in different cultures. A proper comparative understanding of the relationship between science and bureaucracy, therefore, cannot be achieved by merely checking for the presence or absence, in one culture, of certain bureaucratic institutions, practices, ideas and groups that were important in the scientific development of a different culture. One should be especially careful not to yield to the temptation to bring in the situation in one culture as a standard or criterion to judge the situation in another.

In regards to this issue, I find it helpful to draw upon the study of the relationship between science and another cultural element: religion. This relationship has been studied intensively, at least in the case of the West.[19] The relationship between scientific development and bureaucracy in China may well be compared to that between science and the Christian religion in the West, because the degree of dominance of the bureaucracy in Chinese society was comparable to that of Christianity in the West. Ever since it took root in Western society in late antiquity, the Christian religion has been a dominant cultural force throughout Western history, just as bureaucracy was in China throughout the imperial period. Bureaucracy and Christianity also had similarly powerful influences upon the intellectuals of China and the West respectively: most of the intellectuals in traditional China were, at least potentially, bureaucrats dedicated to Confucian values and beliefs, just as the intellectuals in the medieval West were potential clerics devoted to Christian values and ideology.

As the relationship between science and religion – which can be, and have often been, seen as forces opposing each other – has been studied in depth in the past several decades, we have come to learn about the extremely complex and multifaceted nature of the interaction between the two in the course of Western history. Religion did not simply play an 'inhibiting' role, but did in fact frequently aid Western scientific development in many different ways over different periods and in different regions. Studies of the Scientific Revolution, for example, have shown various formative roles played by the Christian religion in the rise of many strands of modern science: experimentalism, mechanical

19 John H. Brooke, *Science and Religion: Some Historical Perspectives* (Cambridge: Cambridge University Press, 1991); David C. Lindberg and Ronald L. Numbers, eds., *God and Nature: Historical Essays on the Encounter between Christianity and Science* (Berkeley: University of California Press, 1986). For the relationship between science and religion in China, see Yung Sik Kim, "Some Reflections on Science and Religion in Traditional China," *Han'guk Kwahak-sa Hakhoe-ji* [Journal of the Korean History of Science Society] 7 (1985): 40-49 (Chapter 7 of this book).

philosophy, 'Puritan ethics,' and the like.[20] Indeed, how could it have been otherwise? Is it even conceivable that such a dominant element as religion could have played no role in a key event that occurred in Western society? And turning our attention back to China and to the matter at hand, is it not obvious that bureaucracy, a similarly dominant element in traditional Chinese society, will turn out to have played a similarly complex and multifaceted role in scientific development in China?[21] At the very least, is it not clear that making an 'either/or' type of judgment regarding whether its role was 'stimulating' or 'inhibiting' will not lead to an adequate understanding of the complex relation between the two?

3 New Perspectives for the Study of Science and Bureaucracy in China

There are a few new, potentially fruitful, perspectives for the study of the relationship between science and bureaucracy in China.

To begin with, a salient feature of Chinese bureaucracy was that it was thoroughly Confucian. All Chinese bureaucrats, at least after the state adopted Confucianism in the Han (206 BC-AD 220), subscribed to Confucian values and beliefs. Their education was primarily based on Confucian texts; the examinations taken by young scholars aspiring to become bureaucrats centered on testing the candidates' knowledge of Confucian classics; frequently, the outcomes of the examinations were based on how faithful the answers were to what the examiners considered to be the true Confucian spirit. In this respect, the subject of the relationship between science and bureaucracy in China can be seen as a part of, or an aspect of, the relationship between science and Confucianism. Much of what can be said about the latter relationship thus bears on our understanding of the interaction between science and bureaucracy.[22]

---

20 No serious student of the Scientific Revolution still holds the simplistic notion of religion opposing and suppressing science. For example, we now have a different, and a good deal more nuanced, understanding of the so-called 'Galileo affair,' which has been largely responsible for such simplistic popular perceptions of the relationship between science and religion.

21 To be sure, there is a clear difference between the two sets of relationships in that science in traditional China did not occupy such an important place as it did in the West. Thus, while the relation between science and religion was the relation between two important elements in the West, this was not the case in China.

22 For the relationship between science and Confucianism, see Wing-tsit Chan, "Neo-Confucianism and Chinese Scientific Thought," *Philosophy East and West*, vol. 6(4), 1957:

Insofar as Confucianism has features that can be considered 'religious,' it turns out to be not so unrelated to the subject of science versus religion after all.[23] An aspect of Chinese bureaucracy that we have already underlined – the fact that Chinese bureaucrats were generalists – can be placed in this context, because it reflects the reluctance to specialize, which was so characteristic of Confucianism from its beginnings, as can be seen from Confucius's famous saying, "The gentleman is not a utensil."[24]

An obvious corollary to this Confucian character of the Chinese bureaucracy is that all Chinese bureaucrats were at the same time scholars, at least in principle. It can thus be expected that Chinese bureaucrats would show a typical scholarly attitude towards technical scientific subjects and those specializing in them. In my own study of Zhu Xi 朱熹 (1130-1200), I found that scholar-bureaucrats like him generally had a condescending attitude toward specialists.[25] True to his doctrine of 'the investigation of things' (*gewu* 格物), which greatly emphasized the need to study knowledge of concrete things and events in all areas of human concerns, Zhu Xi studied and discussed knowledge contained in such specialized traditions as calendrical astronomy, harmonics, geography and alchemy. For some subjects he attained quite a sophisticated level of specialized knowledge, but at the same time he deemed these subjects to be of secondary importance to his principal concern – moral and social problems. He did not have much respect for those specializing in these specialist traditions, taking them to be mere technical experts. He thought, for example, that calendar specialists could only make observations about obvious facts and make practical computations, but could not fathom more profound matters. Sometimes he did not even have adequate respect for their expertise and casually criticized them in matters where their expertise

---

309-32; Yung Sik Kim, "Confucian Scholars and Specialized Scientific and Technical Knowledge in Traditional China, 1000-1700: A Preliminary Overview," *East Asian Science, Technology and Society: An International Journal* 4 (2010): 207-228 (i.e. Chapter 6 of this book).

23 For the religious aspects of Confucianism, see Rodney L. Taylor, *The Religious Dimensions of Confucianism* (Albany: State University of New York Press, 1990).

24 *Analects*, 2.12. Derk Bodde even says that the Chinese bureaucracy lacked 'functional specialization,' which sociologists have traditionally regarded as a necessary feature of bureaucracy: *Chinese Thought*, 215.

25 Yung Sik Kim, "Chu Hsi (1130-1200) on Calendar Specialists and Their Knowledge: A Scholar's Attitude toward Technical Scientific Knowledge in Traditional China," *T'oung Pao*, vol. 78 (1), 1992: 94-115. For a more detailed account, see Yung Sik Kim, *The Natural Philosophy of Chu Hsi (1130-1200)* (Philadelphia: American Philosophical Society, 2000), Chap. 12.

was clearly relevant. For him, they were experts in their own specialized areas, which he did not bother to master himself; he did not hide his confidence that he could have mastered them if he had only tried. The fact that most Chinese bureaucrats were scholars will necessarily have some significance for the relationship between science and technology in China, because scholar-bureaucrats naturally looked differently upon subjects that we would call 'scientific' than they did upon subjects that we would call 'technological.' We should also note that most Chinese bureaucrats were not ideal Confucians. Indeed, at no point in Chinese history were bureaucrats considered acceptable to the high standard of the idealistic Confucians of the time, who always lamented the current state of affairs and wanted to restore the ideal bureaucratic institutions and values that were believed to have existed in the ancient golden age. This was in part the reason why many serious Confucian thinkers, some of them highly original and creative, so often shunned bureaucracy in favor of a private life devoted to scholarship and moral self-cultivation. This tendency became more pronounced from the southern Song onwards, when competition in the civil service examinations became more intense and the odds for successfully embarking on a bureaucratic career became more remote for average scholars.[26] During periods of dynastic change, this tendency became more widespread and frequently took the form of eremitism.[27] This, then, is another reason why it cannot be said simply that the Chinese bureaucracy drained talents from scientific pursuits, for many of the talents, especially the best among them, actually refused to be 'drained' into bureaucracy in this manner. It will be important to find out what kind of intellectual pursuits appealed to those who chose not to, or found themselves unable to, join the bureaucracy. Occasional studies on the subject have shown examples of many scholars who, having been denied a bureaucratic career for various reasons, found an alternative career in the sciences, especially in medicine.[28]

---

26 Peter K. Bol, "The Sung Examination System and the *Shih*," *Asia Major, third series*, 3 (1990): 149-171.

27 Frederick W. Mote, "Confucian Eremitism in the Yuan Period," in *The Confucian Persuasion*, ed. Arthur F. Wright (Stanford: Stanford University Press, 1960), 202-40; Zeng Xiongsheng, "Hiding in Doctoring: The Study of 1364 Medical Practitioners," in *Current Perspectives in the History of Science in East Asia*, ed. Yung Sik Kim and Francesca Bray (Seoul: Seoul National University Press, 1999), 438-448.

28 Joseph Needham, "Social Position of Scientific Men and Physicians in Medieval China," *Proceedings, XIVth International Congress of the History of Science* (Tokyo/Kyoto: Science Council of Japan, 1974), vol. 4: 19-34; Robert P. Hymes, "Not Quite Gentlemen? Doctors in Sung and Yuan," *Chinese Science*, no. 8 (1987): 9-76.

Finally, in considering the relationship between science and bureaucracy in China, note should be taken of the fact that the close connection between the two is one of the key aspects of modern society, in which science has entered the bureaucracy, has become a part of it, and has been thoroughly bureaucratized. In fact, bureaucracy is one of the three major niches where science is found in modern society, the other two being universities and industry. We know that such ties between science and bureaucracy came to exist in the West very late. It can be said to have happened first in France during the last decades of the eighteenth century, when many scientists worked very efficiently in various governmental tasks and projects.[29] This connection has often been considered to have been responsible for the flourishing, both intellectual and institutional, of French science in that period. We cannot but note at this point, however, that science in traditional China was comparatively more bureaucratized, especially so in the fields which Needham has called 'orthodox sciences' – calendrical astronomy, mathematics, harmonics and geography.[30] Does this not provide another reason why it is a mistake to see the bureaucracy's role in the scientific development of China as so one-sided and negative? It must have played a positive role to some degree – and it is this which should be studied more closely. To do so, it will be helpful to analyze what happened during the process of bureaucratization of science in the West.

Meanwhile, it should be mentioned that what was valued at the early stage of this process in the West was the ability, efficiency and creativity displayed by scientists in the practical tasks assigned to them. These were qualities that such scientists came to acquire in the course of their hands-on scientific training and practice, and were not directly derived from their specialized scientific knowledge or expertise. Thus, endowed with these qualities, the French scientists of the late eighteenth and early nineteenth centuries, when science began to be bureaucratized, did not in reality function as specialists confined to their narrow specialties, but more as generalists. They did possess specialized expertise which had helped them gain a bureaucratic post, but that expertise was not directly utilized in their day-to-day activities as bureaucrats. It may be interesting to compare this situation with that of the generalist bureaucrats in the Chinese system.[31]

29 Charles C. Gillispie, *Science and Polity in France at the End of the Old Regime* (Princeton: Princeton University Press, 1980).

30 Needham, "Social Position of Scientific Men and Physicians."

31 We can point out here that, in the West, as the sciences became more specialized, there was a gradual increase in the prestige of specialized knowledge and the practitioners of specialized subjects, whereas in China specialization was accompanied by a decrease in

Finally, the close connection between science and bureaucracy in the modern world suggests an alternative to the Weberian habit, so widespread among students of modern society, of linking science and capitalism together as characteristically modern elements, and as what distinguished the West that achieved 'modernity' from the rest of the world that did not.[32] For those who adopt this Weberian stand, it becomes almost inevitable that science could have developed only in the West, because its partner, capitalism, was a uniquely Western development. The alternative combination I am thinking of is, of course, that of science and bureaucracy, which appears to be at least as natural a candidate for such a linkage in the modern world. One advantage of this non-Weberian and anti-Needhamian[33] combination is that, when one subscribes to it, one is no longer forced to see the West alone as a culture predisposed to the development of modern science; bureaucracy, the new partner of science, had been developed much earlier and to a greater degree in China.

4 Concluding Remarks: New Topics and Questions

The most important subject in the study of the relationship between science and bureaucracy in China should be scientific and technical activities of government offices and the bureaucrats who served in them. However, we need to broaden the scope of our study to ask new questions about them.

Above all, we should study the characteristics of the bureaucrats who staffed the offices assigned to scientific and technological tasks. In particular, the fact that these offices were usually headed by generalist bureaucrats raises a number of questions. Was it only in a nominal sense that they headed these offices? If so, what was the relationship of these generalists with the specialist bureaucrats who did the actual work under their nominal control? And if it was not

---

prestige. This may have something to do with the fact that in the West specialization occurred not simply as a 'narrowing down' of the interest of scientists but as some kind of 'broadening up' in which scientists took over things from the culture at large and incorporated them into their own domain.

32 Such habit underlies the work of Robert K. Merton, for example, who extended Weber's thesis on capitalism and the Protestant ethic into his own celebrated thesis on science and Puritanism: *Science, Technology & Society in Seventeenth Century England* (New York: Harper, 1970). For various critiques of the Merton thesis, see I. Bernard Cohen, ed., *Puritanism and the Rise of Modern Science: The Merton Thesis* (New Brunswick: Rutgers University Press, 1990).

33 It is "anti-Needhamian" in that Needham would have denied the congruence between the two.

so, what kind of scientific and technological tasks did the generalists actually perform and how were they related to, and different from, the works of specialists? What were the positions these generalists occupied in such offices? What did they themselves think of their positions in these offices, especially in relation to their more usual positions in non-scientific and non-technical government offices to which they more naturally belonged? The place – the role, importance and connection – of the scientific and technological offices and the bureaucrats in them, both the generalists and the specialists, within the entire Chinese bureaucratic structure is another subject that needs to be studied in a greater depth. In studying all these subjects, we should look at differences between various scientific and technical branches, and relate such differences with the varying practical functions performed by these different branches for the government. We should also study the historical changes in these aspects over time.

There are, of course, other themes that, if studied, would bring further insight. The first to come to mind is the civil service examination, and the place of science within it. Another potentially fruitful topic is bureaucrats' attitude toward various aspects of science: its utility, political significance, and cultural value. We should also look at the distinction between government as 'patron' (in pre-bureaucratized Western science, for example) and government as 'employer' (in bureaucratized Chinese science) and determine what the resulting differences in the scientific development of China and the West were. The scientific activities of emperors and members of the imperial household, who were not part of the bureaucracy in China, is yet another topic that would have a significant bearing on the subject.

A close study of these issues, and of course of many others which have not been adequately studied so far, will help us better understand the true relationship between scientific development and bureaucracy in China.

PART 3
Comparative Problems in the History of Science in East Asia

CHAPTER 9

The 'Why Not' Question of Chinese Science: The Scientific Revolution and Traditional Chinese Science*

1 Introduction: Various Forms of the 'Why Not' Question

When it comes up, usually in quite casual conversation, that I work in the field of Chinese science and natural philosophy, nearly the first question that people ask me is the so-called 'why not' question, namely 1) "Why did the Scientific Revolution not take place in China?" or 2) "Why did modern science not develop independently in China?" At least, this is my personal experience.

The question was first asked by the Chinese themselves in the late nineteenth and early twentieth centuries, when China fell prey to Western powers. The Chinese of the time felt that this situation was largely due to the successful development in the West of science and technology, which the Chinese themselves had failed to develop, at least not in the 'right' direction. This perception led many Chinese to strive to strengthen and enrich their country by catching up in those areas. But it also made them wonder why science and technology had not developed in China in the first place. Some even asked a more extreme form of the question: Fung Yu-lan, for example, titled one of his articles "Why China Has No Science."[1]

Gradually, however, people came to realize that China did in fact have scientific and technological traditions that were fairly sophisticated and advanced, and as a result of this new awareness, the question has been posed in new terms. Joseph Needham, perhaps the one most responsible for such a shift in the general evaluation of Chinese scientific achievements, has put the question in the following way: Why did China fail "to give rise to distinctively mod-

* Originally published as "The 'Why not' Question of Chinese Science: Scientific Revolution and Traditional Chinese Science," *East Asian Science, Technology, and Medicine* 22 (2004): 96-112.

1 Fung Yu-lan, "Why China Has No Science: An Interpretation of the History and Consequences of Chinese Philosophy," *The International Journal of Ethics* 32 (1922): 237-263. Homer H. Dubs even referred to the Chinese "failure to produce philosophical systems": Homer H. Dubs, "The Failure of the Chinese to Produce Philosophical Systems," *T'oung Pao* 26 (1928): 96-209.

 | DOI 10.1163/9789004265318-011

ern science while having been in many ways ahead of Europe for some fourteen previous centuries"?[2]

There have, of course, been other forms of the question, some of which are quite interesting. Some have asked more specifically why scientific activities in China did not continue or prevail. For example, Robert S. Cohen, a philosopher, phrased the question as: "China amply, confidently and splendidly reached the threshold of modern science and technology but no further, and why not?"[3] For another philosopher, Arnold Koslow, the question is: "Why, given the state of knowledge in Ancient and Medieval China, was there a discontinuance rather than a continuation of the tradition in (say) optics, acoustics, and mathematics?"[4] A historian of the Scientific Revolution, H. Floris Cohen, asked: "Could a Galileo have emerged in China, and if so, could he have prevailed?"[5] Sometimes a potential answer to the question is expressed in the form of a 'why not' question: "Why did scientific activity become functionally differentiated and institutionalized in Western Europe and not in China or elsewhere?"[6] What may be the most neutral, but perhaps the least interesting, form is my own: "Why did the Chinese develop their knowledge of the natural world in the way they did and not in the way the Europeans developed modern science?"[7]

In this paper, rather than asking or trying to answer the 'why not' question again, I shall examine the problems – both theoretical and practical – involved in asking and answering the question. I shall look at various criticisms of the question and examine the validity of such criticisms. I shall take the question as an example of the general question 'Why not an event E in culture A?' and show how the characteristics of event E and culture A influence the way one goes about dealing with the question. It is not just the question itself that is problematic but also the answers that people have looked for and have come

2 E.g., Joseph Needham, *Science and Civilisation in China.* vol. 2 (Cambridge: Cambridge University Press, 1974), xxiii. Of course, Needham formulated the question in many different ways. Sal P. Restivo, for example, identified twelve such 'formulations.' See Sal P. Restivo, "Joseph Needham and the Comparative Sociology of Chinese Science and Modern Science," in *Research in Sociology of Knowledge, Science, and Art,* vol. 2, ed. Robert A. Jones and Henrika Kuklik (Greenwich, Conn.: Aijai, 1979), 43-44 (appendix A).

3 Robert S. Cohen, "The Problem of 19(k)," *Journal of Chinese Philosophy* 1 (1973): 104.

4 Arnold Kowlow, "More on 19(k)." *Journal of Chinese Philosophy* 2 (1975): 185.

5 H. Floris Cohen, *The Scientific Revolution: A Historical Inquiry* (Chicago: University of Chicago Press, 1994), 483.

6 Restivo, "Joseph Needham and the Comparative Sociology," 29.

7 Yung Sik Kim, "Natural Knowledge in a Traditional Culture: Problems in the Study of the History of Chinese Science," *Minerva* 20 (1982): 103.

up with. I shall also examine such problems. Finally, I shall consider the ways in which the question and answers can be, and have been, useful in spite of the many pitfalls. I shall end the paper with a discussion of what can be called 'the Korean why not question.'

My purpose in doing all this is to suggest that we cannot afford to write off the 'why not' question. The widespread hyper awareness of the potential problems arising from the question in the past few decades has discouraged students of the history of Chinese science from asking the question at all. We now hardly see anyone raising the question in serious professional discussions, except for occasional examinations of the legitimacy of asking the question itself that usually end in banning it as an illegitimate question to discuss. An unfortunate consequence of this, however, is that while the scholars in the field – the very ones most qualified to ask the question and most prepared to deal with the dangers involved in posing it – are abstaining from asking it, there are plenty of others out there who are willing to ask, and answer, the question.

2 Targets of the 'Why Not' Question

The first thing we may note about this 'why not' question is that it is not asked with respect to just any culture. Typically, it is asked with respect to a culture 1) when the science of the culture seemed so close to a 'Scientific Revolution' but did not cross the 'threshold'; or 2) when other areas of the culture were so advanced that it makes one wonder why its science was not so. Actually, China, India, and Islam are just about the only cultures for which this grand question is asked, and of these, China is by far the most frequent target of the question. It is obvious that for many people China appears to meet both of the above two criteria. Firstly, for a long period in Chinese history there were traditions of science and technology at quite advanced levels, such as calendrical astronomy, mathematics, medicine, pharmacology, harmonics, and various techniques. Secondly, the Chinese have shown abundant activities and impressive achievements in other cultural areas, like philosophy, arts, government, industry and economy.[8]

On a more restricted scale, the question is also posed regarding other cultures such as seventeenth-century China, the Sejong 世宗 (r. 1418-1450) era of

8 This was especially so in the eleventh century during the Song, the period which some scholars even characterized as the 'early modern' China. See, e.g., Edwin O. Reischauer and John K. Fairbank, *East Asia: The Great Tradition* (Boston: Houghton Mifflin, 1958), 220-225.

Korea, and Tokugawa Japan.[9] Sometimes the 'why not' question turns into a 'why stagnate' or a 'why decline' question: "Why did science stagnate (or decline) in a culture?" This 'why stagnate/decline' question has been aimed at, for example, the sciences of ancient Greece, Islam, Ming China, and the post-Sejong period of Korea, etc.[10]

3 Problems with the Question

Some see problems in the question itself, and wonder whether it can be a meaningful question at all.

3.1 *Inherent Problems and Underlying Assumptions*

Issues can be found concerning all three components of the question: 1) 'why not'; 2) 'Scientific Revolution'; 3) 'in China'. Many are skeptical, for example, about the term 'Scientific Revolution,' and wonder whether it makes any sense to use it even in the West.[11] Some question the validity of the notion of the 'Great Divide' between 'China and the West' as the basis of historical analysis.[12]

---

9 E.g., Nathan Sivin, "Copernicus in China," *Studia Copernicana* 6 (1973): 63-122; Chŏn Sang-un 全相運, *Han'guk kwahak kisulsa* 韓國科學技術史 [History of Science and Technology in Korea] (Seoul: Kwahak segyesa 과학세계사, 1966), 23-25; James R. Bartholomew, "Why Was There No Scientific Revolution in Tokugawa Japan?" *Japanese Studies in the History of Science* 15 (1976): 111-125.

10 For discussions of such questions, see, e.g., H. F. Cohen, *The Scientific Revolution*, Section 4.2; A. I. Sabra, "The Appropriation and Subsequent Naturalization of Greek Science in Medieval Islam: A Preliminary Statement," *History of Science* 25 (1987): 223-243; Roger Hart, *Imagined Civilizations: China, the West, and Their First Encounter* (Baltimore: Johns Hopkins University Press, 2013), Chap. 4; Pak Sŏng-rae 朴星來 (Park Seong-Rae), "Chosŏn sidae kwahaksa-rŭl ŏttŏk'e pol kŏt-in'ga" 조선시대 科學史를 어떻게 볼 것인가 [How Should We View the Science of the Chosŏn Period?], *Han'guksa simin kangjwa* 韓國史市民講座 16 (1995): 145-166, esp. 165-166. It should be noted that, in speaking of these questions, I am not implying that decline and stagnation are the sole alternatives to scientific revolution. The science of a society can continue to grow at a slow pace without leading to revolutionary breakthroughs.

11 Steven Shapin, *The Scientific Revolution* (Chicago: University of Chicago Press, 1996), "Introduction"; David C. Lindberg, "Conceptions of the Scientific Revolution from Bacon to Butterfield: A Preliminary Sketch," in *Reappraisals of the Scientific Revolution*, ed. David C. Lindberg and Robert S. Westman (Cambridge: Cambridge University Press, 1990), 1-26.

12 Roger Hart, "Beyond Science and Civilization: A Post-Needham Critique," *East Asian Science, Technology, and Medicine* 16 (1999): 88-114. Karine Chemla also raised this question while commenting upon an earlier, oral, version of the present chapter.

But even if one sets aside such skepticisms that have become fashionable in recent years, and admits that one can, as I shall do in the remainder of this paper, legitimately speak of the 'Scientific Revolution' and compare 'the West' and 'China,' there is still a problem inherent in the first component of the question, namely whether it is meaningful to ask why something did not happen.

To illustrate the problem, A. C. Graham came up with the striking analogy of a house fire. Pointing out that when a house catches fire, one does not go through all the other houses in the street explaining why they did not catch fire, he asked why one should do a similar thing in the case of the Scientific Revolution. Naturally, he went on to argue that what needs to be asked is "why the Scientific Revolution did take place" where it did, rather than "why it did not take place" elsewhere.[13]

The basic criticism along this line is that in asking why an event E did not take place, the question contains the implied assumption that it is natural that E should have happened. For the case of the 'why not (a) Scientific Revolution' question, for example, the implied assumption is that the Scientific Revolution is something that should occur in every culture at a certain stage of its scientific development. And we know that this assumption is mistaken, for, as Graham points out, nowhere was the Scientific Revolution an event that was bound to occur, not even in Western Europe, where it did occur.

3.2 *'Why E' = 'Why Not (Not E)'*

Nonetheless, the above criticisms do not constitute a sufficient reason for us to avoid the 'why not' question altogether. We can still ask the question without making the above assumptions. Furthermore, if anyone, based on such a theoretical criticism of the 'why not' type question, insists that only the 'why' question should be asked, I have an equally theoretical counter-argument, namely that there cannot be an absolute distinction between the 'why' and the 'why not' questions, for one can take any 'why' question and turn it into a 'why not' question:

'why E' = 'why not (not E)'.

Should we, then, criticize those asking 'why E' for asking a 'why not' question, namely 'why not (not E)', and thus for assuming it natural that 'not E' should have happened (or that E should not have happened)?

13 A. C. Graham, "China, Europe, and the Origins of Modern Science," in *Chinese Science: Explorations of an Ancient Tradition*, ed. Shigeru Nakayama and Nathan Sivin (Cambridge, MA: MIT Press, 1973), 45-69. The 'house fire' analogy appears on p. 51.

The above equation, although it may appear strange at first sight, is quite plausible in the cases where 'not E' is more natural than E, for then one will want to ask about the causes for the less natural event of 'not (not E)', which is E. The 'why the Scientific Revolution' question could be considered such a case because, as we have noted, the Scientific Revolution is not an event that can be expected to occur naturally in any culture: indeed, it can be said that it is more natural for a culture not to have a Scientific Revolution. Thus, one may naturally ask a 'why not (not E)' question: "Why did Western Europe not enjoy the natural state of not having the Scientific Revolution?" An interesting example of asking this form of 'why not (not E)' is William Theodore de Bary's following remark, which, though it might at first appear odd, contains an entirely natural question to ask, and thus nicely supports my point here.

> ... it could be asked, in slightly modified non-Western terms (in this case, the normative Neo-Confucian language of East Asia): why could the West not measure up to the civilized behavior of Confucian East Asia? Why did it not act in a mature, responsible way by getting its own house in order and staying home, as China, Japan, and Korea were doing, rather than chasing around the world to make trouble for others?[14]

In fact, many events in history can be viewed from such a perspective. Unless one has sufficient reason to consider any given historical event 'E1' as a natural, or even an inevitable, event to occur in every culture, 'not E1' should be considered more likely. Therefore, assuming 'not E1' and asking 'why not (not E1)' is a more plausible thing to do in such cases, rather than assuming 'E1' and asking 'why not E1.' Nevertheless, asking 'why E1' can in principle be considered as asking a 'why not' type question (why not [not E1]) and assuming that 'not E' should have happened.

3.3 *The Nature of the Scientific Revolution*

From the above discussion, it should be sufficiently clear that in asking and answering 'why not E,' a great deal depends on the nature of E. One's notion of the nature of an event that did not happen influences greatly the way one goes about dealing with the question of 'why it did not happen.'

Yet E is not an event that can be discussed in isolation, that is, in general, abstract terms: it is a concrete event in a particular cultural setting. In other

14 William Theodore de Bary, *East Asian Civilizations: A Dialogue in Five Stages* (Cambridge, Mass.: Harvard University Press, 1988), 68.

words, what we refer to as 'why not E' in reality is 'why not E in culture A.' It thus depends very much on the 'culture A' about which the question is asked.

For example, in asking 'why not E' one can distinguish a culture in which E did not happen but still could have happened on the one hand, and a culture in which E could not have happened at all, on the other. Some think that Islam and China were different in this respect: when E is the Scientific Revolution, Islam is the former kind of culture, and China the latter. Among other things, it is noted that Islamic culture shared some of the intellectual foundations with the West, which did undergo the Scientific Revolution.[15] If we accept this, we can conclude that, of the two non-Western cultures, considerations of Islamic culture will yield intercultural understanding more readily transferable to Western history. However, this argument involves the same kind of assumptions that have been criticized in the previous subsection.

Perhaps one may be tempted to consider the Scientific Revolution an event that could have occurred only in Western Europe of the sixteenth and seventeenth centuries, and then the question will not need to be asked with respect to other cultures. But if we do not wish to go to that extreme, and instead admit that it can happen in cultures other than early modern Europe, what kind of 'Scientific Revolution' are we talking about when we ask the 'why not (a) Scientific Revolution' question? Are we speaking of the Scientific Revolution (E) that actually happened in Western Europe (B)? When we know that E happened in B and ask 'why not E in A,' are we asking 'why not [E in B] in A'? In other words, is there only one particular form of E ('E in B,' namely the E which actually happened in B) that can happen anywhere in the world? And if so, can it happen in other cultures – can 'E in B' conceivably happen in A? Or is it possible that there are more than one forms of E? For example, can there be a form of E which is different from the actual 'E in B' and that can still happen somewhere in the world (although it actually did not happen anywhere)?

One aspect of the Scientific Revolution that can be noted in this regard is that the peaks of the developments that constituted the Scientific Revolution in Europe were not followed by consolidation or stagnation, let alone decline. That was what did happen, however, in nearly all other cases of scientific upsurge in history – ancient Greece, medieval Islam, and Song China, for example. Indeed, this pattern of intermittent growths between periods of stagnation, and even decline, is much more common than what happened in Europe following the Scientific Revolution. It is noteworthy, then, that unlike those other instances where flourishing scientific development reached a peak after a while and then slowed down after a few hundred years, the develop-

15 H.F. Cohen, *The Scientific Revolution*, 143.

ments that had begun during the European Scientific Revolution continued and even accelerated. And they spread across the rest of the world.[16]

3.4 *Misunderstanding the Scientific Revolution*

Naturally, a misunderstanding of E in the question 'why not E' will pose problems. For example, an inaccurate understanding of what happened in Europe during and before the Scientific Revolution has led to all sorts of wrongheaded speculations concerning the 'why not (a) Scientific Revolution' question. The most frequently found such misunderstanding is the assumption that the sciences of all traditional cultures were more or less similar up to the fourteenth and fifteenth centuries, just before the Scientific Revolution, and that after this only European science took off. We know that this is wrong; we know that there were in fact a lot of differences – both in content and in context – between medieval Western science and traditional Chinese science.

One notable such difference was that in the medieval West natural philosophy was accepted by the universities and formed an essential part of their curriculum. Natural phenomena were thus frequently discussed, and often became subjects of heated debates among the medieval schoolmen. The situation was vastly different in traditional China. It is not to say that subjects dealing with the natural world were not accepted by the intellectual mainstream in China. They were tolerated; or at least, they were not excluded. Nor is it my point that objects and phenomena in the natural world were unimportant to traditional Chinese thinkers. They were important too. The point is, rather, that they were unproblematic. Natural objects and phenomena were not excluded, but were taken for granted, and hence simply accepted as they were observed, reported, or recorded. It is for this reason that, while there were heated discussions and continuing debates on moral and social problems among traditional Chinese thinkers, natural phenomena were not very often discussed or debated in this manner. Natural philosophy did not belong to the list of subjects for which they showed much enthusiasm.[17]

---

16 To be sure, if we look at individual countries like Italy or England, there were indeed periods of stagnation and decline. However, at the same time science was flourishing in other parts of Europe. One may ask whether it was in fact the pluralism of Europe that helped maintain the pace, or even accelerate it.

17 This is evident, for example, in the manner in which Zhu Xi 朱熹 (1130-1200), the author of the so-called 'neo-Confucian synthesis,' discussed natural phenomena. Frequently, he spoke of them not for their own intrinsic interest, but for some external purposes. He alluded to familiar and obvious natural phenomena mainly in dealing with moral and social problems, drawing analogies between these natural phenomena and moral and social problems, which were usually more complicated and challenging. Only rarely did

A corollary of the misunderstanding of taking all traditional sciences to be more or less the same is the belief that modern science came about due to new scientific methodology. For Needham and for many others, for example, modern science was the result of the combination of mathematics and experimentation.[18] We know, of course, that the real situation was far more complicated. Another ensuing misunderstanding is the belief that continuation of scientific development would have eventually produced a Scientific Revolution. Concerning this we may ask, as Arnold Koslow has done, "Why should Chinese [scientific] traditions, even if they continued, broadened, and deepened, lead to what has been called modern science?"[19] Still another misunderstanding is confusing scientific development and technological development. Many factors have been put forward as potential explanations for the development of technology, but not that of science, which in traditional China, as in most other traditional cultures, was an activity largely separate from technology. Benjamin Nelson has noted, for example, that it is easy to see some of the roles played by capitalistic interest in technical utilization of scientific knowledge, but not so to see its role in science per se.[20]

These are only the more frequent examples of misunderstandings concerning what happened in the Scientific Revolution. Such misunderstandings have misled numerous authors asking and answering the 'why not' question, as well as their readers. Nevertheless, this is not a good enough reason to ban the question outright; it only emphasizes the importance of having an accurate understanding of the nature of E when asking 'why not E.'

3.5 *'Why Stagnate/Decline' Question*

In principle, the 'why stagnate' or 'why decline' questions suffer from the same problems as the 'why not' question. The main focus of these types of questions in particular is the apparent correlation between the progress and the decline of a culture at large and that of the science within that culture. We may need to ask whether science can continue to flourish while all other areas are in decline. And if it can, this may mean that science is a unique branch of culture,

---

he mention such common natural phenomena for their own sake. For more detailed discussion of this point, see Yung Sik Kim, *The Natural Philosophy of Chu Hsi (1130-1200)* (Philadelphia: American Philosophical Society, 2000), 3-4.

18 E.g., Joseph Needham, *Science and Civilisation in China*, vol. 3 (Cambridge: Cambridge University Press, 1959), 150-168.

19 Koslow, "More on 19(k)," 185.

20 Benjamin Nelson, "Sciences and Civilizations, 'East' and 'West': Joseph Needham and Max Weber," in *Philosophical Foundations of Science*, ed. R. S. Cohen and M. Wartofsky (Dordrecht: Reidel, 1974), 463.

different from other branches. But then we can also ask the reverse question: can science alone decline in a culture while other areas of the culture flourish? As we have noted in Section 2, asking the 'why not' question presupposes not only that it can but also that it actually did in China.

Another point to be noted is that while the 'stagnation' or 'decline' of science may often be the result of rejection and suppression, it is not always so. Frequently, stagnation or decline is the result of simple neglect, or of something being simply ignored, rather than active rejection or suppression. Sometimes it can result from acceptance followed by isolation and stabilization, as in Islam and post-Song China. In the former instance, science was welcomed by isolated, marginal intellectual circles of society; in the latter, scientific knowledge entered, and became part of, the comprehensive corpus of learning, and was taken for granted by the intellectual mainstream.[21]

4 Problems in the Answers

It is not only the question that is problematic. The answers that have been suggested suffer from their own set of problems.

4.1 *Answers Required for the 'Why Not' Question*

There have been problems in the ideas people have about the kind of answers the 'why not' question should satisfy. Some think that the answers required to the 'why not' question are 'causes' which can provide a rigorous causal explanation. H. F. Cohen illustrates this problem by rephrasing the 'why not' question of ancient Greek science in the following manner: "What, in other words, was absent from Greek science that was indispensable for bringing forth early modern science?"[22]

However, this is usually not what we are looking for when we ask the 'why not' question. For example, in asking 'why not E in A,' we are not looking for necessary and sufficient conditions for E to occur, ones which did not exist in culture A and thus prevented E from occurring in A (or caused 'not E' in A).[23]

---

21 Sabra, "The Appropriation and Subsequent Naturalization of Greek Science in Medieval Islam"; Kim, *The Natural Philosophy of Chu Hsi*, 6-8; Yung Sik Kim, "Confucianism and the Development of Science in East Asia," in *Modern Confucianism and East Asian Civilization: Issues and Perspectives*, ed. Liu Shu-hsien and Lin Yueh-hui (Academia Sinica, Taipei: 2002), 183-213.

22 H.F. Cohen, *The Scientific Revolution*, 242.

23 After considering science in Islam, H. F. Cohen lists four features that affect the development of science in general, of which three are given in the form of necessary conditions

After all, we do not seek such 'causes' when we ask the usual 'why' questions in history, as in 'Why the French Revolution?' or 'Why the Ming collapse?' And, in fact, the answers that people come up with to questions like these are not necessary and/or sufficient conditions, but rather various other sorts of things: influences, reinforcements, correlations, connections, linkages, and even coincidences, for example. There is no reason, then, that we should insist on finding more rigorously causal 'causes' for the 'why not' question.

4.2 *Problem of 'Inhibiting' Factors*

Many attempts to produce answers to the 'why not' question have suffered from a common problem, namely that of seeking 'inhibiting' factors as answers to the question. It is a problem because this amounts to saying that a factor inhibited something that was not happening, and thus makes one wonder whether it means that the Scientific Revolution was about to happen, or was in progress, in China, and then was inhibited by the inhibiting factor. What is more serious, however, is that the inhibiting factors people have suggested have frequently been the absence, rather than the presence, of certain factors. For example, the inhibiting factors suggested so far have been the absence in China of such ideas as causality, laws of nature, and a creator-god, such methodological elements as proof, formal logic, geometry, symbolic algebra, and experiment, and such social and cultural factors as capitalism, urbanization, merchant class, slavery, universities, organized religion, individualism, toleration, etc.[24] But taking the absence of something as an inhibiting factor really

---

for a scientific revolution to occur: 1) "that science not be perceived by the community at large as being directly opposite to its core values"; 2) the existence of "at least semiprofessional cluster of scientists who exude a deep and abiding belief in the value and dignity of their own life's work"; and 3) that "no Scientific Revolution is feasible...in lands thoroughly destroyed by bands of foreign invaders." The fourth feature speaks of conditions that would give a society "a decisive advantage when it comes to preparing the ground for early modern science." See Cohen, *The Scientific Revolution*, 417.

24 For discussions of these and many other factors, see, e.g., Joseph Needham, "Poverties and Triumphs of the Chinese Scientific Tradition," in Joseph Needham, *The Grand Titration: Science and Society in East and West* (London: George Allen & Unwin, 1969), 11-54; Derk Bodde, *Chinese Thought, Society, and Science: The Intellectual and Social Background of Science and Technology in Pre-modern China* (Honolulu: University of Hawaii Press, 1991); H.F. Cohen, *The Scientific Revolution*, Section 6.5. A rare exception in which the presence of a factor was given as an inhibiting cause is the bureaucracy. For discussions of the role of the bureaucracy in Chinese scientific development, see, in addition to the above cited works, Yung Sik Kim, "Science and Bureaucracy in Traditional China," *The Medieval History Journal* 3 (2000): 363-379 (Chapter 8 of this book).

amounts to saying that the absence of a factor inhibited something which did not happen after all.[25]

4.3 *Mistaken Assumptions and Other Problems*

A still more serious problem that can be found in the answers given so far to the 'why not' question is the one that has plagued comparative historians of science, namely the assumption that a factor, an aspect, or a situation should play the same role in every culture, in particular the assumption that the factors that were significant in Western scientific development are the factors that must have also affected the development of the sciences in China in the same ways. That this assumption is mistaken can easily be seen from the simple fact that many features or situations similar to the ones suggested as answers to the Chinese 'why not' question were present in the West before the Scientific Revolution, but did not prevent the development of modern science there.[26]

Obviously derived from the above sort of assumption is the common trend of attributing the supposed lack of scientific development in China to the different characteristics of some cultural elements or aspects in China from those in the West. Examples are numerous: the Chinese conception of cyclical time versus the Western conception of linear time; the non-inflective Chinese language versus the inflective Western languages; the strength of Chinese mathematics in algebra versus the Western strength in geometry; the Chinese inclination to wave theory versus the Western particulate theory in explaining physical phenomena, and so on.[27]

There are, of course, other problems that can be found in the answers that have been suggested so far. In an interesting philosophical analysis of the problems involved in answering the question, Robert S. Cohen wonders, "What is

---

25 Along this line, Nathan Sivin has identified 'two fallacies' frequently found in intercultural comparisons: 1) taking an earlier state as a 'necessary condition'; 2) taking a later absence as a result of 'inhibiting.' See Nathan Sivin, "Why the Scientific Revolution Did Not Take Place in China – or Didn't It?" *Chinese Science* 5 (1982): 45-66, esp. 56.

26 Taking the example of the supposed inhibiting role of the 'scholar-bureaucrat class,' Nathan Sivin has underpinned this point in the following manner: "But in Europe at the onset of the Scientific Revolution we are faced with the predominance of the Schoolmen and dons, immersed in books, face toward the past, and oriented toward human institutions rather than toward Nature. They did not prevent the great changes that swept over Europe. It would take a more imaginative historian than myself to say whether those changes would have taken place sooner had Scholasticism never existed. (Sivin, "Why the Scientific Revolution Did Not Take Place in China," 57.)"

27 E.g., see Needham, "Poverties and Triumphs of the Chinese Scientific Tradition"; Bodde, *Chinese Thought, Society, and Science*; H. F. Cohen, *The Scientific Revolution*.

the sort of connection which can affect one layer of social action by thought, feeling or action at another layer?"[28] He points out, for example, the difficulty of establishing the comparability of cultural variables in different historical settings. He criticizes the 'fuzziness' of the answers, or their "imprecise characterization and explanatory status of cultural variables". Indeed, some of these variables may be merely "intervening variables, convenient conceptual stand-ins" that have to be replaced later, and some are even themselves candidates for 'why' and 'why not' questions.[29] He also points out the problems in what he calls the 'causal conditional arguments' employed in these discussions.[30]

The problems discussed in this section underline the real dangers faced by those who try to answer the 'why not' question, misunderstandings that have thwarted numerous attempts to answer the question adequately. Yet the existence of such problems should not lead us to reject the question itself. They need only remind us to be mindful of the dangers when answering the question, and to be especially wary of simple, easy answers.

5 Contribution of the Question

There is still another problem, perhaps the most conspicuous one, which is that so many people keep asking this question so riddled with problems. The questioners include not just laymen, but many scholars also; in fact, nearly everyone interested in Chinese culture has asked and answered the question. And this is likely to continue. We might succeed in persuading one another to avoid posing the question, or at least to wait until the time when we have enough knowledge – which, by the way, will never happen! But in the meantime the question will not go away, and still people will go on asking and answering it. But why do people keep asking this question?

It is difficult to deny that many people must have felt that the question has some contribution to make for a better understanding of something, be it scientific development in China (or in other places), the nature of scientific enterprise, Chinese culture in general, or something else. Can we discount their feelings by simply saying that they should not succumb to them because the 'why not' question cannot contribute anything to their understanding? Can we say for sure that the question indeed has made no contribution to the

28 R. S. Cohen, "The Problem of 19(k)," 106.

29 Ibid, 108-109.

30 Ibid., 111. Cf. Koslow's comments on Cohen's paper in Koslow, "More on 19(k)."

field of Chinese science? I do not believe we should be so negative in assessing the contribution of the question.

An undeniable contribution of the 'why not' question is that it has provoked more widespread interest in Chinese science and acted as the spark that propelled many individuals into the field in the first place – whether they recognize it or not. It is only after grappling with the question for some time that these people can realize its inherent problems. For example, only after asking the question, and trying out various answers to it, do they realize the danger involved in assuming that the Scientific Revolution is what should happen at some stage of every culture.

Another important contribution made by the 'why not' question is that it has led many historians of Chinese science to consider, while trying to find answers to the question, not only the contents but the contexts of Chinese science. That is, the question has led many of them to believe that the answer is to be found mainly in the factors external to science itself – intellectual, social, institutional, and other cultural factors. And while this is not entirely correct, such a perspective has encouraged, and will continue to encourage, much good work on the role of such factors in the development of science in China.

The question has also inspired many to engage in the comparative history of science, because the answers to the 'why not' question are usually based upon, and sometimes tested by, comparative analyses of the situations in different cultures – usually comparisons with Europe, where the Scientific Revolution did take place. Such comparisons, when carried out prudently, have contributed greatly to a better understanding of the various differences between the two cultures, and eventually to a better understanding of the two cultures themselves. In fact, the cross-cultural differences we have come to see regarding concepts of time, language, the nature of mathematics, and so on, were the fruits of such comparisons.[31]

Perhaps the most important contribution, however, is to be found in the answers that have been offered so far. And that is the way it should be after all. However problematic they may be as answers to the 'why not' question, they have nonetheless resulted in numerous factors, aspects and situations in traditional Chinese science, society, and culture being put forth, studied and dis-

31 More recently, Geoffrey E. R. Lloyd has underlined the differences in the dominant modes of intellectual arguments between the two cultures, noting the Chinese appeal to authority versus the Western predilection for "polemical argumentation." See G. E. R. Lloyd, *Adversaries and Authorities: Investigations into Ancient Greek and Chinese Science* (Cambridge: Cambridge University Press, 1996).

cussed.[32] Admittedly, any one factor, aspect or situation by itself cannot be the whole answer. But each of these factors can explain, at least to some extent, the lack of scientific development or the absence of the Scientific Revolution. In fact, it would be difficult to dismiss a particular factor as having had no effect on the development of science in China, or elsewhere. Indeed, a given factor did exert an effect within the concrete, particular (and peculiar) context of traditional China. And as the result of those various factors appearing together in such specific historical contexts, and acting together in combination in very complicated ways, they were bound to affect Chinese scientific development in certain tangible ways.

It is quite likely, then, that in trying to answer the 'why not' question we will end up amassing almost all the factors, aspects and situations in China that differed from those in the West. What we do then in presenting our answer – combining all these factors – may be nothing more than pointing to a different 'civilizational setting,' 'cultural orientation,' and 'life-way' of the Chinese, to use Benjamin Nelson's expressions.[33] In other words, the answer will amount to saying essentially that "China was different from the West in numerous ways, and because of these differences the Scientific Revolution did not occur in China." Or if we push it further, it may even become: "The Scientific Revolution did not occur in China because China was different from the West." Of course, we do not need to ask the 'why not' question, and to labor with all those answers, to reach this sort of obvious conclusion. What is important, however, is that we will have learned a lot in the course of reaching this seemingly trivial conclusion. And come to think of it, can our ultimate conclusion be really different in any essential ways from this? Is this not the function of all good, important historical questions after all?

32 Sal P. Restivo has found, in Needham's *Science and Civilisation in China* alone, some 29 "factors inhibiting the emergence of modern science in China and Western Europe," and 33 "factors conducive to the emergence of modern science." See Restivo, "Joseph Needham and the Comparative Sociology," 44-47 (Appendixes B and C).

33 Nelson, "Sciences and Civilizations, 'East' and 'West'," 460-463. Nelson identifies in the cities, universities, professions, estates, etc. of the twelfth and thirteenth centuries the "seedbed of modern European society" (459), and describes the situation in the following terms: "It is … the time when new images and horizons of conscience, self, person, society, the cosmos, action, justice, forms of rule, institutions of law and learning take on a cast that have ever since been distinctive and primary features of the Western European world. (464)"

6 Postscript: The Korean 'Why Not' Question

I will end the paper by considering a related question concerning Korea, a question that can be called the 'Korean why not question': "Why did Korean science, which had reached the highest level in the world in the early fifteenth century, not continue to progress, but stagnate and decline instead?"[34] Of course, this question will have all the problems that we have found for the case of the Chinese 'why not' question. But there is an additional problem that should be considered for the Korean case. In dealing with the 'stagnation' and 'decline' of Korean science and technology in the post-Sejong period, for example, can the historian ignore the situation in China at the time? At what level were the Chinese scientific ideas and techniques adopted by Koreans of the Sejong period? How did these ideas and techniques subsequently progress – or stagnate – in China? If Chinese science and technology did stagnate or decline, was it possible for Korean science and technology to continue to develop independently? There are quite a few Korean scholars who seem to think so, and it is not impossible in principle for science to continue to develop, or even to flourish, in a peripheral area of a culture while it is declining in the center. However, if that is impossible, would that reduce the Korean 'why not' question to being a part of the Chinese 'why not' question?[35]

---

34 Chŏn, *Han'guk kwahak kisulsa*, 23-25; Pak, "Chosŏn sidae kwahaksa-rŭl ŏttŏk'e pol kŏt-in'ga," 159-164.

35 This is really a part of the 'problem of China' in the history of Korean science, which troubles me when I try to think in general terms about the history of traditional Korean science, namely the problem of the significance of the scientific ideas and technical artifacts of China apparently omnipresent in traditional Korean science and technology. What roles and meanings should be assigned to them in the history of Korean science? Would it be permissible for a historian of Korean science to neglect them and to study only what is uniquely Korean? In particular, what should the historian do with those ideas and techniques that were direct importations from China? In other words, how should the historian deal with those theories and practices that were exactly the same as those of China except for the fact that they were discussed and carried out in Korea? Should the history of these theories and practices become a local history of the Chinese cultural world? See Yung Sik Kim, "The 'Problem of China' in the Study of the History of Korean Science: Korean Science, Chinese Science, and East Asian Science," *Disquisitions on the Past and Present*, no. 18 (2008): 185-198 (Chapter 12 of this book).

CHAPTER 10

The Ideas of the Earth's Rotation in Seventeenth- and Eighteenth-Century East Asia: Independent Development, Transmission from the West, and Chinese Forerunners*

The idea of the earth's rotation is one of the most salient elements of the modern world-picture that differed fundamentally from traditional East Asian notions, according to which the earth was more or less flat, more or less at rest, more or less at the center of the heavens.[1] When the idea of a rotating earth entered the East Asian world in the seventeenth century, it made waves. The idea underwent a long period of rejection, confusion, misunderstanding, hesitation, adjustment, and modification, before it was eventually accepted by most of the educated population. Various aspects and events in this long, complicated process have been studied, but we do not yet have the complete picture.[2] In this paper, I will try to shed some light by sorting out various events and aspects of the process, and examining some of them more closely.

---

* Originally published as "Independent Development, Transmission from the West, and Chinese Forerunners: Ideas about the Earth's Rotation in Seventeenth- and Eighteenth-Century East Asia," *Aisa Major*, third Series 22 (2009): 101-120.

1 I am using the expression 'East Asia' in a very limited manner in this paper. It does not cover the whole of East Asia. All my examples are from China and Korea, and may not represent the situation in other parts of East Asia, particularly in Japan.

2 Boleslaw Szczesniak, "Notes on the Penetration of the Copernican Theory into China (Seventeenth-Nineteenth Centuries)," *Journal of the Royal Asiatic Society* (April 1945): 30-38; Nathan Sivin, "Copernicus in China," *Studia Copernicana* 6 (1973): 63-122; Park Seong-Rae, "Hong Tae-yong's Idea of the Rotating Earth," *Han'guk Kwahak-sa Hakhoe-ji* [Journal of the Korean History of Science Society] 1 (1979): 39-49; Ogawa Haruhisa 小川晴久, "Higashi Ajia ni okeru chiten(dō)setsu no seiritsu" 東アシアにおける地轉(動)説の成立 [Development of the Theory of the Earth's Rotation (Motion) in East Asia], *Tongbang hakchi* 東方學志 23-24 (Seoul: Yŏnse daehakkyo Kuk'hak yŏn'guwon 延世大學校 國學研究院, 1980), 375-387; Shi Yunli 石云里, "Shiqi shiji Zhongguo de zhun Gebaini xueshuo" 十七世紀中國的準哥白尼學說, *Da ziran tansuo* 大自然探索 14 (1995): 122-127; Yang Xiaoming 楊小明, "Gebaini rixin didongshuo zai Zhongguo de zuizao jieshao" 哥白尼日心地動說在中國的最早介紹, *Zhongguo keji shiliao* 中國科技史料 20 (1999): 67-73; Shi Yunli, "*Tianbu zhenyuan* yu Gebaini tianwenxue zai Zhongguo de zaoqi chuanbo" 天步眞原與哥白尼天文學在中國的早期傳播, *Zhongguo keji shiliao* 21 (2000): 83-91.

 | DOI 10.1163/9789004265318-012

I begin by comparing how the idea of the earth's rotation was received with how the concept of the earth's sphericity was. While both ideas were novel, and out of place, in the established worldviews of East Asia, the extents to which they were new were different. The situations surrounding the two ideas in the West at the time of transmission to the East Asian world also differed. This in turn led to differences in the way East Asian scholars responded to the two ideas. The existence of 'forerunners' of the idea of the earth's rotation in earlier Chinese sources further complicated the situation. I will examine the case of Zhang Zai 張載 (1020-1077), mentioned most frequently as such a forerunner, and consider the possible contexts in which such forerunners were 'discovered' and 'used' by some seventeenth- and eighteenth-century East Asian thinkers. In light of such factors, modern scholars have raised both possibilities regarding the concept of a rotating earth: independent development in East Asia and transmission from the West. I will consider both of these possibilities, taking a Korean scholar, Kim Sŏk-mun 金錫文 (1658-1735) as an example. Finally, I will look at another Korean scholar, Sŏ Myŏng-ŭng 徐命膺 (1716-1787), and suggest that the so-called theory of 'the Chinese origin of Western learning' provided a context for East Asian thinkers to 'discover' and 'use' such 'forerunners.'

1 The Earth's Sphericity and the Earth's Rotation in the West and in East Asia

The circumstances surrounding the idea of a spherical earth (*diqiu* 地球) was more or less clear, both in the West and in East Asia. Its origin was the West, where the idea of a spherical earth at the center of the heavens had been firmly embedded in Western natural philosophy from ancient times.

In East Asia, in contrast, the idea was new and directly in conflict with some basic components of traditional cosmology. For example, there was the widely held ancient idea of the 'circular heaven and square earth' (*tianyuan difang* 天圓地方).[3] We know, of course, that this expression did not literally mean a square earth, but it is difficult to deny that the idea made a round earth difficult to imagine. And although there were some expressions in ancient Chinese texts that seemed to imply a spherical earth, some of which were even used as evidence of Chinese forerunners, such interpretations do not survive close

3 For a brief account of this idea in ancient China, see John S. Major, *Heaven and Earth in Early Han Thought: Chapters Three, Four, and Five of the* Huainanzi (Albany: State University of New York Press, 1993), 32-35.

scrutiny. For instance, there was the egg analogy in which the heaven and the earth are compared to an eggshell and yolk respectively.[4] Yet the emphasis in this analogy was on the spherical and shell-like shape of heaven's vault and on the complete enclosure of the earth by the heavens, and not on the actual shape of the earth. Another example was the ancient knowledge of the variation of the height of the North Pole, that is, that the elevation of the North Star from the horizon becomes higher as one goes northward. This well-known fact is a sort that can be easily explained in terms of a spherical earth, but it was not impossible to explain it in terms of a flat earth, which in fact was what the traditional Chinese did. Although the above two ideas were among the main pieces of evidence used by the Jesuits and their Chinese associates in arguing for a spherical earth,[5] no Chinese thinker seems to have done so before them.

What was in more serious conflict with the notion of a spherical earth in East Asia was the idea – or ideology – of '*Zhonghua* 中華' ('Central Splendor'). The concept of a round earth upset the accepted world order of the Chinese, namely that China, the 'Central Kingdom' (*Zhongguo* 中國), was at the center of the world, which entitled the Chinese to enjoy the 'Central Splendor.' The central position of China on the surface of the flat earth was so important an element of the Chinese – East Asian – worldview that it can only be compared to the central position of man in the traditional earth-centered world-picture of the West. If the earth was to be found spherical, it would have a devastating implication, similar to that of the Copernican shattering of the man-centered universe in the West, because there could be no single center on the surface of a spherical earth. Naturally, many Confucian thinkers opposed the idea of a spherical earth when it entered East Asia.[6] There were continuing debates and criticisms of the new calendars based on Western astronomy, and the criti-

4 This analogy first came up in Zhang Heng 張衡 (78-139) and Wang Fan 王蕃 (228-266), and quoted repeatedly by subsequent writers, including Zhu Xi 朱熹 (1130-1200). See, e.g., Cheng Wenguang 鄭文光 and Xi Zezong 席澤宗, *Zhongguo lishi shang de yuzhou lilun* 中國歷史上的宇宙理論 (Beijing: Renmin chubanshe, 1975), 69-70.

5 Chen Meidong 陳美東 and Chen Hui 陳暉, "Mingmo Qingchu xifang diyuanshuo zai Zhongguode chuanbo yu fanxiang" 明末清初西方地圓說在中國的傳播與反響, *Zhongguo keji shiliao* 中國科技史料 21 (2000): 6-12.

6 Pingyi Chu, "Trust, Instruments, and Cross-Cultural Scientific Exchanges: Chinese Debate over the Shape of the Earth, 1600-1800," *Science in Context* 12 (1999): 385-411; Ku Man-ok 具萬玉, "Chosŏn hugi 'chigu'sŏl suyong-ŭi sasangsajŏk ŭiŭi" 朝鮮後期 '地球'說 受容의 思想史的 意義 [The Intellectual Historical Significance of the Acceptance of the Theory of the Spherical Earth in Late Chosŏn], *Han'guksa-ŭi kujo-wa chŏn'gae – Ha Hyŏn'gang kyosu chŏngnyŏn kinyŏm nonch'ong* 韓國史의 構造와 展開 – 河炫綱教授定年記念論叢 (Seoul: Hyean, 2000), 717-747; Jongtae Lim, "Locating a Center on the Surface of a Globe:

cisms were often on political grounds, because the *Zhonghua* world order, now threatened by the idea of a spherical earth, was as much political and cultural as it was cosmological.[7]

The situation involving the idea of the earth's rotation was different from that of the earth's sphericity in both the West and East Asia. It was a new idea in the West, one that conflicted with the established worldview. Naturally, it had already created great controversy in the West before coming to East Asia. As is well known, the idea was banned by the Church and was suppressed; sometimes those holding the idea were persecuted.

The idea of the earth's rotation was new in East Asia as well. However, the possibility of the earth's motion had not necessarily been ruled out by the cosmological discussions of traditional East Asia. Unlike with the idea of a spherical earth, there was no essential component of the established cosmology of East Asia that would exclude the idea of the earth's motions. To be sure, the above-mentioned idea of the 'circular heaven and square earth' could have potentially clashed with the concept of the rotation of the earth. But unless the character '*fang* 方' literally meant 'square,' it did not make rotation of the earth impossible. Furthermore, there was the *Yijing* 易經 expression that "heaven's motion is strong" (*Tianxing jian* 天行健). Although this phrase could suggest the movement of heaven, its counterpart expression that "the earth's tendency is *kun*" (*Dishi kun* 地勢坤) did not indicate an earth at rest. Also, while the common association of movement (*dong* 動) with yang and stillness (*jing* 靜) with yin could imply a movement of the heaven (associated with yang) and the stillness of the earth (associated with yin), this correlation does not seem to have actually hindered the idea of the earth's rotation. If anything could have gotten in the way of accepting the idea, it would have been commonsense objections, such as the fact that man on earth does not feel the earth's movement. But from very early times, the Chinese knew how to handle such objections to the possibility of the earth's movement, arguing from the analogy of a man on a boat not feeling the movement of the boat.[8]

Negotiating China's Position on the Spherical Earth in Seventeenth and Eighteenth-Century China and Korea," *Historia Scientiarum* 17 (2008): 175-188.

7 For the complex nature of these debates and criticisms, see Yi-long Huang, "Court Divination and Christianity in the K'ang-hsi Era," *Chinese Science* 10 (1991): 1-20; Pingyi Chu, "Scientific Dispute in the Imperial Court: The 1664 Calendar Case," *Chinese Science* 14 (1997): 7-34.

8 See, e.g., the Han 漢 apocryphal text, *Gaolingyao* 考靈曜. In explaining the 'four wanderings' (*siyou* 四遊) of the earth (discussed below), it says: "The earth constantly moves and does not stop. [It is] like a man sitting in a boat. The boat moves, but the man does not feel [the movement]." Quoted in *Bowuzhi* 博物志, *Sibu beiyao* 四部備要 ed., 1.1b. Zhu Xi also quoted the passage: *Zhuzi yulei* 朱子語類 (1270 edition, reprinted in 1473), 96.3a.

Thus, for the Chinese before the arrival of the Jesuits, the earth did not have to be at absolute rest. Movement of the earth along an orbit was entirely possible, and the rotation on its own axis was by no means ruled out. Indeed, various ancient sources have passages that speak of, or appear to speak of, the earth's motions. The foremost example is the so-called 'four wanderings' (*siyou* 四遊), the yearly movement of the earth – the true nature of which is not clear – that referred sometimes to ascending and descending motions of the earth, and sometimes to motions of the earth in the four compass directions.[9] There are other expressions of the earth's movement in various ancient sources.[10] Although such expressions began to disappear from about the fifth century AD with the general waning of interest in cosmological speculations, Zhu Xi 朱熹 (1130-1200) in the twelfth century still raised the possibility of the earth's motion by asking: "How can we know for sure that [while] heaven moves outside, the earth does not follow it and rotate?"[11]

Thus, it is not surprising that the idea of the earth's rotation did not create much controversy when it entered East Asia. The response to the idea was relative silence, compared with the noisy controversy stirred by the idea that the earth was round. Some aspects of the actual historical situation surrounding the transmission of the idea of the earth's rotation complicated the situation and reinforced the 'silence.' The timing of the transmission of the idea is one such aspect. It fell precisely in the period during which the idea was officially banned by the Catholic authorities, that is, between 1616 (when the Copernican idea was placed in the Index) and 1757 (when the ban was lifted). The fact that the carriers of the idea were the members of the Society of Jesus also complicated the situation. While the Jesuits in China in much of this period were actively appropriating Western scientific knowledge in their efforts to attract the Chinese scholars' attention, they were forbidden to openly discuss this novel, and potentially interesting, idea. The presence of earlier Chinese assertions about the earth's motions further complicated matters.

Yet another complicating factor is that the earth had not one but two principal motions: the daily rotation on its own axis and the yearly revolution in its orbit around the sun. The acceptability of the two motions was different, both

9 The idea of the 'four wanderings' appeared in various ancient sources, including the *Zhoubi suanjing* 周髀算經 (Chap. 5), and were variously interpreted as movements of the earth, heaven or the luminaries. For brief discussions of 'the four wanderings,' see Zhongguo tianwenxueshi zhengli yanjiu xiaozu 中國天文學史整理研究小組, *Zhongguo tianwenxueshi* 中國天文學史 [History of Chinese Astronomy] (Beijing: Kexue chubanshe 科學出版社, 1987), 171-173.

10 Cheng and Xi, *Zhongguo lishi shang de yuzhou lilun*, 105-113.

11 *Zhuzi yulei*, 86.9a. For further discussion on this remark, see below.

in the West and in East Asia. In the West, the idea of an orbital revolution of the earth was a more serious problem because it disrupted the man-centered universe by displacing the earth from the center of the world. In contrast, the possibility of an axial rotation could be raised, as long as it was not accepted as truth. The situation was more or less the opposite in East Asia. While the possibility of an orbital revolution clearly existed in traditional cosmological discussions (for example, in the above-mentioned theory of the 'four wanderings'), the possibility of an axial rotation did not. Thus, we are faced with a paradoxical situation in which the Jesuit transmitters were more strictly forbidden to speak of the earth's orbital revolution, towards which the East Asian world would have been more receptive.

It is thus very difficult to document the spread of the idea of the earth's rotation from the West to East Asia. Many things remain unclear. When and how did the Western idea actually enter the East Asian world? To what extent could the Jesuits discuss the idea – or did they want to discuss it, or were they willing to discuss it? To what extent were the East Asians aware of the existence of – and did they have access to – the idea, one that was forbidden to the Western transmitters themselves? Was it possible that the East Asian intellectual world was completely barred from it until the Catholic ban was lifted? Perhaps it is impossible to give a clear answer to any of these questions. By contrast, the introduction, controversies and eventual acceptance of the idea of a spherical earth in East Asia can be more or less fully documented, as we have seen.

It is clear, however, that some East Asian scholars began to speak of the rotation of the earth soon after the arrival of the Jesuits. Although it has often been said that the idea did not enter China until 1760, when Michel Benoist (Jiang Youren 蔣友人, 1715-1774) wrote about it in the margins of the world map that he presented to Qing emperor Qianlong 乾隆 (r. 1735-1796), it had previously found its way into the writings of a number of seventeenth- and eighteenth-century Chinese and Korean thinkers.[12]

Therefore, some modern scholars have suggested that the idea of the earth's rotation was independently discovered (or developed) by East Asian thinkers in the seventeenth and eighteenth centuries. This has given rise to debates involving the originality of their ideas, leading to questions of whether, and how, they arrived at the ideas of the earth's rotation. While names of Chinese scholars like Wang Fuzhi 王夫之 (1619-1692) and Huang Baijia 黃百家 (1643-

---

12 See the references cited in note 1. Benoist's account of the earth's rotation became public after it was discussed in Ruan Yuan's 阮元 (1764-1849) *Chouren zhuan* 疇人傳, published in the early nineteenth century.

1709) have been mentioned in this connection,[13] more frequently noted were the seventeenth- and eighteenth-century Korean scholars Kim Sŏk-mun and Hong Tae-yong 洪大容 (1731-1783).

2 A Forerunner of the Idea of the Earth's Rotation in East Asia? – Zhang Zai

In considering the possible East Asian origins of the idea of the earth's rotation, we will first examine the fact that many of these seventeenth- and eighteenth-century East Asian scholars fingered Zhang Zai 張載 (1020-1077) as an East Asian forerunner. Scholars such as Wang Fuzhi, Huang Baijia, Kim Sŏk-mun, and Hong Tae-yong saw the idea of the earth's rotation in the following two very ambiguous passages from the "*Sanliang*" 參兩 chapter of Zhang Zai's *Zhengmeng* 正蒙, which I present first in Chinese characters.

> (Passage I) 地純陰凝聚於中, 天浮陽運旋於外, 此天地之常體也. 恒星不動, 純繫乎天, 與浮陽運旋而不窮者也. 日月五星, 逆天而行, 幷包乎地者也. 地在其中, 雖順天左旋, 其所繫辰象隨之, 稍遲則反移徙而右爾. 間有緩速不齊者, 七政之性殊也. ...

> (Passage II) 凡圜轉之物, 動必有機. 旣謂之機, 則動非自外也. 古今謂天左旋, 此直至粗之論爾, 不考日月出沒, 恒星昏曉之變. 愚謂在天而運者, 惟七曜而已. 恒星所以爲晝夜者, 直以地氣乘機左旋於中, 故使恒星河漢, 因北爲南, 日月因天隱見. 太虛無體, 則無以驗其遷動於外也.

Not only seventeenth- and eighteenth-century thinkers, but also modern scholars have seen the idea of the earth's rotation in these passages and considered Zhang Zai as a real forerunner.[14] Let us first translate the passages following one such modern scholar, Yamada Keiji 山田慶兒, perhaps the most enthusiastic in seeing in these passages the notion of a rotating earth. Yamada interprets the underlined portion of Passage I as follows:

---

13 Ogawa, "Higashi Ajia ni okeru chiten(dō)setsu no seiritsu"; Shi, "Shiqi shiji Zhongguo de Gebaini xueshuo"; Yang, "Gebaini rixin didongshuo zai Zhongguo."

14 Yamada Keiji 山田慶兒, *Shushi no shizen-gaku* 朱子の自然學 [Master Chu's Study of Nature] (Tokyo: Iwanami shoten, 1978), 40-42. Ogawa, "Higashi Ajia ni okeru chiten(dō) setsu no seiritsu"; Cheng and Xi, *Zhongguo lishi shang de yuzhou lilun*, 111-112.

> The sun, the moon, and the five stars move in the direction opposite to heaven's rotation, and they also surround the earth. The earth is inside them, and though [it, i.e., the earth,] follows the sky and rotates leftward,...

For Passage II, Yamada interprets the underlined portion as speaking of the earth's rightward rotation instead of the heaven's leftward rotation. However, he had to change "*zuoxuan*" 左旋 (leftward rotation) to "*youxuan*" 右旋 (rightward rotation) to do so.

> People have spoken about heaven's leftward rotation. This is an extremely crude theory, [not knowing that it is the earth that rotates rightward rather than heaven rotating leftward, and] not considering the rise and setting of the sun and the moon, and the change of the morning and evening of the fixed stars. I would say that what moves in heaven are only the seven luminaries. The reason that the fixed stars have day and night is that the earth's *qi* rotates *rightward* inside relying on the *ji* 機 [of motion]. Thus it makes the fixed stars and the Milky Way [appear to be rotating] from north to south, and the sun and the moon appear and disappear following heaven.[15]

It is not possible to determine for sure exactly what Zhang Zai meant in these ambiguous passages. What needs to be underlined, however, is that reading the earth's rotation in these passages does not make Zhang Zai appear creative or imaginative; rather, such a reading makes him appear strange and difficult, if not confused and even ridiculous. For nowhere in Zhang Zai's sayings and writings does he actually show any awareness of the possibility of a rotation of

15 Kim Sŏk-mun made a further change: "地乘氣右旋於中" instead of "地氣乘機右旋於中." Some modern scholars like Ogawa and Yi Yong-bŏm 李龍範 read Kim Sŏk-mun's sentence as "地機乘氣右旋於中," but what they read as "機" appears to me to be a mark left as the result of crossing out a miswritten character in Kim Sŏk-mun's manuscript. See Ogawa Haruhisa 小川晴久, "Chijŏn(dong)sŏl-esŏ uju muhannon-ŭro – Kim Sŏk-mun-kwa Hong Tae-yong-ŭi segye" 地轉(動)說에서 宇宙無限論으로 – 金錫文과 洪大容의 世界 [From the Theory of the Earth's Rotation (Motion) to the Theory of Infinite Universe: The World of Kim Sŏk-mun and Hong Tae-yong], *Tongbang hakchi* 21 (1979): 55-90; Yi Yong-bŏm, "Kim Sŏk-mun-ŭi chijŏnnon-kwa kŭ sasangjŏk paegyŏng" 金錫文의 地轉論과 그 思想的 背景 [Kim Sŏk-mun's Theory of the Earth's Rotation and Its Intellectual Background], *Chindan hakpo* 震檀學報 41 (1976): 81-107. In another edition of the *Yŏkhak isipsado ch'onghae* 易學二十四圖總解, believed to be Hwang Yun-sŏk's 黃胤錫 (1729-1791) personal copy, the sentence is corrected to "地氣乘機右旋於中."

the earth. Indeed, in the passage immediately following Passage II, which Yamada and others interpret as referring to the earth's rightward rotation, Zhang Zai speaks of heaven rotating leftward. It is quite unlikely that if Zhang Zai really had the idea of the earth's rotation, he would have spoken about it in such a casual and confusing manner.

On the other hand, there is a perfectly reasonable way to make sense of the passages without reading the earth's rotation into them. For this new interpretation, we should adopt a different punctuation in one place in Passage I: making the phrase "*di zai qi zhong*" 地在其中 as part of the previous sentence, rather than beginning the next sentence with it. Then we take the subject of the previous sentence, "*riyue wuxing*" 日月五星, to be the subject of the next sentence too: it is "the sun, the moon and the five stars" that "follow the heaven and rotate leftward."[16] This is perfectly in line with Zhang Zai's "theory of leftward rotation" (*zuoxuanshuo* 左旋說), namely that the sun, the moon and the five stars all rotate leftward along with heaven, which is exactly what the next sentences say. The result is:

> The sun, the moon, and the five stars move in the direction opposite to heaven's rotation, and they also surround the earth, which is inside them. Although [they, i.e., the sun, the moon, and the five stars,] follow heaven and rotate leftward, ... 日月五星, 逆天而行, 并包乎地者也, 地在其中. [日月五星]雖順天左旋, ...

For Passage II, what we need to do is to take "*diqi*" 地氣 to be Zhang Zai's, or more likely the scribe's, mistake for "*tianqi*" 天氣. True, attributing mistakes and changing the text is not a choice to be made lightly. Yet if we have to attribute a mistake to the passage anyway, this choice appears far more likely than having Zhang Zai make the mistake of saying "leftward rotation" while meaning "rightward rotation," and at a crucial place at that. If we accept the new alternative, Passage II becomes:

> People have spoken about heaven's leftward rotation. This is an extremely crude theory, [only speaking about heaven while] not considering the rise and setting of the sun and the moon...I would say that what moves in heaven are only the seven luminaries. The reason that the fixed stars have day and night is that the *qi* of heaven rotates *leftward* inside relying on the *ji* [of the movement of heaven]. Thus it makes the fixed stars and the

16 This is exactly what Zhu Xi did in his commentary to the passage: *Zhangzi quanshu* 張子全書 (*Sibu beiyao* edition) 2.6b.

> Milky Way [rotate along with it] from north to south... 古今謂天左旋, 此直至粗之論爾, 不考日月出沒 ... 愚謂在天而運者, 惟七曜而已. 恒星所以爲晝夜者, 直以天氣乘機左旋於中, 故使恒星河漢, 因北爲南 ...

In this way we can avoid the necessity of putting into the thought of Zhang Zai the unlikely idea of the earth's rotation, which did not match with the rest of his thought.

Does this reading rule out Zhang Zai as a possible source of the idea of the earth's rotation as held by those seventeenth- and eighteenth-century thinkers mentioned above? Or is it still possible that they found the idea in Zhang Zai whether the latter had intended it or not?

What should be noted in this connection is that all readings – by Wang Fuzhi, Huang Baijia and Kim Sŏk-mun, as well as by the modern scholars – of the concept of the earth's rotation in the above-quoted passages of Zhang Zai's came after the arrival of the Jesuits. While Zhang Zai's *Zhengmeng* had been a key philosophical text studied by nearly all Confucian scholars, and while his theory of the leftward rotation contained in it had been frequently discussed and praised, no one until the seventeenth century had interpreted the above passages as suggesting the earth's rotation. Zhu Xi, for example, did not mention them while speaking of the possibility of the earth's rotation in the remark seen earlier. That such a reading was never made before the coming of the Jesuits bespeaks the remoteness of the possibility of the earth's rotation in the traditional East Asian mind. It was only after they were introduced to the idea by the Jesuits that East Asian thinkers could make such a reading of a well-known text.

In fact, Zhu Xi's remark is also worth closer examination because it was considered by some Korean scholars of the late Chosŏn as another 'forerunner' concerning the earth's rotation.[17] What Zhu Xi actually said was the following:

> I think that the movement of heaven has deviations, and the earth rotates following heaven, and [its movement also] has deviations. It is simply that now we sit like this and think only that the earth does not move. How can we know for sure that [while] heaven moves outside, the earth

17 Yi Ik, for example, spoke of this remark by Zhu Xi while considering the possibility of the earth's rotation. *Sŏngoho sasŏl* 星湖僿說, "Tiandimen" 天地門, Chap. 3 (Seoul: Kyŏnghi ch'ulp'ansa 慶熙出版社, 1967), vol. 1, 76-77.

> does not follow it and rotate? 想是天運有差. 地隨天轉而差. 今坐於此但知地之不動耳. 安知天運於外而地不隨之以轉耶.[18]

In this remark, Zhu Xi does appear to have in mind a rotation of the earth around its own axis, which to him was also the axis of heaven's rotation. But it is not clear what kind of rotation he had in mind for the earth to make as it 'followed' heaven's rotation.[19] Nor is it clear how seriously he was committed to the idea of the earth rotating. It may very well have been merely a passing thought about a possibility. His followers seem to have taken it as just that. Even these passages were not given much attention before the arrival of the Jesuits.

3 Independent Development of the Idea of the Earth's Rotation? – Kim Sŏk-mun

After the Jesuits' arrival, the sudden appearance of scholars who read such an unlikely idea of the earth's rotation from the above passages suggests that a shift must have occurred. Something seems to have predisposed East Asian thinkers to suddenly perceive the idea of the earth's rotation in these passages.[20] Naturally, one is tempted to think that they might have come across the idea from Western sources. Still, one cannot rule out the possibility that East Asian thinkers came up with the idea independently in the course of their cosmological speculations. Let us look at the case of Kim Sŏk-mun, perhaps

18 *Zhuzi yulei*, 86.9a.

19 In another conversation, Zhu Xi spoke of the earth's motion in connection with the theory of the 'four wanderings,' which, according to him, "refers to the four wanderings and ascending and descending of the earth…In spring, [the earth] wanders to the east for thirty thousand *li* 里; in summer it wanders to the south for thirty thousand *li*; in autumn it wanders to the west for thirty thousand *li*; in winter it wanders to the north for thirty thousand *li*." 謂地之四遊升降 … 春游過東三萬里, 夏游過南三萬里, 秋游過西三萬里, 冬游過北三萬里. (*Zhuzi yulei*, 86.10a) Later in the same conversation, when a disciple illustrated these motions with the motion of an empty vessel floating around on water, Zhu Xi approved. In yet another passage he said: "The motion of the earth now is a motion in just one place, and the motion also does not reach far" 今之地動只是一處動動亦不至遠也. (*Zhuzi yulei*, 100.6a) So he seems to have been thinking about yet another kind of motion, smaller than the above two.

20 Ogawa has also suggested this: "Higashi Ajia ni okeru chiten(dō)setsu no seiritsu," 380.

the first East Asian thinker who not only pronounced the idea of a rotating earth but constructed a world system based on it.[21]

Kim Sŏk-mun's cosmological theory in his *Yŏkhak isipsado ch'onghae* 易學二十四圖總解 (General Interpretation of Twenty-four Charts in *Yijing* Studies) was composed of the following three basic constituents:[22]

1) the cosmological structure of the nine more of less concentric heavenly spheres;
2) the cosmogonic remarks at the beginning of Zhou Dunyi's 周敦頤 (1017-1073) *Taiji tushuo* 太極圖說;
3) Shao Yong's 邵雍 (1011-1077) theory of the cosmic cycles of *yuan-hui-yun-shi* 元會運世.

In Kim Sŏk-mun's cosmological structure of the 'nine heavens' (*jiutian* 九天), laid out in his *Hwanggŭk kuch'ŏn do* (*Huangji jiutiantu*) 黃極九天圖,[23] the outermost heavenly sphere, 'Heaven of the Supreme Ultimate' (*taiji tian* 太極天), is motionless. The next sphere, 'Heaven of the Supreme Void' (*taixu tian* 太虛天), moves extremely slowly, which Kim Sŏk-mun referred to as 'the slightest motion' (*weidong* 微動). Further inward in the spheres, the motions become gradually faster: 'Heaven of the Fixed Stars' (*jingxing tian* 經星天), 'Heaven of Saturn' (*zhenxing tian* 鎮星天), 'Heaven of Jupiter' (*suixing tian* 歲星天), 'Heaven of Mars' (*yinghuo tian* 熒惑天), 'Heaven of the Sun's Wheel' (*rilun tian* 日輪天, which coincides with the Heavens of Mercury and of

21 E.g., Yi, "Kim Sŏk-mun-ŭi chijŏnnon-kwa kŭ sasangjŏk paegyŏng"; Yabuuchi Kiyosi 藪内清, "Richo gakusha no chikyu kaitensetsu" 李朝科學者の地球回轉說 [The Chosŏn Scholars' Theory of the Earth's Rotation], *Chōsen gakuhō* 朝鮮學報 49 (1968): 427-434; Ogawa, "Chijŏn(dong)sŏl-esŏ uju muhannon-ŭro"; Ogawa, "Higashi Ajia ni okeru chiten(dō)setsu no seiritsu"; Park, "Hong Tae-yong's Idea of the Rotating Earth."

22 The standard text for the *Yŏkhak isipsado ch'onghae*, the one used in the present paper, is appended to Min Yŏng-kyu 閔泳珪, "Sipchil segi Yijo hag'in-ŭi chidongsŏl – Kim Sŏk-mun-ŭi Yŏkhak isipsadohae" 十七世紀 李朝學人의 地動說 – 金錫文의 易學二十四圖解 [A Seventeenth-Century Chosŏn Scholar's Theory of the Earth's Motion: Kim Sŏk-mun's *Yŏkhak isipsadohae*], *Tongbang hakchi* 16 (1975): 1-44. For discussions of Kim Sŏk-mun's cosmological theory, see Yi, "Kim Sŏk-mun-ŭi chijŏnnon-kwa kŭ sasangjŏk paegyŏng"; Jun Yong Hoon, "Transformation of the Western Science and Its Role in the Cosmological Discourse in the Late 17th Century Korea," in *Current Perspectives in the History of Science in East Asia*, ed. Yung Sik Kim and Francesca Bray (Seoul: Seoul National University Press, 1999), 497-503; Moon Joong-yang, "Traditional Cosmology Associated with the *I-ching* and Anti-Cosmological Discourses in 18th-Century Korea," *Seoul Journal of Korean Studies* 12 (1999): 177-227, esp. 183-197.

23 *Yŏkhak isipsado ch'onghae*, 5.

Venus), and 'Heaven of the Moon's Wheel' (*yuelun tian* 月輪天). The innermost is the 'Heaven of the Earth's Wheel' (*dilun tian* 地輪天), the motion of which is the fastest at one rotation a day. And its center is not the center of heaven, but it is removed from it by 180,000 *li* 里 (1 *li* is approximately 400 meters).

Kim Sŏk-mun used the *Taiji tushuo* to explain the cosmological structure outlined above. He interpreted the *Taiji tushuo* phrase "The supreme ultimate moves and generates the yang" (*taiji dong er sheng yang* 太極動而生陽), as referring to the slightest motion of the Heaven of the Supreme Void emerging from the motionless outermost Heaven of the Supreme Ultimate. The next phrase, "Movement becomes extreme and becomes still" (*dong ji er jing* 動極而靜), was interpreted to mean that the earth, which moves the fastest, nevertheless appears to people on it as motionless. The phrase "It is still and generates the yin" (*jing er sheng yin* 靜而生陰), then, meant to him that the motionless earth produces shadow (which is yin) on its surface on the opposite side of the sun. Finally, he took the phrase "Stillness becomes extreme, and it moves again" (*jing ji fu dong* 靜極復動) to mean that people on earth, apparently at rest, feel that heaven moves.

Based on the principle (which he adopted from Italian Jesuit Giacomo Rho's [Luo Yagu 羅雅谷] *Wuwei lizhi* 五緯曆指) that the distance covered by one day's movement of each heaven should be the same (*zhutian nengli bi deng* 諸天能力必等), and using the known value of the circumference of the earth (the daily movement of the earth), 90,000 *li*, Kim Sŏk-mun was able to compute the circumference of all the nine heavens, and thus their distances from the center of the heaven. (For example, if the period of rotation of a heaven is *n* days, the circumference of that heaven is *n* times 90,000 *li*.) For the period of rotation of the Heaven of the Fixed Stars, he used 25,440 years, the period of the precession of equinoxes. And for 'the slightest motion' of the Heaven of the Supreme Void, he took the angular speed of 1 *xu* 虛 per day, where the '*xu*' is the smallest unit of angle (1 *du* 度 = 1 degree = 60^9 *xu*). The period of rotation for the Heaven of the Supreme Void then becomes:

$$60^9(xu/du) \times 360(du/circumference) \times 1\,(day/xu) \times 1/365.2425(year/day) = 9.93 \times 10^{15}(years).$$

The outermost Heaven of the Supreme Ultimate does not move, and its circumference cannot be measured.

Having accounted for the structure of the world in this manner, Kim Sŏk-mun went on to deal with cosmological time. He took up Shao Yong's theory of cosmic cycles (1 *yuan* = 12 *hui* = 12 × 30 *yun* = 12 × 30 × 12 *shi* = 12 × 30 × 12 × 30 years), but gave it his own interpretation and went further than Shao Yong. He associated different periodic motions of the heavens, of the ecliptic (the sun's apparent path), and of the earth with these cycles. One day, to begin with, is associated with the daily rotation of the earth. One year is the period of the yearly revolution of the Heaven of the Sun's Wheel. The next time unit for which Kim Sŏk-mun found cosmological significance is 1 *yun* = 360 years, the period of oscillation of the size of the sun's orbit, the ecliptic. His next cosmic period is that of the revolution of the earth along the orbit around the center of the heaven. He took it to be equal to 2 *hui*. But the number he used for *hui* was 12,720 years, and was thus different from the one used by Shao Yong, 10,800 years. He chose it to be equal to half the period of the precession of equinoxes (25,440 years). Kim Sŏk-mun had said earlier that, because Shao Yong's numbers did not fit the calendars, he had chosen his own numbers for *yuan* and *hui* that differed from Shao Yong's. Finally, one *yuan* for Kim Sŏk-mun was 60 *hui* = 763,200 years, and is the period of the variation of the angle between the equator and the ecliptic from 0 degree to 45 degrees. He noted that his *yuan* is approximately six times Shao Yong's *yuan*, which is 129,600 years. Kim Sŏk-mun did not stop here, but went on to speak of still greater cosmic periods, even mentioning a period of $763{,}200^8$ years, which is greater than 10^{48} years.

This was a system based more on numerological speculations than on observations of the actual world. Nonetheless, Kim Sŏk-mun seems to have believed that in this manner he had constructed a coherent cosmological system. In the course of this construction, he assigned two different motions to the earth. The earth's rotation constituted an essential part of this system, in which the motion is the fastest at the center, and slower outside, coming to complete rest at the outermost. Another kind of motion, namely the revolution around heaven's center (from which the center of the earth is removed by 180,000 *li*) with the period of 25,440 years, was assigned to the earth to take account of the cosmological cycles. This made some modern scholars consider that the idea of a rotating earth was a result of Kim Sŏk-mun's own creative imagination, which he came up with as he tried to create a coherent cosmological theory.[24] Yet does this mean that Kim Sŏk-mun came to the idea independently from Western ideas about the earth's rotation?

Again, it is not possible to reach a conclusive answer to this question. An argument for independent development can be made based on the absence of a fundamental conflict of the idea with the basic elements of the traditional

24 E.g., Jun, "Transformation of the Western Science and Its Role"; Moon, "Traditional Cosmology Associated with the *I-ching*."

East Asian cosmological system, and on the commonsense idea that it is natural for round things to turn. For once the sphericity of the earth is accepted, there would be nothing to prevent an East Asian thinker like Kim Sŏk-mun from imagining the possibility of a rotation of that spherical earth.[25]

4 Transmission from the West?

However, in my opinion, it is extremely difficult to imagine that Kim Sŏk-mun did not know that the idea of the earth's rotation came from the West. It is far more likely – although I do not have any direct evidence for it – that he had, in some way or other, heard or read about the idea of the earth's rotation and incorporated the idea into his cosmological system, or built his cosmological system around it. Copernican ideas were being circulated in China before the above-mentioned Chosŏn scholars discussed the earth's rotation.[26] Indeed, the Jesuits were more or less free to discuss the possibility that the earth rotated on its own axis if only they refuted it afterwards. For example, in the *Wuwei lizhi*, Giacomo Rho presented the following argument for the possibility of the earth's rotation, which he rejected afterwards:

> Now above the earth's surface we see the stars move leftward. But they are not real movements of the stars. The stars do not have the motion of one rotation a day. It is simply that the earth becomes one sphere together

25 Jongtae Lim, in his comment on an earlier version of the present paper, persuaded me to think about the connection between the idea of the round earth and that of its rotation. Indeed, Hong Tae-yong said that the earth's "body is a perfect circle and rotates without stop." 其體正圓旋轉不休停. *Ŭisan mundap* 毉山問答, in *Tamhŏnsŏ* 湛軒書, *inner collection* (*naejip* 內集), supplements (*poyu* 補遺) 4.19a. It has to be emphasized, however, that the notion of the spherical earth did not exist either in East Asia before the arrival of the Jesuits (see the first section of this paper). It is only after the Jesuits introduced the notion of the earth's sphericity that we find those who inferred the earth's rotation based on the spherical shape of the earth.

26 See the references cited in note 1. In particular, Shi Yunli has shown that Huang Daozhou 黃道周 (1585-1646) and You Yi 游藝 (1614-1684) had already spoken of the earth's rotation, and has suggested that Fang Yizhi 方以智 (1611-1671) and Huang Baijia read about it in the *Tianbu zhenyuan* 天步眞原 of Jean-Nicolas Smogulecki (Mu Nige 穆尼閣, 1611-1656). Shi Yunli has further suggested the possibility that these scholars' ideas were the sources of the Chosŏn scholars' theory of the earth's rotation. See Shi, "Shiqi shiji Zhongguo de zhun Gebaini xueshuo"; Shi, "Cong Huang Daozhou dao Hong Darong: shiqi shiji Zhong Chao didongshuo de bijiao yanjiu" 從黃道周到洪大容 – 十七世紀中朝地動說的比較研究, *Ziran bianzhengfa tongxun* 自然辨證法通訊 19 (1997): 60-65.

> with the [surrounding] *qi* and fire, and moves from west to east, one rotation a day. It is like a man traveling on a boat looking at trees and other things on the shore; he does not feel that he moves but feels as though the shore moved. The reason (*li* 理) that people on the earth see the westward motion of the stars is the same. In this manner [by accepting the earth's rotation], one can avoid many motions in the heavens by the one motion of the earth, and large rotations in the heavens by the small rotation of the earth.[27]

This, then, gives rise to further questions. Why did Kim Sŏk-mun, if he borrowed the idea of the rotation of the earth from some Western source, keep silent about the source? The question may push us back to the possibility of an independent development. But if one is unable to accept that possibility, what was the reason for Kim Sŏk-mun's silence? Was he hiding the source of his idea? And if so, why would he want to do so? Not only Kim Sŏk-mun but Hong Tae-yong too was silent about the source of his concept of a rotating earth. It is likely that the latter read or heard about it in Kim Sŏk-mun's work,[28] but he did not say so. Why did he act in this way?

These questions, in turn, lead to another set of questions. How important, for Kim Sŏk-mun and for his readers, was the cosmological issue of whether the earth moves or not? We have pointed out that the earth's motion was not an idea in direct conflict with the basic components of East Asian cosmology, and it is thus possible that the issue was not as serious as the one involving the idea of a spherical earth.[29] But could it also have been unimportant for Kim Sŏk-

27 *Xinfa suanshu* 新法算書, *Siku quanshu* 四庫全書 edition, 36.7b-8a. Kim Sŏk-mun quoted Rho's argument almost verbatim in his *Yŏkhak isipsado ch'onghae*, 32a.

28 For example, Hong Tae-yong was in close contact with Huang Yun-sŏk 黃胤錫 (1729-1791), who was deeply interested in Kim Sŏk-mun's theory. See Kim Tae-jun 金泰俊, *Hong Tae-yong P'yŏngjŏn* 洪大容評傳 [The Life and Work of Hong Tae-yong] (Seoul: Min'ŭmsa 民音社, 1987), 107-108. Moon Joong-yang has suggested that Hong Tae-yong's critical attitude towards the correlative mode of thinking made him reluctant to speak of Kim Sŏk-mun, whose cosmological system was rooted in correlative thinking. See Mun Chung-yang 文重亮, "Chŏnt'ongjŏk chayŏn insik ch'egye-ŭi sajŏk pyŏnhua: Sangsuhakchŏk, kironjŏk ujuron-ŭi simhua, chŏn'gae" 傳統的 自然認識 體系의 史的 變化: 象數學的, 氣論的 宇宙論의 深化, 展開, in *Han'guk sil'hak sasang yŏn'gu* 韓國實學思想研究, vol. 4, *Kwahak kisul p'yŏn* 科學技術篇, ed. Yŏnse daehakkyo Kuk'hak yŏn'guwon 延世大學校國學研究院. (Seoul: Hyean 慧眼, 2005), 90.

29 Not only the idea of the earth's rotation, but even ideas about an infinite universe and plural worlds, etc., for example, which could be found in Hong Tae-yong's grand cosmological theory were not in any fundamental conflict with the traditional East Asian worldview. And the source of those ideas could also be traced back to such earlier East Asian

mun, who made it an essential ingredient of his cosmological system? And if so, was that the reason why he was silent about where it came from? In other words, was the issue so trivial that he did not care to mention its source? Or on the contrary, did he consider it important enough to want to hide it?

Still another question can be asked in this connection. How conscious was Kim Sŏk-mun about the problem, and consequence, of accepting an idea from the West? Also, did the fact that he was Korean, and not Chinese, make it easier for him to accept that Western idea? Or were his feelings towards the ideas of Chinese forerunners like Zhang Zai and Zhu Xi different from his feelings towards Western ideas? Regarding this, it is worthwhile to point out that Kim Sŏk-mun read about Western concepts from books written in classical Chinese with the European authors' names Sinicized. He used the expression 'the calendar specialists' (*lijia* 曆家) to refer to Western astronomers just as he did to their Chinese counterparts, so he may not have even made much of a distinction between the two.[30]

5 The Theory of 'the Chinese Origin of Western Learning' and the Search for Chinese Forerunners – Sŏ Myŏng-ŭng

One aspect of the intellectual climate of East Asia in the seventeenth and the eighteenth centuries can help us make sense of this complicated situation. That East Asian thinkers pointed to earlier Chinese sources – 'discovering' or 'inventing' Chinese 'forerunners' – could be understood in the context of the so-called theory of 'the Chinese origin of Western learning' (*xixue zhongyuan* 西學中源) that was widespread among seventeenth- and the eighteenth-century Chinese thinkers. According to this theory, Western scientific ideas that came to China had their origins in ancient China: Chinese had such knowledge in their ancient golden age, but later the knowledge disappeared and fell into the hands of the barbarians, who developed it further and brought it back to China.[31] This theory was held also by many Korean scholars of the late Chosŏn, among whom there was a strong tendency to try to make sense of the new

sources as Shao Yong, Sŏ Kyŏng-dŏk 徐敬德 (1489-1546), and Choi Sŏk-chŏng 崔錫鼎 (1646-1715). On Hong Tae-yong's cosmological theory, see, e.g., Moon, "Traditional Cosmology Associated with the *I-ching*," 212-219.

30 E.g., *Yŏkhak isipsado ch'onghae*, 32a.

31 On the theory of "the Chinese origin of Western learning," see, e.g., John B. Henderson, "Ch'ing Scholars' Views of Western Astronomy," *Harvard Journal of Asiatic Studies* 46 (1986): 139-143; Jiang Xiaoyuan 江曉原, "Shilun Qingdai xixue zhongyuanshuo" 試論清代西學中源說, *Ziran kexueshi yanjiu* 自然科學史研究 7 (1988): 101-108; Pingyi Chu,

Western knowledge in terms of ancient Chinese ideas. A typical attitude of many of these Korean thinkers when faced with a novel Western cosmological idea was to try to find precedents or prototypes in ancient Chinese sources.

The best example is that of Sŏ Myŏng-ŭng, whose perceptions of the incoming Western astronomical and cosmological knowledge and of his own natural philosophical system based on *Yijing* studies (*yixue* 易學) were set in this sort of 'Chinese origin' theory.[32] Even the idea underlying the nineteenth-century slogan 'Chinese substance and Western function' (*zhongti xiyong* 中體西用), which advocated adopting Western science and technology as 'functions' (*yong* 用) while holding onto Chinese values and culture as 'substance' (*ti* 體),[33] could be found in Sŏ Myŏng-ŭng's attitude toward Western scientific knowledge. He took Western astronomical knowledge as 'function,' and *Yijing* studies as 'substance.' In particular, he saw Fu Xi's *Xiantian yi* 先天易 (*Yi* Prior to Heaven) as the source, or the substance (*ti*), while the functions (*yong*) were the computational method of *gougu* 句股[34] and calendrical astronomy. This was what Sŏ Myŏng-ŭng meant when he said: "The calendars and the *Yijing* 'mutually [form] the inside and the outside' (*xiang wei biaoli* 相爲表裏)."[35]

Thus, according to Sŏ Myŏng-ŭng, both the substance and the function existed in ancient China; afterwards, function disappeared from China and went to the West, where it was developed further before coming back to China. However, Western astronomy had deficiencies because it lacked the proper substance, that is, a simple, systematic theory. Sŏ Myŏng-ŭng wrote:

> After the burning [of books] in the Qin 秦, the *Xiantian yi* disappeared into the elixir kitchens; the *gougu* lost its substance, while the *Xiantian yi* lost its functions…The *gougu* of the Western countries…is like water without origin, and thus does not see its simplicity truly reaching to the

---

"Remembering Our Grand Tradition: The Historical Memory of the Scientific Exchanges between China and Europe, 1600-1800," *History of Science* 41 (2003): 193-215.

32 Pak Kwŏn-su 朴權壽, "Sŏ Myŏng-ŭng-ŭi yŏkhakchŏk ch'ŏnmun'gwan" 徐命膺의 易學的 天文觀 [Sŏ Myŏng-ŭng's Interpretation of Astronomy in the Framework of the *Yijing* Studies], *Han'guk Kwahak-sa Hakhoe-ji* 20 (1998): 57-101.

33 On the theory of 'the Chinese substance and Western function,' see Wang Ermin 王爾民, "Qingji zhishi fenzi de zhongti xiyong lun" 清季知識分子的中體西用論, in his *Wan Qing zhengzhi sixiangshilun* 晚清政治思想史論 (Taipei: Huashi chubanshe, 1969).

34 For a brief discussion of the *gougu* method, see Joseph Needham, *Science and Civilisation*, vol. 3 (Cambridge: Cambridge University Press, 1959), 22ff.

35 *Sŏnch'ŏn sayŏn* 先天四演, *Pomanjae ch'ongsŏ* 保晚齋叢書 (Kyujanggak 奎章閣 ed.), 32.

Xiantian yi. It is like a tree whose branches are cut off and thus does not see its great flow.[36]

In this way, he could accept, as many of his contemporaries had to, knowledge of Western astronomy that proved more accurate, while pointing out its deficiency in 'substance.' To supplement this and to retrieve its proper substance, Western astronomical knowledge had to be incorporated into the system based on *Yijing* studies (*yixue* 易學). And that was what Sŏ Myŏng-ŭng did. In fact, his system of *Yijing* studies accommodated not only Western astronomical knowledge but traditional Chinese astronomical knowledge as well.[37]

Sŏ Myŏng-ŭng seems to have felt more keenly than others about the inevitability to accept – at least not to avoid – Western knowledge. He said, for example, "How can one dislike something [just] because it has come from a foreign country?"[38] Naturally, he felt more strongly about 'the Chinese origin of Western learning,' which is quite understandable if one considers the intellectual climate surrounding him. After the humiliating defeat to the Manchus in the early seventeenth century, a strong anti-barbarian sentiment became widespread among Korean Confucians, along with what can be called the 'Korean *Zhonghua*' (Chosŏn chunghwa 朝鮮中華) theory – the notion according to which Korea was now the true bearer of the *Zhonghua* (Central Splendor) since China itself was ruled by the barbarian Manchus.[39] There was animosity towards anything barbarian, including the scientific knowledge of the Western 'barbarians.' Chosŏn scholars were furthermore swept up in what was called '*chonjuron*' 尊周論, admiration for ancient Zhou culture, the culture that was considered ideal by many neo-Confucians. Naturally, the need to legitimize the barbarian knowledge of Western astronomy for the *Zhonghua* was especially keen in this climate.

Thus, it is easy to understand how, in this kind of atmosphere, another Chosŏn thinker, Kim Sŏk-mun, could have made an appeal to Zhang Zai as the originator of the idea of the earth's rotation without mentioning any Western sources. When he did cite a Western source in connection with the idea, it was to criticize Westerners for *not* accepting the possibility of the earth's rotation.

36 *Sŏn'guje* 先句齊, *Pomanjae ch'ongsŏ*, Preface.

37 Pak, "Sŏ Myŏng-ŭng-ŭi yŏkhakchŏk ch'ŏnmun'gwan," 74-89.

38 *Piryejun* 髀禮準, *Pomanjae ch'ongsŏ*, Preface.

39 On the Korean *Zhonghua* theory, see Chŏng Ok-cha 鄭玉子, *Chosŏn hugi Chosŏn Chunghwa sasang yŏn'gu* 朝鮮後期 朝鮮中華思想 研究 [A Study of the Idea of the Korean *Zhonghua* in Late Chosŏn] (Seoul: Ilchisa 一志社, 1998).

For example, Kim Sŏk-mun condemned Johann Schreck's (Deng Yuhan 鄧玉函) argument in the *Cetian yueshuo* 測天約說 against the earth's rotation:

> The Western theory says the following. "The body of the earth never moves. It does not leave its own place; nor does it rotate. If what does not leave its own place were to leave [its own place], then it would not be at the exact center of the heavens. If what does not rotate were to rotate, man should feel it. Also, if it rotated, it must rotate once every day. Its motion is extremely fast. All the moving birds [will appear to] move slower if they move in the same direction as the earth's rotation, and faster if they move in the opposite direction. If a man throws a thing upward from the earth, its return to the earth cannot be at its original place. Now [we know] all these are not so, which suffices to make it clear that the earth does not rotate." What the Western theory takes as evidence that the earth does not rotate is nothing more than this. It can be said to be an extremely crude argument.[40]

6 Concluding Remarks

The title of this paper lists three different possibilities concerning the origin of the idea of the earth's rotation in East Asia: 'independent development,' 'transmission from the West,' and 'Chinese forerunners.' The first two are mutually exclusive: the idea was either transmitted from the West or developed independently in East Asia. The third complicates the picture: the existence of Chinese forerunners of the idea could have facilitated both of the two possibilities. We have examined grounds for arguing for, and speculating about, each of the above possibilities, but none are compelling enough for us to reach a definitive conclusion. Between the mutually exclusive alternatives, however, I am inclined toward 'transmission from the West.' I believe that, given the availability of Copernican ideas after the arrival of the Jesuits and the sudden appearance of the idea of the earth's rotation afterwards, the 'burden of the proof' rests with those who argue for independent development, and no such proof has yet come forth.

Regarding the situation surrounding the idea of the earth's rotation, it is not only the origin of the idea that is unclear, but in fact many other things as well. Perhaps the only thing we can be sure of is that the idea was eventually

40 *Yŏkhak isipsado ch'onghae*, 33a. The quotation is from *Cetian yueshuo*: in *Xinfa suanshu*, 11.14a.

accepted in the latter half of the nineteenth century by East Asian intellectuals. Except for someone like Kim Sŏk-mun, who recorded his views more or less explicitly, it is difficult even to tell exactly what kind of motion these East Asian thinkers meant by 'rotation' (*zhuan* 轉) of the earth. Frequently, they did not have a clear, fixed conception about the earth's rotation: some could still have doubts while accepting it by and large; others could still have suspicions while rejecting it at the same time.

CHAPTER 11

Problem of Meiji Japan in the History of Science in East Asia*

Japan achieved great success in the latter half of the nineteenth century in the adoption of Western science and industrialization. This success was so impressive, and so different from what happened in other East Asian countries, that a great deal of attention has been given to understanding and explaining it. In fact, Japanese scientific and technological development in the Meiji 明治 period (1867-1912) is considered such a success story that it is seen as an exception. It is, however, precisely the exceptional nature of this case that suggests the possibility of a different interpretation. And this will be the subject of the present paper.[1]

We are accustomed to viewing this phenomenon simply as part of 'early modern' East Asian history in which an East Asian country was more successful than its neighbors in adopting and assimilating science and technology from the West. The emphasis here is on the notion that the Japanese development of science and technology took place *after* the Western countries had already developed them. But is this view correct? Couldn't Japanese development have been a part of the development of science and technology that was going on at the time in a few leading countries of the world? This is the question I will attempt to tackle in various forms in the remainder of this paper.

* Originally published as "Problem of Early Modern Japan in the History of Science in East Asia," *Historia Scientiarum* 18 (2008): 49-57.

1 The examples I will mention in this paper are fairly well known and available in many secondary sources. The useful English materials, from which most of my examples have been drawn, are: James R. Bartholomew, *The Formation of Science in Japan: Building a Research Tradition* (New Haven: Yale University Press, 1989); Tessa Morris-Suzuki, *The Technological Transformation of Japan: From the Seventeenth to the Twenty-first Century* (Cambridge University Press, 1994); Scott L. Montgomery, *Science in Translation: Movements of Knowledge through Cultures and Time* (Chicago: University of Chicago Press, 2000), part 2; Morris Low, *Science and the Building of a New Japan* (New York: Palgrave Macmillan, 2005). These works contain citations to many references in the Japanese language. I will not try to identify the original Japanese sources for the information I use in the present paper.

 | DOI 10.1163/9789004265318-013

1 Japan as One of the Major World Powers in the Late Nineteenth-Century

A source of the problem is that those holding this conventional view see the West as one single unit, and Japan on the outside, and not part of it. This makes us look at the transmission of Western science to Japan in a manner entirely different from the ways we view scientific transmissions from Europe to the United States or among European countries themselves. It forces us to treat the Japanese case strictly as an 'inter-cultural' transmission between the West and East Asia and precludes the possibility of viewing it together with the other cases of 'intra-cultural' transmission among Western countries. But is the distinction really so clear-cut? Was the transmission of science and technology from the West to Japan so very different from the transmission of such knowledge from, say, Germany to England, or from England to the United States?

Now, if we look at the scientific and industrial development in the Western countries of the period more closely – and especially if we take note of the dates when the various events took place – what happened in Japan could, and I think should, be seen as part of the single development of science and industrialization of the leading Western countries that was going on in the latter half of the nineteenth century. Japan was taking part in this competitive process. To look back now at the scientific and technological development of various countries at the time, we cannot but notice that in many respects Japan was either on a par with, or even ahead of, major Western countries, and was doing similar things in similar ways. Whether the Japanese at the time realized it or not, much of what Japan did was not vastly different from what the leading European countries or the United States did in the late decades of the nineteenth century when they found themselves lagging behind others in some aspects of science and technology and made an effort to catch up.

This was very unlike what other East Asian countries were doing: China and Korea were not consciously participating in that competitive process, but were trying to import and adopt what they thought had already been accomplished in the West. The Chinese and Korean people's sense of the situation was not so much that they were 'lagging behind' in the competition with the Western powers, whom they could hope to catch up with, as that they were confronted with Western powers, whom they had to cope with in order to survive.

2 Japanese Scientific and Technological Achievements in the Meiji Period

To illustrate the point I am trying to make, I would like to mention some of the remarkable scientific and technological developments that took place in Japan from the mid-nineteenth-century onward – in universities, academic societies, government, and industries.

In 1862, a Shogunate navy officer was sent to Leiden. He was the first Japanese person to be sent to Europe for formal technical study, and many followed afterwards.[2] Learning science directly in Europe enabled these early Japanese students to learn not only Western scientific knowledge, but also scientific values and practice.[3] Science became an important subject in compulsory elementary education instituted in the 1870's. Courses named *'kyūri'* 窮理 (investigation of principles) taught about such things as vacuums, electricity and magnetism, thermometers and barometers. Fukuzawa Yukichi's 福澤諭吉 (1834-1901) textbook on introductory physics was published in 1868 and was adopted as a primary school text in 1872.[4]

In 1873, the College of Engineering (*Kōgakuryō* 工學寮, *Kōbu daigakkō* 工部大學校) was established by the Ministry of Engineering (*Kōbushō* 工部省) under the directorship of Henry Dyer (1848-1918), a British engineer, and staffed by eight other young engineers.[5] The college consisted of six departments: civil engineering, mechanical engineering, telegraphy, architecture, applied chemistry and metallurgy, and mining. It was "one of the world's first comprehensive institutions for basic science and engineering."[6] One of its teachers, William E. Ayrton (1847-1908), was the first professor of electrical engineering in the world.[7] By 1885 the college employed 49 Western professors and experts, and enrolled 478 students, of whom 211 graduated.[8] The engineering program of

2 Bartholomew, *Formation of Science in Japan*, 41.

3 Ibid., 78.

4 Morris-Suzuki, *Technological Transformation of Japan*, 80.

5 The charter for the College was granted in 1871. This and most other information in this paragraph is derived from Eikoh Shimao, "Some Aspects of Japanese Science," *Annals of Science* 46 (1989): 72-74.

6 Bartholomew, *Formation of Science in Japan*, 91. Indeed, the only precedent seems to have been the Polytechnic Institute in Zurich (ETH). See Shimao, "Some Aspects of Japanese Science," 73.

7 Morris-Suzuki, *Technological Transformation of Japan*, 82.

8 Shimao, "Some Aspects of Japanese Science," 74.

the college was later to be emulated by the British, who did not have such a polytechnic institute at the time.[9]

Engineering and science were important in the newly established Tokyo University (東京大學).[10] When it was founded in 1877, its college of science had departments of chemistry, mathematics, physics and astronomy, biology, geology, and engineering. It would be difficult to find many similar examples even among the Western universities at this early date. These science programs of the new university showed a steady growth. For example, the first chair in physics was established in 1879; the first three students graduated majoring in physics in 1882, and this number reached 104 by 1900. When Tokyo Imperial University (東京帝國大學) was inaugurated in 1886, the College of Engineering became part of it. In 1890, the Tokyo School of Agriculture was absorbed by the university, making it "the world's first university to include a School of Agriculture."[11]

There were a great number of Western teachers teaching science to Japanese students in colleges and other schools. Some 8,000 of them came from Europe and the United States between 1870 and 1900.[12] Again, I cannot think of many other countries which could have boasted that many science teachers in that period. Some Western scientists of a very high caliber spent a significant part of their early career in Japan; some began their career in Japan and established their names there. In addition to Dyer and Ayrton mentioned above, we can list James A. Ewing, John Milne, Robert E. Atkinson, and E. S. Morse. What is more, they were hired on the recommendation of first-rate European scientists, such as W. J. M. Rankine, Alexander W. Williamson, and even Lord Kelvin.[13]

It is significant, however, that the Japanese government had employed these foreigners to educate the Japanese so that they could eventually be replaced by Japanese teachers. And this ambitious goal was achieved quickly, the switch being more or less complete by about 1900. While Westerners occupied 12 of the 15 chairs in the Faculty of Science in Tokyo University in 1877, the number

9 See also W. H. Brock, "The Japanese Connexion: Engineering in Tokyo, London, and Glasgow at the End of the Nineteenth Century," *The British Journal for the History of Science* 14 (1981): 227-243.

10 For the information in this paragraph, see Shimao, "Some Aspects of Japanese Science," 74-77.

11 Ibid., 75.

12 Montgomery, *Science in Translation*, 219.

13 Bartholomew, *Formation of Science in Japan*, 64-66; Shimao, "Some Aspects of Japanese Science," 76-77.

decreased to only two of the 13 chairs in 1886.[14] Already in the 1880s, and increasingly in the 1890s, international-level research was being carried out in Japan in physics, chemistry and in medicine.[15] Kitasato Shibasaburo (北里柴三郎, 1853-1931), the world-class bacteriologist and founding director of the Research Institute for Infectious Diseases (founded in 1892) is a good example. Industrial research laboratories began to be established in the 1890s: Onoda 小野田 Cement Laboratory and Mazda Lamp Laboratory are good examples.[16]

Scientific societies were established in Japan in the 1870's.[17] The first was the Tokyo Mathematical and Physical Society (東京數學物理學會), which was founded in 1877. The Tokyo Chemical Society (東京化學會) followed the next year, and began publication in 1880. The year 1880 also saw the foundation of the world's first seismological society (地震學會) in Tokyo. These early dates are comparable to the dates of the foundation of major scientific societies in the United States: the American Chemical Society (1876); the American Mathematical Society (1888); the American Physical Society (1899).

Industrialization of the country proceeded relatively early and rather rapidly. The construction of the main electrical communication lines began in 1869, and was completed by 1876. It may be noted in this connection that Japan had an advantage in that it embarked on industrialization at a time when electrical power was beginning to be put to practical use.[18]

3 Japanese Awareness of Their Own Status as a Major World Power

Still more significant for my point, however, was the awareness among the Japanese themselves that their country was part of a keen competition with Western powers. The Japanese at the time were making comparisons of various aspects of their science and technology with those of the leading European countries. In 1900 a House representative complained that Japan, with its 40 million people, had only two universities whereas Germany had over 20, and Britain had more than five, without hiding his feeling that Japan should match these leading European countries. Two years later another representative

14 Shimao, "Some Aspects of Japanese Science," 76.

15 Bartholomew, *Formation of Science in Japan*, 176ff.

16 By the 1930s the number of industrial research laboratories reached about 300. Morris-Suzuki, *Technological Transformation of Japan*, 110.

17 Bartholomew, *Formation of Science in Japan*, 86.

18 Morris-Suzuki, *Technological Transformation of Japan*, 111. Showing the great pace of the electrification of Japan, in 1935 more Japanese households (89 percent) had access to electrical lighting than in Germany or the United States. Ibid., 130.

lamented the small number of Japanese university students compared with European countries.[19] In 1900 in the House of Peers, a former vice-minister of education contrasted Japan's educational budget with those of France and Germany: while these Western countries were spending about 8 percent of their national budget on public education, Japan was spending only 5 percent, which he found unacceptable.[20]

Japanese of the period did not simply copy science from a single European country, from, say, Germany. Until the Meiji era, Dutch science was the most important, but from the 1860's, the "Japanese began to notice English and American naval power, French chemistry and astronomy, German physics and medicine."[21] The early Japanese universities were not modeled exclusively on German universities, as is often assumed; the Japanese saw certain disadvantages in the German model, and were also attracted to the centralized French model.[22] For their patent system they even adopted the Spanish model because they thought it was less restrictive.[23] Also, they sought a connection with the foremost scientists of Europe, without confining their search to any one field or to any one country. Indeed, the role models of early Japanese scientists were not the foreign scientists working in Japan, but rather the first-rate Western scientists themselves.[24]

By the late nineteenth century, there was a feeling among many Japanese that Japan had now become self-sufficient in its technological power. Already in 1860, Fukuzawa Yukichi found "really nothing new" in San Francisco, which he visited as a member of an official delegation.[25] This awareness may have been behind his later scheme of 'Leaving Asia' (*datsu-a* 脱亞).[26] In the final years of the nineteenth century, some Japanese even argued for a 'Japanization' as opposed to the previous trend of 'Westernization,' revealing their confidence that Japan had now become more or less self-sufficient in terms of technology.[27] Although, on the whole, Japanese still clung to the slogan 'Westernization of Japan,' this was merely because such a position was

19 Bartholomew, *Formation of Science in Japan*, 111.

20 Ibid., 150-151.

21 Montgomery, *Science in Translation*, 216.

22 Bartholomew, *Formation of Science in Japan*, 97-98.

23 Ibid., 132.

24 Ibid., 75ff.

25 Morris-Suzuki, *The Technological Transformation of Japan*, 67.

26 Fukuzawa's influential essay on 'Leaving Asia' (*Datsu-a ron* 脱亞論) is translated in *Japan: A Documentary History: The Late Tokugawa Period to the Present*, ed. David John Lu (Armonk, New York: M.E. Sharpe, 1996), 351-353.

27 Montgomery, *Science in Translation*, 223.

diplomatically convenient, as well as a way to prevent the Western powers from feeling threatened.[28]

This growing awareness of Japan's status was not restricted to the Japanese, either. Europeans also began to take Japan seriously as a competitor in scientific development, and as an industrialized imperial power on a par with them. Already in the 1870's such feelings began to be expressed. In 1878 a British journalist referred to Ayrton's laboratory in Tokyo as "the finest physical laboratory which exists perhaps in the world."[29] In the same year we also find James Maxwell's remark concerning Ayrton and Perry's (John Perry, 1850-1920) collaborative work that it was "threatening the centre of electrical development."[30] These are only a few of many contemporary examples of admiration expressed regarding Japan's achievement.[31] In the 1900's, after Japan's victory in the war with Russia, there were attempts by many Westerners to understand the extraordinary military and economic success of Japan.[32] The Japanese themselves also felt that access to European industrial research facilities became more difficult after Japan's victory in the Sino-Japanese war, as the Europeans were on guard against the competition represented by an industrializing, and already industrialized, Japan.[33]

4. Historical Background

We may speculate about the factors that could provide explanations for this difference between Japan and other East Asian countries, although my purpose in this essay is not to explain, or find the causes behind, this difference, but to point out that such a difference existed and that one should have this difference in mind in trying to understand what actually happened. Perhaps

28 Graeme J. N. Gooday and Morris F. Low, "Technology Transfer and Cultural Exchange: Western Scientists and Engineers Encounter Late Tokugawa and Meiji Japan," *Osiris* 13 (1998): 102.

29 [Anonymous], "A Visit to Professor Ayrton's Laboratory," *Japan Weekly Mail* (October 1878): 1129, quoted in Gooday and Low, "Technology Transfer and Cultural Exchange," 117.

30 Quoted in ibid., 118.

31 Brock, "The Japanese Connexion." In 1897, a British journal, *Engineer*, published a feature article titled "Pioneers of Modern Engineering Education in Japan," and the journal noted the contribution of the Japanese themselves, and not only Western experts, to the remarkable achievement of engineering education in Japan in these decades. Gooday and Low, "Technology Transfer and Cultural Exchange," 99.

32 Ibid., 114.

33 Bartholomew, *Formation of Science in Japan*, 144.

to find what readied the soil for Japan's scientific and technological progress, we could turn to the developments that had already occurred in the Tokugawa 徳川 period (1603-1867): urbanization, industrial development, agriculture and seagoing technology, for example. The result was "the relatively narrow gap between imported technologies and endogenous [Japanese] know-how," owing to the developments inside Japan before its doors opened to the outside world.[34] Furthermore, the foundations had been laid in the second half of the eighteenth century for the translation of Western scientific works, especially after Shogun Yoshimune (吉宗, r. 1716-1745) lifted the ban on technical literature from Europe.[35] The presence, and the awareness, of the threat of Western imperialists acted as a stimulus for the sense of competition also. One can even point to the presence in Japan of an alternative leadership that was able to step in when the inability of the ruling regime to defend the country discredited it, as W. T. de Bary noted in connection with the Meiji Restoration leaders.[36] Also, one is tempted to find some significance of earlier periods in the history of Japan, which went down paths that were considerably different from those of the other East Asian countries and in many ways similar to the West. Feudalism and the tradition of military rulers are among the examples that come up most frequently.

Along with these, one can also point to differences in the ways the Japanese adopted Western science. For example, Western astronomy in the seventeenth century did not infiltrate government in Japan as it did in China and Korea. Also, Japanese intellectuals first studied Western science not through physics but through medicine. But what is more important are the differences in the Japanese perception of Western science. For example, whereas the Chinese and the Koreans tried to incorporate Western science into the traditional philosophical system, the *Yijing* 易經 system for example,[37] the Japanese took Western science into a new context. In other words, they created a new system around, and based on, Western science.

---

34 Morris-Suzuki, *The Technological Transformation of Japan*, 60.

35 Montgomery, *Science in Translation*, Chap. 6.

36 William Theodore de Bary, *East Asian Civilizations: A Dialogue in Five Stages* (Cambridge, MA: Harvard University Press, 1988).

37 Willard J. Peterson, "Fang I-chih: Western Learning and the 'Investigation of Things,'" in *The Unfolding of Neo-Confucianism*, ed. William Theodore de Bary (New York: Columbia University Press, 1975), 369-411; Yung Sik Kim, "Western Science, Cosmological Ideas, and the *Yijing* Studies in Seventeenth- and Eighteenth-Century Korea," *Seoul Journal of Korean Studies* 14 (2001): 299-334; Benjamin A. Elman, *On Their Own Terms: Science in China, 1550-1900* (Cambridge: Harvard University Press, 2005).

Undeniably, there was an earlier Japanese attempt to incorporate Western science into a traditional Chinese framework.[38] But in the second half of the nineteenth century, especially after the Meiji Restoration, this changed. On the whole, the Chinese philosophical system had largely lost its appeal among the Japanese. There was even a growing contempt among many Japanese for what they considered "the antiquated barbarism of China."[39] We may also highlight the general attitude around the 1800's that already "saw Western science as significantly superior to Chinese 'technique,'" and the so-called 'National Learning' (*kokugaku* 國學) movement of the eighteenth century with its tendency away from China in favor of Japanese native learning.[40] Such an attitude did not come up in China or in Korea till much later.[41] It is of symbolic significance that the modern term for science, '*kagaku*' 科學, was coined in Japan in the early 1870's, and was quickly adopted both in Japan and in China, replacing the traditional Chinese term '*kakuchi*' 格致.[42] Granted, later, during World War II, there was a realization that Japan should follow a different path of scientific development from the West.[43] However, this came after a long period of all-out efforts at Westernization, which even included a campaign for beef-eating as a mark of civilized status among the Japanese.[44]

5 Reflections on the Present Situation

The mentality of many Japanese today, who seem to be reluctant to consider themselves as part of East Asia, may reflect their unconscious, intuitive under-

38 Wai-ming Ng, "The *I Ching* and the Adaptation of Western Science in Tokugawa Japan," *Chinese Science*, no. 15 (1998): 94-117.

39 Prasenjit Duara, *Rescuing History from the Nation: Questions and Narratives of Modern China* (Chicago: University of Chicago Press, 1995), 37.

40 Montgomery, *Science in Translation*, 213-214.

41 One can also note in this regard that the kind of attitude expressed in the slogan "Chinese substance and Western function" (*zhongti xiyong* 中體西用) – "*wakon yōsai*" 和魂洋才 (Japanese spirit and foreign talents) in Japanese – was less prominent in Japan compared to China and Korea, where the slogan was "*tongdo sŏgi*" 東道西器 (Eastern Way and Western tools). See Yung Sik Kim, "Some Reflections on Science and Technology in Contemporary Korean Society," *Korea Journal* 28.8 (1988): 6.

42 It is not clear exactly when and by whom this term was coined. But many, including Eikoh Shimao, trace the first use of the term to a magazine article by Nishi Amane (西周, 1829-1897) in 1874: Shimao, "Some Aspects of Japanese Science," 71.

43 Morris-Suzuki, *The Technological Transformation of Japan*, 145.

44 Duara, *Rescuing History from the Nation*, 22.

standing of this feature of their own early modern situation, which I have been discussing. But, on the whole, the Japanese do not seem to grasp this situation consciously and fully. Many Japanese believe that early modern, Westernized Japan was doing more or less the same thing as its neighbors, China and Korea, competing with them but doing it better, faster, and more efficiently than them. They do not realize that actually early modern Japan was doing something completely different.

Admittedly, Japan continued to accept and assimilate Western science and culture. However, after a while, Japan was no longer simply absorbing the successes the Western powers had already achieved. Japan itself had become one of the 'Western' powers, competing with them, and imposing itself upon – and threatening – its own neighbors in East Asia.

The failure of many Japanese to comprehend their own early modern background even seems to underlie the issues that some Japanese people have with their East Asian neighbors. For they fail to see the reality of the situation – that, far from being a part of the community of East Asian countries, Japan was actually playing the role of a 'Western' power imposing upon their East Asian neighbors. Earlier, this lack of understanding had given rise to the notion of 'the Great East Asia' (*Daitōa* 大東亞), or 'the Greater East Asian Co-Prosperity Sphere' (*Daitōa kyōeiken* 大東亞共榮圈),[45] with Japan as part of it, and had justified their invasion and exploitation of other East Asian – and then Southeast Asian – countries; some Japanese had even viewed their acts of aggression as acts of protection of their neighbors against the intrusion of Western powers.

Nor did Chinese and Koreans have a clearer grasp of this situation. They did not realize that Japan in the second half of the nineteenth century was not doing the same thing as they were; they did not recognize that Japan was competing not with the other East Asian countries but really with the West. Failure of the Chinese and the Koreans to see early modern Japan for what it really was – a newly emerged 'Western' power – aggravated an already troublesome situation. This lack of understanding has created among Japan's East Asian neighbors a peculiar kind of animosity directed at Japan, but not at the Western powers, for the threat and imposition of Japan must have been all the more frustrating for Koreans and Chinese coming as they did not from a 'foreign' power, but from their own neighbor. This has also blinded them to a major factor that lay behind the real situation: the development in the Tokugawa period

45 On the "the Greater East Asian Co-Prosperity Sphere," see Joyce C. Lebra, *Japan's Greater East Asia Co-Prosperity Sphere in World War II: Selected Readings and Documents* (New York: Oxford University Press, 1975).

that prepared Japan for its subsequent success. Instead, many Koreans and Chinese have tended to ignore this, considering the Japanese success to be the result of a lucky, even accidental, case of rapid adoption and imitation.

6 Concluding Remarks

Finally, I have some words of qualification and caution before ending this essay. I am not saying that Japan is not East Asia. It is. However, it is a particular kind of East Asia. And it provides yet another reason why we should not accept the notion of 'the Great Divide' of the West versus East Asia, the habit of considering everything in East Asia (irrespective of the period or geographical differences) on one side and putting everything in the West on the other side.[46] As we have seen, the notion of this divide, which had its origin in the West, most notably in the 'Orientalisms' of the Europeans, seems to have prevented East Asians – and not only the Chinese and Koreans, but the Japanese people themselves – from seeing the truth of the scientific and industrial development in their own part of the world in the late nineteenth century.

46 For a critique of the idea of the 'Great Divide,'" see Roger Hart, "Beyond Science and Civilization: A Post-Needham Critique," *East Asian Science, Technology, and Medicine* 16 (1999): 88-114.

CHAPTER 12

The Problem of China in the Study of the History of Korean Science: Korean Science, Chinese Science, and East Asian Science*

Because of the Chinese origin of many ideas and artifacts of traditional Korean science, and the overwhelming Chinese influence on its development, the history of Korean science suffers from a kind of identity crisis, based on the issue of whether traditional Korean science should be considered simply a part of Chinese science. Jeon Sang-woon, for example, begins his account of traditional Korean science by referring to its history as "a tributary of the mainstream of scientific developments in traditional China."[1] But Jeon quickly adds that Koreans did not merely adopt ideas and techniques from China, but modified them and adapted them to their own needs and conditions. This led to many 'independent' developments of Korean science, and often to new discoveries and inventions. Much of the previous work – and indeed most of the early works by modern scholars – on traditional Korean science have focused on the new contributions by Koreans and their differences from the Chinese developments.[2]

Yet this raises a problem, which has been troubling me when I attempt to think in general terms about the history of traditional Korean science. The problem, which I call "the problem of China in the study of the history of Korean science," is basically about the significance of the scientific ideas and technical artifacts from China that are apparently pervasive throughout traditional Korean science and technology.[3] What roles and meanings should be assigned to such ideas and techniques in the history of Korean science? Would it be reasonable, or permissible, for a historian of Korean science to neglect

---

* Originally published as "The 'Problem of China' in the Study of the History of Korean Science: Korean Science, Chinese Science, and East Asian Science," *Disquisitions on the Past and Present,* no. 18 (2008): 185-198.

1 Sang-woon Jeon, *Science and Technology in Korea: Traditional Instruments and Techniques* (Cambridge, Mass.: MIT Press, 1974), 1.

2 Yung Sik Kim, "Problems and Possibilities in the Study of the History of Korean Science," *Osiris* 13 (1998): 48-79.

3 Although I have come to realize that the problem is more or less common to all the other branches of Korean history, to me it appears more pronounced in the history of science. I am not sure, however, whether that is indeed the case, and, if so, why.

 | DOI 10.1163/9789004265318-014

them and to study only what is uniquely Korean? It would of course be understandable if a foreigner were to be interested in knowing mainly those elements and aspects that were unique to Korea. But should – or could – the Korean historian be the same? These are the questions I was first confronted with as I began to think about this 'problem of China.' In discussing these questions in the present paper, however, I will be posing a good number of other questions. Indeed, rather than trying to come up with solutions for them, I will present the problem and share my concerns about these issues.

1 Some Key Questions

I became aware of this problem while supervising the dissertation research of my graduate students on topics in Korean science. I found very often that large parts of the writings of traditional Korean thinkers or scientists which my students were reading consisted of excerpts and collations of earlier Chinese writings. Of course, this does not discredit the scholarship of these traditional Korean thinkers who wrote them. What they were doing was essentially the same kind of thing that many of their contemporaries in traditional China were doing. But it posed problems for my students, and then for myself.

Expressed in a blunt manner, the problem is as follows: How should the historian of Korean science treat those ideas and techniques that were straightforward importations from China? In other words, how should the historian deal with the theories and practices that were exactly the same as those of China except for the fact that they were discussed and carried out in Korea geographically? If the historian restricts himself to what is uniquely Korean, he will have to leave out much that is actually Korean, but not uniquely so. Conversely, if the historian includes all that took place in Korea, he will have to deal with much that is no different from what was happening in China. Is this a legitimate way to study and understand traditional Korean science?[4]

One might agree with Nakayama Shigeru, who, while speaking of "the primary sources of science that were...written in classical Chinese by people in Korea, Vietnam, and Japan," once said: "One can speak of all these sources as Chinese science."[5] However, even if we follow Nakayama and refer to these

4 Language cannot be used as a distinguishing factor for the historian, because traditional Korean scholars/scientists invariably used classical Chinese for their writings. The situation was more or less like Arabic in medieval Islam, or Latin in the Christian medieval West.

5 Shigeru Nakayama, "History of East Asian Science: Needs and Opportunities," *Osiris* 10 (1995): 86.

sources as "Chinese science," there are still fruitful questions which can be asked by the historian of Korean science. Through what channels were those scientific ideas and technical artifacts transmitted to Korea? To what extent did Koreans master – understand, adopt, and assimilate – the ideas, techniques, and artifacts that had arrived from China? Did Koreans accept them individually or as parts of a whole? And how much were they committed to that whole body, system, or corpus of Chinese science? Frequently, it appears that Koreans adopted Chinese systems in their entirety. But was this always the case? Which among them did Koreans prefer, and why? Indeed, these are questions that can shed more light on the nature of scientific and technical activities in traditional Korea, and their place in traditional Korean culture and society.

But what about the actual content of these ideas and techniques that were the same as those of China? Should the historian of Korean science ignore them? These ideas and techniques were of course mostly a continuation, duplication, or repetition of those of China. But can we conclude that they are therefore of no importance to the historian of Korean science? To him such materials are clearly different – different from the similar ones written or used by the Chinese. But what is the significance of this difference?

Sometimes a historian of Korean science keenly studies such material, written by a given Korean scholar, without realizing that the same content had already been written by an earlier Chinese scholar. Now what happens, or what should happen, when the historian finds out that what he thought was Korean was originally Chinese? Clearly, the significance of the material to the historian will change the moment he finds that out. What is the nature of this change and how should the historian deal with the material?

Now, we can turn the question by 180 degrees and ask what the significance of such materials is to the historian of Chinese science. Should he treat the same contents differently (either as more important or as less important) because they were written by a non-Chinese person outside China, and had been written by a Chinese earlier? Or should he treat them in the same way as he would treat similar contents written somewhere in China by a Chinese person? And at this point, one is led to ask whether the history of Korean science becomes a local history, or peripheral history, of Chinese science.[6] Or one may

6 A very interesting case comes to my mind in this connection. For some years, Wann-sheng Horng's (Hong Wansheng 洪萬生) research group on the history of mathematics at National Taiwan Normal University have been studying mathematical works written by eighteenth- and nineteenth-century Koreans. And since so few Korean historians study such materials, his group is the largest group in the world working on traditional Korean mathematics. Now

try to evade these questions altogether by wondering whether the term 'East Asian science' might resolve the matter.

2 The Problem of 'Center versus Periphery'

The problem mentioned above can be seen as an example of the issue of 'center versus periphery' in the history of science, in which China is viewed as the center and Korea is a periphery. A few things come to mind when we look at the history of Korean science from this perspective.

2.1 *Chinese 'Center' versus Korean 'Periphery'*

Firstly, China may be too large geographically to call it a 'center.' Indeed, there may be centers and peripheries within China itself. Various localities in China can be regarded as centers and peripheries. For example, Beijing and the Jiangnan 江南 region can both be considered as centers during various times in history. In the early fifteenth century, and the second half of the eighteenth century, Korea could also be considered as one such center of 'Chinese' science. Even within Korea, one might argue that there were centers and peripheries: the court of King Sejong 世宗 (r. 1418-1450) during the first half of the fifteenth century, and the metropolitan Seoul area in the eighteenth century are two examples.

When we think of periods like these, my problem – the 'problem of China in the history of Korean science' – can be turned upside down and become the 'problem of Korea in the history of Chinese science.' When a historian of Chinese science finds an original, or unique, development in Korea, how should he deal with it? Should he deal with it differently from when a similar original or unique development occurred in a particular locality of China? In other words, should unique developments in science be given special attention when they arise in Korea but should simply be taken for granted when they appear in some locality in China? This last question often arises both when I encounter overenthusiastic responses from Korean historians concern-

do they view their work as the history of Korean mathematics, Chinese mathematics, or East Asian mathematics? See, e.g., Wann-Sheng Horng, "Sino-Korean Transmission of Mathematical Texts in the 19th Century: A Case Study of Nam Pyong-gil's *Kugo Sulyo Tohae*," *Historia Scientiarum* 12 (2002): 87-99; Hong Wansheng (Horng Wann-Sheng 洪萬生), "Shiba shiji dongsuan yu zhongsuan de yiduan duihua: Hong Zhengxia vs. He Guozhu" 十八世紀東算與中算的一段對話: 洪正夏 vs. 何國柱 [A Dialogue between Korean and Chinese Mathematics in the Eighteenth Century: Hong Chong-ha versus Ha Kuk-chu], *Hanxue yanjiu* 漢學研究 20 (2002): 57-80.

ing a Korean priority (a discovery or invention made first in Korea before anywhere in the world, and especially before China) and when I discern difficulties that some Chinese historians have in accepting such Korean priority.[7]

Although such cases of Korean priorities have attracted more attention from Korean historians, the more common instances, in which developments that had first occurred in China occurred in Korea at a later time, are equally interesting, and at times yield even more interesting insights into the Korean, and the Chinese, situations. Examination of these cases will show, for example, how differently (or similarly) the Koreans responded to a situation while knowing what the Chinese had done in the same situation earlier. Lim Jongtae's study of the surge, among eighteenth-century Korean scholars, of the sort of cosmological speculations found among the Chinese scholars of the Ming-Qing transition period enhanced our understanding of the intellectual climates of both seventeenth-century China and eighteenth-century Korea.[8] We may also note that many seventeenth- and eighteenth-century Korean scholars quoted and discussed the views of Chinese scholars of the Song, Yuan, and Ming periods in the same way as they would have discussed the views of their contemporaries.

There is yet another aspect of the 'problem of China.' It surfaced during a dissertation committee meeting for a student of mine, who wrote an excellent dissertation, entitled "Assimilating the Jesuits' Geographical Knowledge in Seventeenth- and Eighteenth- Century Korea." As the student delved more deeply into the Chinese background of the Korean sources, his dissertation expanded to cover a substantial amount of Chinese materials, including those of earlier periods. However, a colleague on his dissertation committee, a member of the Korean History department, protested by saying, "This is not Korean history." A historian of China on the committee then rebutted by saying, "This

7 I might add, in this connection, that Korean historians' tendency to search for something uniquely Korean arose mainly as a reaction to the historiographical tendency of the Japanese colonial period, which tended to stress the negative aspects of Korean history and criticize in particular its lack of originality and its dependence on China. But part of the reason for the persistence of this tendency among Korean historians is the widespread feeling that Korean contributions are still being ignored or neglected by contemporary Chinese and Japanese historians. See Seong-Rae Park, "Pride and Prejudice in the Hitoriography of Science in East Asia," in *Current Perspectives in the History of Science in East Asia*, ed. Yung Sik Kim and Francesca Bray (Seoul: Seoul National University Press, 1999), 3-12.

8 Im Chong-t'ae 林宗台 (Lim Jongtae), *Sipch'il sipp'al segi Chungguk-kwa Chosŏn-ŭi sŏgu chirihak ihae* 17,18世紀 中國과 朝鮮의 西歐 地理學 理解 [Seventeenth- and Eighteenth-Century Korean and Chinese Scholars' Interpretation of the Western Geographical Knowledge] (Seoul: Changbi 創批, 2012), Chap. 3, esp. 162-199.

is exactly how we should study Korean history." In the end the dissertation was approved with much praise, but a word was added to the title of the dissertation: "Assimilating the Jesuits' Geographical Knowledge in Seventeenth- and Eighteenth- Century Korea *and China*." Thus, what the student wrote was proclaimed as not just 'Korean' history but as 'Korean and Chinese' history. To be sure, this did not mean that Korean history should not deal with what was happening in China. But it certainly carried an implicit message that the Korean historian does not have to look into what was happening in China during, or before, the events in Korea that he studies.

2.2 *Dominance of the Chinese 'Center'*

On the whole, however, the Chinese dominance in East Asian science was almost absolute. As Nakayama has pointed out, China was the 'center' where significant developments almost always took place first and subsequently spread to the 'peripheries.'[9] Granted, one may point to different geographic areas of China where there did appear some degree of shifting of centers and peripheries. However, these shifts were never as frequent or oscillatory as they were in the West, for example. And in no field was this dominance as complete as in science. While a few other countries in East Asia could gain temporary superiority over China in political, military or economic aspects, Chinese dominance in science was hardly ever challenged.

Due to this absolute supremacy, the Chinese centrality was accepted by people living at the peripheries of the Chinese cultural world. This is best illustrated by the acceptance, among eighteenth-century Koreans, of the so-called theory of 'the Chinese origin of Western learning' (*xixue zhongyuan* 西學中源), the theory that the Western scientific ideas that came to China had their origins in ancient China. According to this theory, the Chinese already had knowledge of them in the ancient golden ages, but such knowledge was lost later and fell into the hands of the Western barbarians, who developed it further and brought it back to China.[10] Many eighteenth-century Korean scholars

9 Shigeru Nakayama, "The Spread of Chinese Science into East Asia," Kim and Bray, *Current Perspectives in the History of Science in East Asia*, 13-20.

10 On the theory of 'the Chinese origin of Western learning,' see John B. Henderson, "Ch'ing Scholars' Views of Western Astronomy," *Harvard Journal of Asiatic Studies* 46 (1986): 139-143; Jiang Xiaoyuan 江曉原, "Shilun Qingdai xixue zhongyuanshuo" 試論清代西學中源說 [An Essay on the Theory of the Chinese Origin of Western Learning in the Qing Period], *Ziran kexueshi yanjiu* 自然科學史研究 [Studies in the History of Natural Sciences] 7 (1988): 101-108; Wang Yangzong 王揚宗, "Xixue zhongyuan shuo zai MingQing zhi ji de youlai ji qi yanbian" '西學中源"說在明清之際的由來及其演變,' *Dalu zazhi* 大陸雜誌 (Taipei 臺北) 90 (1995): 39-45.

accepted this theory, and due to the historical circumstances they faced, many of them became ardent advocates of it. After the humiliating defeat of their country to the Manchus in early seventeenth century, anti-barbarian sentiments became widespread among Korean Confucians. Since the Manchu barbarians had already conquered the Chinese heartland, many Korean Confucians of the time came to consider themselves to be the true bearers of the 'Central Splendor' (*Zhonghua* 中華, *chunghwa* in Korean), or the 'Korean Central Splendor' (*Chosŏn chunghwa* 朝鮮中華). The '*chonjuron*' 尊周論, or the tendency to admire the ancient Zhou 周 culture – the culture that was considered ideal by many neo-Confucians – became predominant among Korean scholars.[11] In this atmosphere, the need to rationalize the barbarian knowledge of Western science for the 'Central Splendor' became a concern of particular urgency: the theory of 'Chinese origin' provided a good rationale for such legitimization.[12]

Here the 'problem of China' manifests itself to an extreme degree. The mentality of these Korean scholars was exactly the same as that of the Chinese, to such an extent that they would identify themselves with Chinese culture and attribute Chinese, and not Korean, origin to Western ideas. In some cases Koreans showed an even stronger adherence to the Chinese tradition than the Chinese themselves. And as a result, an interesting departure from the situation in China took place. Whereas for the seventeenth-century Chinese the similar military humiliation gave rise to some critical, soul-searching attempts at a re-examination and purification of their own tradition through critical studies of the classics, many of their Korean contemporaries tended to cling even more strongly to the classical tradition. While in China there was a growing rejection of the Cheng-Zhu 程朱 school of Song-Ming neo-Confucianism,

11 On the Korean *Zhonghua* theory and the '*chonjuron*,' see Chŏng Ok-cha 鄭玉子, *Chosŏn hugi Chosŏn Chunghwa sasang yŏn'gu* 朝鮮後期 朝鮮中華思想硏究 [A Study of the Idea of the Korean *Zhonghua* in Late Choson] (Seoul: Ilchisa 一志社, 1998).

12 Pak Kwŏn-su 朴權壽, "Sŏ Myŏng-ŭng-ŭi yŏkhakchŏk ch'ŏnmun'gwan" 徐命膺의 易學的 天文觀 [Sŏ Myŏng-ŭng's Interpretation of Astronomy in the Framework of the *Yijing* Studies], *Han'guk Kwahak-sa Hakhoe-ji* 韓國科學史學會誌, [Journal of the Korean History of Science Society] 20 (1998): 93. Even the way Qing culture was accepted could be seen as a manifestation of Chinese origin theory: a number of Korean scholars, especially those who belonged to the 'Northern Learning school' (*pukhakp'a* 北學派), saw no problem in accepting Qing culture since they considered it as the 'remaining institution' (*yuje* 遺制) of the ancient 'Chinese Splendor.' On the 'Northern Learning school,' see Yu Ponghak 劉奉學, *Yŏnam ilp'a pukhak sasang yŏn'gu* 燕巖一派 北學思想 硏究 [Studies of the Ideology of 'Northern Learning' of the School of Yŏnam (Pak Chi-wŏn)] (Seoul: Ilchisa 一志社, 1995).

in Korea development of the orthodox Zhu Xi 朱熹 learning continued and deepened, at least for some time. Even more remarkable is the difference in the Chinese and Korean scholars' attitudes toward the study of the *Yijing* 易經 (*yixue* 易學). Whereas the Qing Chinese were attempting to purge it of some of its excessive, contaminating elements – such as the numerological speculations of Shao Yong 邵雍 (1011-1077) and Cai Yuanding 蔡元定 (1135-1198) – Koreans like Kim Sŏk-mun 金錫文 (1658-1735) and So Myŏng-ŭng 徐命膺 (1716-1787) proceeded to develop these elements further.[13]

It is interesting to note, however, that even under such circumstances Koreans did not seem to feel that they were full members of Chinese learned circles. Chinese scholars could read the writings of Korean scholars of the Chosŏn period because they were written in classical Chinese. Yet in reality, the Chinese rarely paid attention to Korean texts. Even Korean authors always seemed to be writing with only Korean readers in mind. Occasional recognition by the Chinese did make Koreans proud of their achievements, but there were few such instances.[14] Some Korean scholars who visited Beijing tried to take advantage of the occasion to impress Chinese scholars. Pak Chi-wŏn 朴趾源 (1737-1805), for example, in preparing his meetings with Chinese colleagues, chose topics of conversation which he thought would impress them. What he chose, after some deliberation, was the idea of the earth's rotation and the question of what a man on the moon would see if he looked at the earth from the moon.[15] Would provincial – and therefore also peripheral – Chinese scholars have shown similar attitudes towards the scholars in Beijing or other Chinese centers of learning?

While the Korean and Chinese situations display an overwhelming amount of parallels in the traditional period, differences become more pronounced during the period of the transmission of Western science to Korea. Perhaps it

---

13 John B. Henderson, *The Development and Decline of Chinese Cosmology* (New York: Columbia University Press, 1984), 184-193; Moon Joong Yang, "Traditional Cosmology Associated with the *I-ching* and Anti-Cosmological Discourses in 18th-Century Korea," *Seoul Journal of Korean Studies* 12 (1999): 177-227; Yung Sik Kim, "Western Science, Cosmological Ideas, and the *Yijing* Studies in Seventeenth- and Eighteenth-Century Korea," *Seoul Journal of Korean Studies* 14 (2001): 299-334.

14 Hong Tae-yong 洪大容 (1731-1783) and Kim Chŏng-hŭi 金正喜 (1786-1856) might have been a couple of such exceptions. See Kim T'ae-jun 金泰俊, *Hong Tae-yong p'yŏngjŏn* 洪大容評傳 (A Critical Biography of Hong Tae-yong) (Seoul: Min'ŭmsa 民音社, 1987), Chapters 5-6; Benjamin A. Elman, *From Philosophy to Philology; Intellectual and Social Aspect of Change in Late Imperial China* (Cambridge, Mass., Harvard University Press, 1984), 155-156.

15 Kim T'ae-jun, *Hong Tae-yong p'yŏngjŏn*, 252-259.

is for this reason that recently more work is being done on the late Chosŏn period, when the influence of Western science became prevalent and divergence from China became evident. Nevertheless, China played the role of center even as a channel for the introduction of Western science to the East Asian world. For example, until well into the nineteenth century, Koreans relied almost entirely on Chinese books to learn about Western science. In fact, the Korean dependence on China for books – not just Western books, but all books – grew in the late Chosŏn period. In the eighteenth century, when the books imported from China increased rapidly in number and became more widely available, historians note a considerable decrease in interest in the collected works of Korean scholars, which continued to be printed in Korea.[16]

The notion of the Chinese centrality must have been so deeply ingrained in Korean minds that Koreans rarely tried to learn Western languages.[17] They must have felt that what they were receiving from China was enough for them. Also, some Koreans might have accepted Western scientific knowledge largely because the Chinese had already approved it. For instance, when the Korean government decided to adopt the *Shixianli* 時憲曆 (*Sihŏn-ryŏk* in Korean), it was not so much a decision to adopt a Western calendar as a decision to adopt the new calendar that had already been adopted by the Chinese government.[18] In this sense, the adoption of the *Shixianli* can be seen as a mere continuation of the long-term practice of the Chosŏn government adopting a newly established official Chinese calendar. On the other hand, one may wonder whether it was easier for Koreans than for the Chinese to accept Western ideas anyway because Koreans were already very much accustomed to accepting Chinese

16 Kang Myŏng-gwan 姜明官, "Chosŏn hugi sŏjŏk-ŭi suip yut'ong-kwa changsŏga-ŭi ch'ulhyŏn: sipp'al sipku segi Kyŏnghwa sejok munhwa-ŭi han tanmyŏn" 朝鮮後期 書籍의 輸入·流通과 藏書家의 出現 – 18,19世紀 京華世族 文化의 한 斷面 [Importation and Distribution of Books and Emergence of Book Collectors in Late Choson: An Aspect of the Culture of Major Scholarly Families in Eighteenth- and Nineteenth-Century Seoul], *Minjok munhaksa yŏn'gu* 民族文學史研究 9 (1996): 171-194.

17 Pak Sŏng-rae 朴星來 (Seong-Rae Park), *Han'guksa-edo kwahak-i innŭn'ga* 韓國史에도 科學이 있는가 [Is There Science in Korean History Also?] (Seoul: Kyobo mun'go 教保文庫, 1998): 204-206.

18 On the adoption of the *Shixianli* by the Chosŏn government, see Chŏn Yong-hun 全勇勳 (Yong Hoon Jeon), "Sipch'il-sipp'al segi sŏyang kwahak-ŭi toip-kwa kaltŭng: *Sihŏn-ryŏk* sihaeng-kwa chŏlgi paech'ipŏp-e taehan nollan-ŭl chungsim-ŭro" 17-18 世紀 西洋科學의 導入과 葛藤 – 時憲曆 施行과 節氣配置法에 對한 論難을 中心으로 [Conflict over the Introduction of Western Science in the Seventeenth and Eighteenth Centuries: Adoption of the *Shixianli* and Controversies over the Arrangement of the (Twenty-four) Seasonal Points], *Tongbang hakchi* 東方學志 no. 117 (2002): 1-49.

ideas, which were also 'foreign.' Indeed, Korean scholars of the time used the same expressions – 'calendrical books' (*lishu, yŏksŏ* 曆書) and 'calendar specialists' (*lijia, yŏkka* 曆家) – in referring both to Chinese and to Western astronomical books translated into Chinese and to their authors with Sinicized names.

2.3 *Initiatives at the Korean 'Periphery'*

Admittedly, there were times when Koreans exerted concentrated efforts to find, study, develop, and establish what was uniquely their own. In particular, during the reign of King Sejong in the early fifteenth century, such efforts were made in many areas.[19] The project of devising the system of *han'gŭl* 한글, the Korean alphabet originally named '*hunmin chŏngŭm*' 訓民正音 (Correct Sounds for Edifying People), is the most important and characteristic fruit of such attempts. Projects of a similar nature were carried out in many scientific subjects. For example, in the field of calendrical astronomy, Sejong's astronomers succeeded for the first time in Korean history in preparing a calendrical system based on the latitude of Seoul, which replaced the one based on the latitude of the Chinese capital that had been used until then. With this new system, named *Ch'ilchŏngsan* 七政算, they were able to predict the solar and lunar eclipses of 1447. There were many important achievements in medicine, of which the compilations of two medical treatises – the *Hyangyak chipsŏngbang* 鄉藥集成方 (Collected Prescriptions of [Korean] Native Medicines) and the *Ŭibang yuch'w i* 醫方類聚 (Classified Compilation of Medical Prescriptions) – are generally considered as representing the culmination of Koreans' efforts to develop their own system of medical knowledge, distinct from that of China. There was also remarkable progress made in the field of agriculture: new theories, techniques, and other practices were developed, most of them geared towards the needs of the newly adopted intensive and continuous cultivation. These efforts are well documented in the *Nongsa chiksŏl* 農事直說 (Straight Accounts of Agriculture). Other areas that showed similar progress in the period were music, geography, and cartography. Techniques in such areas as printing, paper-making, and firearms, and even forensic medicine, also improved greatly during this period.

---

19 Chŏn Sang-un 全相運 (Sang-woon Jeon), "Chosŏn chŏn'gi-ŭi kwahak-kwa kisul – Sibosegi kwahakkisulsa yŏn'gu chaeron" 朝鮮前期의 科學과 技術 – 15세기 科學技術史 研究 再論 [Science and Technology in Early Choson: Another Discussion on the Study of the History of the Fifteenth-Century Science and Technology], *Han'guk Kwahak-sa Hakhoe-ji*, 14 (1992): 141-170.

These activities are generally viewed as indicative of the independent spirit of Korean science of the period, and a testimony to the fact that Koreans were beginning to recognize the differences between Korea and China in terms of language, latitude, resources, climate, and so on. Naturally, the subjects of such projects were the ones in which significant local differences could be expected – "the winds and earths [of China and Korea] are not the same" (*p'ungt'o pul il* 風土不一), to use their own terms. Recently, however, Moon Joong Yang has pointed out that even these efforts can be seen instead as attempts by Sejong and the scholars of his court to reach the same level of scientific achievement as the Chinese.[20] According to Moon, when the Koreans of the Sejong period endeavored to construct a calendrical system based on the latitude of Seoul, their intense efforts reflected not so much an independent spirit as a desire to achieve a level of competence in calendrical astronomy that was on a par with the Chinese. In other words, their aim was to become truly equal to the Chinese, by having a calendrical system based on their own capital. In the end, they could not call their calendrical system a *li* 曆 (calendar, *yŏk* in Korean), but called it a *suan* 算 (computation, *san*). This, of course, was to avoid provoking the Ming government, which would have become suspicious if it had found out that the Korean king had established an independent calendrical system, the exclusive right of the Ming emperor. Indeed, this example illustrates yet another facet of Korean acceptance of Chinese centrality.

Another period of interest is the reign of King Chŏngjo (正祖, r. 1776-1800).[21] The astronomers of his court not only based the calendar on the latitude of Seoul, but also tried to provide the calendrical numbers for the latitudes and longitudes of all the provincial capitals of Korea, just as the Chinese had been doing in their annual calendars. While Chŏngjo's efforts too pointed to an independent spirit (and, having experienced the collapse of the Ming 'Central Splendor,' this spirit could have been that much more genuine), in a sense this again shows how eager Koreans were to upgrade their calendars to the level of the Chinese calendars. It is thus significant that the numbers they attempted to put in those calendars were not obtained from observations; they were

20 Mun Chung-yang 文重亮 (Joong-Yang Moon), "Sejongdae kwahakkisul-ŭi chajusŏng tasi pogi" 世宗代 科學技術의 '自主性' 다시보기 [Another Look at the Science and Technology of the Sejong Period], *Yŏksahakpo* 歷史學報 no. 189 (2006): 39-72.

21 Mun Chung-yang, "Sipp'al segi huban Chosŏn kwahak kisul-ŭi ch'ui-wa sŏngkyŏk: Chŏngjodae chŏngbu bumun-ŭi ch'ŏnmun yŏksan hwaltong-ŭl chungsim-ŭro" 18世紀 後半 朝鮮 科學技術의 推移와 性格 – 正祖代 政府 部門의 天文曆算 活動을 中心으로 [Trends and Characteristics of Science and Technology in Late Eighteenth-Century Choson: The Royal Astronomical Projects in the Period of Chŏngjo], *Yŏksa-wa Hyŏnsil* 歷史와 現實 39 (2001): 199-231.

obtained from computation. What was important for them was to have numbers placed in their calendars so that they appeared as complete as the Chinese calendars.

King Chŏngjo's endeavors did produce some results. Toward the end of the eighteenth century, when his astronomers mastered the detailed computational methods of the *Shixianli* – and especially after Sŏ Ho-su 徐浩修 (1736-1799) became in charge of the Astronomical Bureau (*kwansanggam* 觀象監) and compiled the *Kukcho yŏksang go* 國朝曆象考 (Examination of the Calendrical Astronomy of the [Chosŏn] Dynasty) – the court was confident enough to choose the Korean calendar over the Chinese when differences occurred between the two calendars in the computations of the seasonal points. (We know that earlier when such cases occurred, the court would punish the astronomers for making mistakes.) Such confidence led Chŏngjo to compile a Korean equivalent of the *Lüli yuanyuan* 律曆淵源 (Sources of Harmonics and Calendrical Astronomy). Similar effort was also put into agricultural books.[22]

Finally, concerning the issue of 'center versus periphery,' it has to be pointed out that there can be differences in the degree of 'peripheral-ness.' For example, the situation in Japan was different from in Korea: Japan was more peripheral, but accepted their peripheral-ness and the centrality of China to a lesser extent. That may be why the 'problem of China' seems to be less pressing in the study of the history of Japanese science.[23]

3 Comparison with the West: Local versus Universal Science

In dealing with these issues, it would be useful to look at comparable situations in the West. Unfortunately, however, there does not seem to have been a situation in Europe comparable to the one-sided relationship between Chinese science and Korean science. Undoubtedly, there were periods in Western history when the science of one individual country was dominant. Yet the center of dominance frequently shifted from one country to another.

Perhaps the example of Greek science in the ancient Mediterranean world and in the medieval West shows some parallels. However, while the over-

22 Mun Chung-yang, *Chosŏn hugi surihak-kwa suri tamnon* 朝鮮後期 水利學과 水利 談論 [Hydraulics and Hydraulical Discourses in Late Chosŏn] (Seoul: chimmundang 集文堂, 2000), 134-144.

23 Is this really the case? I wonder whether my Japanese colleagues would agree with me on this point.

whelming degree of dominance was comparable, the location of the center in these two cases shifted continuously from one place to another. Another case similar to the East Asian situation would be that of eighteenth- and early nineteenth-century Europe, where the leading French science was transmitted to the rest of Europe. But again, compared with the Chinese case, the period of French dominance was fairly brief. Moreover, the contents of this prevailing science of the French were not solely of their own making, but contained important elements from elsewhere. Nevertheless, it may be worthwhile to draw comparisons between French-dominated European science of the time and Chinese-dominated East Asian science. It would be helpful to analyze the ways in which the historians of Western science deal with the French-style sciences being practiced in other parts of Europe at the time. Also, how should the historian regard the overwhelming prevalence of the United States in science after World War Two? This of course is also relevant to Korea, where we see a dominant influence of the United States, not only in scientific content, but also in terms of institutions, education, and even personnel (mostly having received American training).[24] Are we then faced with a problem of essentially the same nature – the 'problem of the United States in the study of modern Korean science'?

One can look at the 'problem of China' from a different perspective by looking at it as a problem of a 'universal' science versus a 'local' science. Comparable situations in the West can shed some light on this. For example, when we speak of nineteenth-century 'English,' 'French,' 'German,' or 'Italian' science, or 'American' science, we are talking about such 'local' sciences. Now, is the relationship between Korean science and Chinese science similar to that between one of these local sciences and a universal European science? And if so, should so-called 'Chinese science' be considered the universal science of the East Asian world, or 'East Asian science'? To provide another example, how do historians of European science regard the periods in which the sciences of Spain, Italy or the Netherlands no longer contributed significantly to European science, or concerning the periods in which Russian or American science had not yet reached the level of European science at the center? Historians of East Asian science may benefit from studying how historians of Western science approach these questions.

---

24 Yung Sik Kim, "Some Reflections on Science and Technology in Contemporary Korean Society," *Korea Journal*, vol. 28, no. 8 (August 1988): 4-15.

4 The Problem and the Korean 'Why Not' Question

The 'problem of China' can be raised in connection with what can be called the Korean 'why not' question.[25] For example, in dealing with the 'stagnation' and 'decline' of Korean science and technology after the Sejong period, during which some areas of Korean science and technology had reached a peak, indeed even the highest level then existing in the world, can the historian ignore the situation in China at the time? Can he ignore questions such as: At what stage of advancement were the Chinese scientific ideas and techniques adopted by Koreans of the Sejong period? How did these ideas and techniques subsequently progress – or decline – in China?[26] If Chinese science and technology did stagnate or decline, would it have been possible for Korean science and technology to continue to develop independently? The importance of Islamic influence, which Jeon Sang-woon has alluded to, but not elaborated upon,[27] is an indication that the situation in Korea could have been different from, or at least more complicated than, that of China. Nevertheless, if we were to conclude in the end that it was not possible for Korean science to prosper alone, would that mean that the Korean 'why not' question should be reduced to a part of the Chinese 'why not' question?

5 Concluding Remarks

As I have stated in the beginning of this paper, I have raised questions rather than provided answers or explanations about the 'problem of China' in the study of the history of Korean science, Chinese science, and East Asian science. How prevalent are these concerns I mention among historians of science focusing on Korea, China or East Asia in general? Studies by historians of science working on scientific developments in different regions of the world might provide interesting comparisons.

25 Yung Sik Kim, "The 'Why not' Question of Chinese Science: Scientific Revolution and Traditional Chinese Science," *East Asian Science, Technology, and Medicine*, no. 22 (2004): 96-112 (Chapter 9 of this book).

26 We read about the supposed decline of many sciences in the Ming, although there are some debates concerning the reality of such decline. On the decline of science in the Ming, see, e.g., Joseph Needham, *Science and Civilisation in China*, vol. 3 (Cambridge: Cambridge University Press, 1959), 50-52; Li Yan and Du Shiran, *Chinese Mathematics: A Concise History* (Oxford: Clarendon Press, 1987), 175; Liu Dun, "400 Years of the History of Mathematics in China – An Introduction to the Major Historians of Mathematics since 1592," *Historia Scientiarum* 4 (1994): 103-104. See also Willard J. Peterson, "Calendar Reform prior to the Arrival of Missionaries at the Ming Court," *Ming Studies* 21 (1986): 45-61.

27 Chŏn Sang-un, "Chosŏn chŏn'gi-ŭi kwahak-kwa kisul."

Bibliography

Abraham, Gary A. "Misunderstanding the Merton Thesis." *Isis* 74 (1983): 368-387.

Analects (*Lunyu* 論語).

Azuma Jūji 吾妻重二. "Shu Ki 'Shueki sandōkei kōi' ni tsuite." 朱熹周易參同契考異について, *Nihon Chūgokugakukaihō* 日本中國學會報, no. 36 (1984): 175-190.

Baker, Don. "A Different Thread: Orthodoxy, Heterodoxy, and Catholicism in a Confucian World." In *Culture and the State in Late Chosŏn Korea*, edited by Jahyun Kim Haboush and Martina Deuchler, 199-230. Cambridge, Mass.: Harvard University Press, 1999.

Balazs, Etienne. *Chinese Civilization and Bureaucracy.* Translated by H.M. Wright. New Haven: Yale University Press, 1964.

Barbour, I. A. *Issues in Science and Religion.* Englewood-Cliffs, NJ: Prentice-Hall, 1966.

Bartholomew, James R. "Why Was There No Scientific Revolution in Tokugawa Japan?" *Japanese Studies in the History of Science* 15 (1976): 111-125.

Bartholomew, James R. *The Formation of Science in Japan: Building a Research Tradition.* New Haven: Yale University Press, 1989.

Berggren, J. L. "Islamic Acquisition of the Foreign Sciences: A Cultural Perspective." In *Tradition, Transmission, Transformation,* edited by Ragep and Ragep, 263-283.

Biagioli, Mario. *Galileo, Courtier: The Practice of Science in the Culture of Absolutism.* Chicago: Chicago University Press, 1993.

Bodde, Derk. "Chinese 'Law of Nature': A Reconsideration." *Harvard Journal of Asiatic Studies* 39 (1979): 139-155.

Bodde, Derk. "Evidence for 'Laws of Nature' in Chinese Thought." *Harvard Journal of Asiatic Studies* 20 (1957): 709-727.

Bodde, Derk. "The Attitude toward Science and Scientific Method in Ancient China." *T'ien Hsia Monthly*, 2 (1936): 139-160.

Bodde, Derk. *Chinese Thought, Society, and Science: The Intellectual and Social Background of Science and Technology in Pre-modern China.* Honolulu: University of Hawaii Press, 1991.

Bodde, Derk. *Essays in Chinese Civilization.* Princeton: Princeton University Press, 1981.

Bol, Peter K. "Chu Hsi's Redefinition of Literati Learning." In *Neo-Confucian Education: The Formative Stage*, edited by William Theodore de Bary and John W. Chaffee, 151-185. Berkeley: University of California Press, 1989.

Bol, Peter K. "The Sung Examination System and the *Shih*." *Asia Major, third series*, 3 (1990): 149-171.

Bray, Francesca and Georges Métailié. "Who Was the Author of the *Nongzheng Quanshu*?" In *Statecraft & Intellectual Renewal in Late Ming China*, edited by Catherine Jami, Peter Engelfriet, and Gregory Blue, 322-359. Leiden: Brill, 2001.

 | DOI 10.1163/9789004265318-015

Bray, Francesca, Vera Dorofeeva-Lichtmann, and Georges Métailié, eds. *Graphics and Text in the Production of Technical Knowledge in China: The Warp and the Weft.* Leiden: Brill, 2007.

Bray, Francesca. "Agricultural Illustrations: Blueprint or Icon?" In *Graphics and Text in the Production of Technical Knowledge in China,* edited by Francesca Bray, Vera Dorofeeva-Lichtmann and Georges Métailié, 521-567. Leiden: Brill, 2007.

Bray, Francesca. "Introduction: The Powers of *Tu.*" In *Graphics and Text in the Production of Technical Knowledge in China,* edited by Francesca Bray, Vera Dorofeeva-Lichtmann and Georges Métailié, 1-78. Leiden: Brill, 2007.

Bray, Francesca. "Science, Technique, Technology: Passages between Matter and Knowledge in Imperial Chinese Agriculture." *British Journal for the History of Science,* 41(2008): 319-344.

Brock, W. H. "The Japanese Connexion: Engineering in Tokyo, London, and Glasgow at the End of the Nineteenth Century." *The British Journal for the History of Science* 14 (1981): 227-243.

Brook, Timothy. "Xu Guangqi in His Context: The World of the Shanghai Gentry." In *Statecraft & Intellectual Renewal in Late Ming China,* edited by Catherine Jami, Peter Engelfriet, and Gregory Blue, 72-98. Leiden: Brill, 2001.

Brooke, John H. *Science and Religion: Some Historical Perspectives.* Cambridge: Cambridge University Press, 1991.

Busch, Heinrich. "The *Tung-Lin* Academy and Its Political and Philosophical Significance." *Monumenta Serica* 14 (l949-1955): 1-163.

Ch'en, Kenneth K. S. *Buddhism in China.* Princeton: Princeton University Press, 1964.

Ch'oe Chu, trans. 崔炷 註譯. *Ch'ŏn'gong kaemul* 天工開物. Seoul: Chŏnt'ong Munhuasa 傳統文化社, 1997.

Chan, A. K. C., G. K. Clancey, and H Loy, eds. *Historical Perspectives on East Asian Science, Technology and Medicine.* Singapore: World Scientific Publishers, 2002.

Chan, Hok-lam and William Theodore de Bary, eds. *Yüan Thought: Chinese Thought and Religion under Mongols.* New York: Columbia University Press, 1982.

Chan, Wing-tsit, ed. *Chu Hsi and Neo-Confucianism.* Honolulu: University of Hawaii Press, 1986.

Chan, Wing-tsit. *A Source Book in Chinese Philosophy.* Princeton: Princeton University Press, 1963.

Chan, Wing-tsit. "Chu Hsi's Completion of Neo-Confucianism." *Sung Studies, In Memoriam Etienne Balazs,* second series, vol. 1, 59-90. Paris: Mouton, 1973.

Chan, Wing-tsit. "Neo-Confucianism and Chinese Scientific Thought." *Philosophy East and West* 6 (1957): 309-332.

Chan, Wing-tsit. "The Ch'eng-Chu School of Early Ming." In *Self and Society in Ming Thought,* edited by William Theodore de Bary, 29-51. New York: Columbia University Press, 1970.

Chan, Wing-tsit. "The Evolution of the Neo-Confucian Concept *Li* as Principle." *Tsing Hua Journal of Chinese Studies*, new series, 4 (1964): 123-149.

Chao, Yüan-ling. "The Ideal Physician in Late Imperial China: The Question of *Sanshi*." *East Asian Science, Technology, and Medicine* 17 (2000): 66-93.

Chen Chun 陳淳, *Beixi ziyi* 北溪字義.

Chen Lai 陳來. *Zhu Xi zhexue yanjiu* 朱熹哲學研究, 2nd ed. Shanghai: Huadong shifan daxue chubanshe 華東師範大學出版社, 1987.

Chen Meidong 陳美東 and Chen Hui 陳暉. "Mingmo Qingchu xifang diyuanshuo zai Zhongguode chuanbo yu fanxiang" 明末清初西方地圓說在中國的傳播與反響. *Zhongguo keji shiliao* 中國科技史料 21 (2000): 6-12.

Cheng, Chung-ying. "Chu Hsi's Methodology and Theory of Understanding." In *Chu Hsi and Neo-Confucianism*, edited by Wing-tsit Chan, 169-196. Honolulu: University of Hawaii Press, 1986.

Cheng Hao 程顥 and Cheng Yi 程頤. *Er Cheng ji* 二程集, modern punctuated ed. Beijing: Zhonghua shuju 中華書局, 1981.

Cheng Hao 程顥 and Cheng Yi 程頤. *Henan Chengshi yishu* 河南程氏遺書.

Cheng Wenguang 鄭文光 and Xi Zezong 席澤宗. *Zhongguo lishi shang de yuzhou lilun* 中國歷史上的宇宙理論. Beijing: Renmin chubanshe 人民出版社, 1975.

Cheng Yi 程頤, *Yichuan wenji* 伊川文集.

Ching, Julia. *The Religious Thought of Chu Hsi.* Oxford: Oxford University Press, 2000.

Ching, Julia. *To Acquire Wisdom: The Way of Wang Yang-ming.* New York: Columbia University Press, 1976.

Choadam 座談. "Tasan, Chujahak, kŭrigo Sŏhak" 茶山, 朱子學, 그리고 西學. *Tasanhak* 茶山學, no. 2 (2001): 210-274.

Chŏn Sang-un 全相運 (Sang-woon Jeon). "Chosŏn chŏn'gi-ŭi kwahak-kwa kisul – Sibosegi kwahakkisulsa yŏn'gu chaeron" 朝鮮前期의 科學과 技術 – 15世紀 科學技術史 研究 再論. *Han'guk Kwahak-sa Hakhoe-ji* 韓國科學史學會誌 14 (1992): 141-170.

Chŏn Sang-un 全相運 (Sang-woon Jeon). *Han'guk kwahak kisulsa* 韓國科學技術史 (History of Science and Technology in Korea). Seoul: Kwahak segyesa 科學世界社, 1966.

Chŏn Yong-hun 全勇勳 (Yong Hoon Jeon). "Sipch'il-sipp'al segi sŏyang kwahak-ŭi toip-kwa kaltŭng: *Sihŏn-ryŏk* sihaeng-kwa chŏlgi paech'ipŏp-e taehan nollan-ŭl chungsim-ŭro" 17 - 18世紀 西洋科學의 導入과 葛藤 – 時憲曆 施行과 節氣配置法에 對한 論難을 中心으로, *Tongbang hakchi* 東方學志 117 (2002): 1-49.

Chŏng Kyu-yŏng 丁奎英. *Saam sŏnsaeng yŏnbo* 俟菴先生年譜. Seoul: Chŏngmunsa 正文社, 1984.

Chŏng Ok-cha 鄭玉子. *Chosŏn hugi Chosŏn Chunghwa sasang yŏn'gu* 朝鮮後期 朝鮮中華思想 研究. Seoul: Ilchisa 一志社, 1998.

Chŏng Ok-cha 鄭玉子. *Chosŏn hugi yŏksa-ŭl ihae* 朝鮮 後期 歷史의 理解. Seoul: Ilchisa 一志社, 1993.

Chŏng Yak-yong 丁若鏞. *Yŏyudang chŏnsŏ* 與猶堂全書. Seoul: Yŏgang ch'ulp'ansa 驪江出版社, 1985.

Chu, Ping-yi. "Technical Knowledge, Cultural Practices and Social Boundaries: Wan-nan Scholars and the Recasting of Jesuit Astronomy, 1600-1800." Ph.D. diss., University of California – Los Angeles, 1994.

Chu, Pingyi. "Remembering Our Grand Tradition: The Historical Memory of the Scientific Exchanges between China and Europe, 1600-1800." *History of Science* 41 (2003): 193-215.

Chu, Pingyi. "Scientific Dispute in the Imperial Court: The 1664 Calendar Case." *Chinese Science* 14 (1997): 7-34.

Chu, Pingyi. "Trust, Instruments, and Cross-Cultural Scientific Exchanges: Chinese Debate over the Shape of the Earth, 1600-1800." *Science in Context* 12 (1999): 385-411.

Clagett, Marshall, ed. *Critical Problems in the History of Science.* Madison: University of Wisconsin Press, 1959.

Clagett, Marshall. *Nicole Oresme and the Medieval Geometry of Qualities and Motions.* Madison: University of Wisconsin Press, 1968.

Clagett, Marshall. *The Science of Mechanics in the Middle Ages.* Madison: University of Wisconsin Press, 1959.

Clunas, Craig. *Superfluous Things: Material Culture and Social Status in Early Modern China.* Honolulu: University of Hawai'i Press, 2004.

Cohen, H. Floris. *The Scientific Revolution: A Historical Inquiry.* Chicago: University of Chicago Press, 1994.

Cohen, I. Bernard, ed. *Puritanism and the Rise of Modern Science: The Merton Thesis.* New Brunswick: Rutgers University Press, 1990.

Cohen, R. S. and M. Wartofsky, eds. *Philosophical Foundations of Science.* Dordrecht: Reidel, 1974.

Cohen, Robert S. "The Problem of 19(k)." *Journal of Chinese Philosophy* 1 (1973): 103-117.

Cooper, William C. and Nathan Sivin. "Man as a Medicine: Pharmacological and Ritual Aspects of Traditional Therapy Using Drugs Derived from the Human Body." In *Chinese Science: Explorations of an Ancient Tradition*, edited by Shigeru Nakayama and Nathan Sivin, 203-272. Cambridge, Mass.: MIT Press, 1973.

Cullen, Christopher. "A Chinese Eratosthenes of the Flat Earth." *Bulletin of the School of Oriental and African Studies (London)* 39 (1976): 106-127.

Cullen, Christopher. "An Eighth Century Chinese Table of Tangents." *Chinese Science*, no. 5 (1982): 1-33.

Cullen, Christopher. "The Science/Technology Interface in Seventeenth-Century China: Song Yingxing on *qi* and the *wu xing*." *Bulletin of the School of Oriental and African Studies* 53 (1990): 295-318.

Dardess, John W. *Confucianism and Autocracy: Professional Elites in the Founding of the Ming Dynasty.* Berkeley: University of California Press, 1983.

Daston, Lorraine. "The Nature of Nature in Early Modern Europe." *Configurations: A Journal of Literature, Science, and Technology* 6 (1998): 149-172.

De Bary, William Theodore, ed. *Self and Society in Ming Thought.* New York: Columbia University Press, 1970.

De Bary, William Theodore, ed. *The Unfolding of Neo-Confucianism.* New York: Columbia University Press, 1975.

De Bary, William Theodore and Irene Bloom, eds. *Principle and Practicality: Essays in Neo-Confucianism and Practical Learning*. New York: Columbia University Press, 1979.

De Bary, William Theodore and Ja Hyun Kim Haboush, eds. *The Rise of Neo-Confucianism in Korea.* New York: Columbia University Press, 1985.

De Bary, William Theodore and John W. Chaffee, eds. *Neo-Confucian Education: The Formative Stage.* Berkeley: University of California Press, 1989.

De Bary, William Theodore. "Chu Hsi's Aims as an Educator." In *Neo-Confucian Education: The Formative Stage,* edited by William Theodore de Bary and John W. Chaffee, 186-218. Berkeley: University of California Press, 1989.

De Bary, William Theodore. *East Asian Civilizations: A Dialogue in Five Stages.* Cambridge, MA: Harvard University Press, 1988.

De Bary, William Theodore. *Neo-Confucian Orthodoxy and the Learning of the Mind-and-Heart.* New York: Columbia University Press, 1981.

Deane, Thatcher E. "The Chinese Imperial Astronomical Bureau: Form and Function of the Ming Dynasty Qintianjian from 1365 to 1627." Ph.D. diss., University of Washington – Seattle, 1989.

Dear, Peter. *The Intelligibility of Nature: How Science Makes Sense of the World.* Chicago: University of Chicago Press, 2006.

Dorn, Harold. *The Geography of Science.* Baltimore: Johns Hopkins University Press, 1991.

Duara, Prasenjit. *Rescuing History from the Nation: Questions and Narratives of Modern China.* Chicago: University of Chicago Press, 1995.

Dubs, Homer H. "The Failure of the Chinese to Produce Philosophical Systems." *T'oung Pao* 26 (1928): 96-209.

Elkana, Yehuda, ed. *The Interaction between Science and Philosophy.* Atlantic Highlands, N.J.: Humanities Press, 1974.

Elman, Benjamin A. "Collecting and Classifying: Ming Dynasty Compendia and Encyclopedias (*Leishu*)." *Extrême-Orient, Extrême-Occident hors série* (2007): 131-157.

Elman, Benjamin A. *A Cultural History of Civil Examinations in Late Imperial China.* Berkeley: University of California Press, 2000.

Elman, Benjamin A. *From Philosophy to Philology; Intellectual and Social Aspect of Change in Late Imperial China.* Cambridge, Mass., Harvard University Press, 1984.

Elman, Benjamin A. *On Their Own Terms: Science in China, 1550-1900.* Cambridge: Harvard University Press, 2005.

Elvin, Mark. *The Pattern of the Chinese Past: A Social and Economic Interpretation.* Stanford: Stanford University Press, 1973.

Fukuzawa Yukichi 福澤諭吉. "Leaving Asia" (*Datsu-a ron* 脫亞論). In *Japan: A Documentary History: The Late Tokugawa Period to the Present*, edited by David John Lu, 351-353. Armonk, New York: M.E. Sharpe, 1996.

Fung, Yu-lan. *A History of Chinese Philosophy*, 2 vols. Princeton: Princeton University Press, 1937, 1952.

Fung, Yu-lan. "Why China Has No Science: An Interpretation of the History and Consequences of Chinese Philosophy." *The International Journal of Ethics* 32 (1922): 237-263.

Furth, Charlotte, Judith T. Zeitlin, and Ping-chen Hsiung, eds. *Thinking with Cases: Specialist Knowledge in Chinese Cultural History.* Honolulu: University of Hawaii Press, 2007.

Furth, Charlotte. "Introduction: Thinking with Cases." In *Specialist Knowledge in Chinese Cultural History,* edited by Charlotte Furth, Judith T. Zeitlin, and Ping-chen Hsiung, 1-27. Honolulu: University of Hawaii Press, 2007. Furth, Charlotte. "The Physician as Philosopher of the Way: Zhu Zhenheng (1282-1358)." *Harvard Journal of Asiatic Studies* 66 (2006): 423-459.

Gallagher, Louis, trans. *China in the Sixteenth Century. the Journals of Matteo Ricci, 1583-1610.* New York: Random House, 1953.

Gardner, Daniel K. *Chu Hsi: Learning to Be a Sage.* Berkeley: University of California Press, 1990.

Gernet, Jacques. "A Note on the Context of Xu Guangqi's Conversion." In *Statecraft and Intellectual Renewal in Late Ming China: The Cross-Cultural Synthesis of Xu Guangqi (1562-1633),* edited by Catherine Jami, Peter Engelfriet, and Gregory Blue, 186-190. Leiden: Brill, 2001.

Gernet, Jacques. *Daily Life in China on the Eve of the Mongol Invasion 1250-1276.* London: Allen & Unwin, 1962.

Gillispie, Charles C., ed. *Dictionary of Scientific Biography*, 16 vols. New York: Charles Scribner's Sons, 1970–1980.

Gillispie, Charles C. "The Natural History of Industry." Isis 48 (1957): 398-407.

Gillispie, Charles C. *Science and Polity in France at the End of the Old Regime.* Princeton: Princeton University Press, 1980.

Girardot, N. J. "The Problem of Creation Mythology in the Study of Chinese Religion." *History of Religions* 15 (1976): 289-318.

Golas, Peter. "'Like Obtaining a Great Treasure': The Illustrations in Song Yingxing's The Exploitation of the Works of Nature." In *Graphics and Text in the Production of*

Technical Knowledge in China, edited by Francesca Bray, Vera Dorofeeva-Lichtmann and Georges Métailié, 569-614. Leiden: Brill, 2007.

Goldschmidt, Asaf. "The Song Discontinuity: Rapid Innovation in Northern Song Dynasty Medicine." *Asian Medicine* 1 (2004): 53-90.

Gooday, Graeme J. N. and Morris F. Low. "Technology Transfer and Cultural Exchange: Western Scientists and Engineers Encounter Late Tokugawa and Meiji Japan." *Osiris* 13 (1998): 99-128.

Goodrich, L. Carrington. "Early Mentions of Fossil Fishes." *Isis* 34 (1942): 25.

Graham, A. C. "China, Europe, and the Origins of Modern Science." In *Chinese Science: Explorations of an Ancient Tradition*, edited by Shigeru Nakayama and Nathan Sivin, 45-69. Cambridge, Mass.: MIT Press, 1973.

Graham, A. C. *Later Mohist Logic, Ethics, and Science.* London: The School of Oriental and African Studies, 1978.

Graham, A. C. *Two Chinese Philosophers: Ch'eng Ming-tao and Ch'eng Yi-ch'üan.* London: Lund Humphries, 1958.

Graham, A. C. *Yin-Yang and the Nature of Correlative Thinking.* Singapore: The Institute of East Asian Philosophies, 1986.

Graham, A. C. and Nathan Sivin. "A Systematic Approach to the Mohist Optics." In *Chinese Science: Explorations of an Ancient Tradition*, edited by Nakayama, Shigeru and Nathan Sivin, 105-152. Cambridge, Mass.: MIT Press, 1973.

Grant, Edward. *Physical Science in the Middle Ages.* New York: Wiley, 1971.

Grant, Edward. "Cosmology." In *Science in the Middle Ages*, edited by David C. Lindberg, 265-302. Chicago: University of Chicago Press, 1978.

Grant, Edward. *Nicole Oresme and the Kinematics of Circular Motion.* Madison: University of Wisconsin Press, 1971.

Guy, R. Kent. *The Emperor's Four Treasuries: Scholars and the State in the Late Ch'ien-lung Era.* Cambridge, Mass: Harvard University Press, 1987.

Haboush, Jahyun Kim and Martina Deuchler, eds. *Culture and the State in Late Chosŏn Korea.* Cambridge, Mass.: Harvard University Press, 1999.

Haeger, John W., ed. *Crisis and Prosperity in Sung China.* Tucson: University of Arizona Press, 1975.

Hall, A. Rupert. "The Scholar and the Craftsman in the Scientific Revolution." In *Critical Problems in the History of Science,* edited by Marshall Clagett, 3-23. Madison: University of Wisconsin Press, 1959.

Han Hyŏng-cho 韓亨祚. "Tasan-kwa Sŏhak: Chosŏn Chujahak-ŭi yŏnsok-kwa tanjŏl" 茶山과 西學: 朝鮮 朱子學의 連續과 斷絶. *Tasanhak* 茶山學, 2 (2001): 128-155.

Han Hyŏng-cho 韓亨祚. *Chu Hi-esŏ Chŏng Yak-yong-ŭro: Chosŏn yuhak-ŭi ch'ŏrhakchŏk p'aerŏdaim yŏn'gu* 朱熹에서 丁若鏞으로 – 朝鮮 儒學의 哲學的 패러다임 硏究. Seoul: Segyesa 世界社, 1996.

Han'guk sasangsa yŏn'guhoe 韓國思想史硏究會. *Insŏng mulsŏng ron* 人性物性論. Seoul: Han'gilsa 한길社, 1994.

Han'guk sasangsa yŏn'guhoe, ed. 韓國思想史硏究會 編. *Chosŏn yuhak-ŭi chayŏn ch'ŏrhak* 朝鮮 儒學의 自然哲學. Seoul: Yemun sŏwŏn 藝文書院, 1998.

Hanshu 漢書.

Hart, Roger. "Beyond Science and Civilization: A Post-Needham Critique." *East Asian Science, Technology, and Medicine* 16 (1999): 88-114.

Hart, Roger. *Imagined Civilizations: China, the West, and Their First Encounter.* Baltimore: Johns Hopkins University Press, 2013.

Hartwell, Robert M. "Demographic, Political, and Social Transformations of China, 750-1550." *Harvard Journal of Asiatic Studies* 42 (1982): 365-442.

Hartwell, Robert M. "Financial Expertise, Examinations, and the Formulation of Economic Policy in Northern Sung China." *Journal of Asian Studies* 30 (1971): 281-314.

Henderson, John B. "Ch'ing Scholars' Views of Western Astronomy." *Harvard Journal of Asiatic Studies* 46 (1986): 121-148.

Henderson, John B. *The Development and Decline of Chinese Cosmology.* New York: Columbia University Press, 1984.

Ho Ping-ti. *The Ladder of Success in Imperial China: Aspects of Social Mobility, 1368-1911.* New York: Columbia University Press, 1962.

Ho, Peng Yoke. *Li, Qi and Shu: An Introduction to Science and Civilization in China.* Hong Kong: Hong Kong University Press, 1985.

Hong Tae-yong 洪大容. *Tamhŏnsŏ* 湛軒書, naejip 內集.

Hong Wansheng (Horng Wann-Sheng 洪萬生). "Shiba shiji dongsuan yu zhongsuan de yiduan duihua: Hong Zhengxia vs. He Guozhu" 十八世紀東算與中算的一段對話: 洪正夏 vs. 何國柱. *Hanxue yanjiu* 漢學硏究 20 (2002): 57-80.

Horng, Wann-Sheng. "Sino-Korean Transmission of Mathematical Texts in the 19th Century: A Case Study of Nam Pyong-gil's *Kugo Sulyo Tohae.*" *Historia Scientiarum* 12 (2002): 87-99.

Hu, Shih. "The Scientific Spirit and Method in Chinese Philosophy." In *The Chinese Mind*, edited by Charles A. Moore, 104-131. Honolulu: University of Hawaii Press, 1967.

Hu, Shih. *The Development of the Logical Method in Ancient China.* Shanghai, 1922.

Huainanzi 淮南子. (*Sibu beiyao* 四部備要 edition).

Huang, Yi-long. "Court Divination and Christianity in the K'ang-hsi Era." *Chinese Science* 10 (1991): 1-20.

Huff, Toby E. *The Rise of Early Modern Science: Islam, China, and the West.* Cambridge: Cambridge University Press, 1993.

Hymes, Robert P. "Not Quite Gentlemen? Doctors in Sung and Yuan." *Chinese Science*, no. 8 (1987): 9-76.

Im Chong-t'ae 林宗台 (Lim Jongtae). *Sipch'il sipp'al segi Chungguk-kwa Chosŏn-ŭi sŏgu chirihak ihae* 17, 18世紀 中國과 朝鮮의 西歐 地理學 理解. Seoul: Changbi 創批, 2012.

Jami, Catherine, Peter Engelfriet, and Gregory Blue, eds. *Statecraft & Intellectual Renewal in Late Ming China: The Cross-Cultural Synthesis of Xu Guangqi (1562-1633)*. Leiden: Brill, 2001.

Jeon, Sang-woon. *Science and Technology in Korea: Traditional Instruments and Techniques.* Cambridge, Mass.: MIT Press, 1974.

Jiang Xiaoyuan 江曉原. "Shilun Qingdai xixue zhongyuanshuo." 試論淸代西學中源說, *Ziran kexueshi yanjiu* 自然科學史硏究 7 (1988): 101-108.

Jin, Qiupeng 金秋鵬. *Zhongguo kexue jishushi: Renwu juan* 中國科學技術史: 人物卷. Beijing: Kexue chubanshe 科學出版社, 1998.

Jones, Robert A. and Henrika Kuklik, eds. *Research in Sociology of Knowledge, Science, and Art*, vol. 2. Greenwich, Conn.: Aijai, 1979.

Jun, Yong Hoon. "Transformation of the Western Science and Its Role in the Cosmological Discourse in the Late 17th Century Korea." In *Current Perspectives in the History of Science in East Asia*, edited by Yung Sik Kim and Francesca Bray, 497-503. Seoul: Seoul National University Press, 1999.

Kaltenmark, Max. *Lao Tzu and Taoism.* Stanford: Stanford University Press, 1969.

Kang Myŏng-gwan 姜明官. "Chosŏn hugi sŏjŏk-ŭi suip yut'ong-kwa changsŏga-ŭi ch'ulhyŏn: sipp'al sipku segi Kyŏnghwa sejok munhwa-ŭi han tanmyŏn" 朝鮮後期書籍의 輸入·流通과 藏書家의 出現 – 18,19世紀 京華世族 文化의 한 斷面. *Minjok munhaksa yŏn'gu* 民族文學史研究 9 (1996): 171-194.

Kibre, Pearl and Nancy G. Siraisi. "The Institutional Setting: The Universities." In *Science in the Middle Ages*, edited by David. C. Lindberg, 120-144. Chicago: University of Chicago Press, 1978.

Kim Hyŏn 金炫. "Chosŏn yuhak-esŏŭi kwisin kaenyŏm" 朝鮮 儒學에서의 鬼神 概念. In *Chosŏn yuhak-ŭi chayŏn ch'ŏrhak*, edited by Han'guk Sasangsa yŏn'gu-hoe, 349-418. Seoul: Yemun sŏwŏn 藝文書院, 1998.

Kim Mun-sik 金文植. *Chosŏn hugi kyŏnghak sasang yŏn'gu* 朝鮮後期經學思想硏究. Seoul: Ilchogak 一潮閣, 1996.

Kim Mun-yong 金文鎔. "Chosŏn sidae Yuhakchadŭl-ŭi Ŭmyang ohaeng-ron" 朝鮮 時代 儒學者들의 陰陽五行論. In *Chosŏn yuhak-ŭi chayŏn ch'ŏrhak*, edited by Han'guk Sasangsa yŏn'gu-hoe, 293-347. Seoul: Yemun sŏwŏn 藝文書院, 1998..

Kim Nak-chin 金洛眞. "Chosŏn yuhakchadŭl-ŭi kyŏngmul ch'iji-ron: Chayŏn insik pangbŏp-kwa kwallyŏnhayŏ" 朝鮮 儒學者들의 格物致知論 – 自然 認識 方法과 關聯하여. In *Chosŏn yuhak-ŭi chayŏn ch'ŏrhak*, edited by Han'guk Sasangsa yŏn'gu-hoe, 71-133. Seoul: Yemun sŏwŏn 藝文書院, 1998.

Kim Sŏk-mun 金錫文. *Yŏkhak isipsadohae* 易學二十四圖解.

Kim Tae-jun 金泰俊. *Hong Tae-yong P'yŏngjŏn* 洪大容評傳. Seoul: Min'ŭmsa 民音社, 1987.

Kim Tae-wŏn 金大原. "Chŏng Yak-yong-ŭi *Ŭiryŏng*. 1" 丁若鏞의 『醫零』. 1, *Han'guk Kwahaksa Hakhoeji* 韓國科學史學會誌 15 (1993): 225-246.

Kim Yŏng-ho 金泳鎬. "Chŏng Tasan-ŭi kwahakkisul sasang" 丁茶山의 科學技術思想. *Tongyanghak* 東洋學 19 (1989): 277-300.

Kim, Yung Sik and Francesca Bray, eds. *Current Perspectives in the History of Science in East Asia.* Seoul: Seoul National University Press, 1999.

Kim, Yung Sik. "'Fossils', 'Organic World-view,' 'The Earth's Movements,' etc.: Problems of Judging East Asian Scientific Achievements from Western Perspectives." In *Historical Perspectives on East Asian Science, Technology and Medicine*, edited by A. K. C. Chan, G. K. Clancey, and H Loy, 14-26. Singapore: World Scientific Publishers, 2002.

Kim, Yung Sik. "The *Ts'an-t'ung-ch'i k'ao-i* and the Place of Internal Alchemy (*Nei-tan*) in Chu Hsi's Thought." *Monumenta Serica* 55 (2007): 99-131.

Kim, Yung Sik. "Western Science, Cosmological Ideas, and the *Yijing* Studies in Seventeenth- and Eighteenth-Century Korea." *Seoul Jouranl of Korean Studies* 14 (2001): 299-334.

Kim, Yung Sik. "Chu Hsi (1130-1200) on Calendar Specialists and Their Knowledge: A Scholar's Attitude toward Technical Scientific Knowledge in Traditional China." *T'oung Pao* 78 (1992): 94-115.

Kim, Yung Sik. "Problems and Possibilities in the Study of the History of Korean Science." *Osiris* 13 (1998): 48-79.

Kim, Yung Sik. *The Natural Philosophy of Chu Hsi (1130-1200).* Philadelphia: American Philosophical Society, 2000.

Kirk, G. S. and J. E. Raven, eds. *The Presocratic Philosophers.* Cambridge: Cambridge University Press, 1957.

Ko Pyŏng-ik 高柄翊. "Tasan-ŭi chinbo gwan" 茶山의 進步觀. In *Chŏng Yak-yong*, edited by Sa-sun Yun, 313-326. Seoul: Koryŏ daehakkyo chulp'anbu, 1990.

Koslow, Arnold. "More on 19(k)." *Journal of Chinese Philosophy* 2 (1975): 181-196.

Kracke, E. A., Jr. *Civil Service in Early Sung China, 960-1067.* Cambridge, Mass.: Harvard University Press, 1953.

Kracke, E. A., Jr. "Sung K'ai-feng: Pragmatic Metropolis and Formalistic Capital." In *Crisis and Prosperity in Sung China,* edited by John W. Haeger, 49-77. Tucson: University of Arizona Press, 1975.

Ku Man-ok 具萬玉. "Chosŏn hugi 'chigu'sŏl suyong-ŭi sasangsajŏk ŭiŭi" 朝鮮後期 '地球'說 受容의 思想史的 意義. *Han'guksa-ŭi kujo-wa chŏn'gae – Ha Hyŏn'gang kyosu chŏngnyŏn kinyŏm nonch'ong* 韓國史의 構造와 展開 – 河炫綱敎授定年記念論叢, 717-747. Seoul: Hyean 慧眼, 2000.

Ku Man-ok 具萬玉. "Chosŏn hugi ch'ŏnche unhaengron-ŭi pyŏnhwa" 朝鮮後期 天體運行論의 變化. *Sirhak sasang yŏn'gu* 實學思想研究 17/18 (2000): 339-368.

Ku Man-ok 具萬玉. *Chosŏn hugi kwahak sasangsa yŏn'gu 1. Chujahakchŏk ujuron-ŭi pyŏndong* 朝鮮 後期 科學思想史 硏究 I. 朱子學的 宇宙論의 變動. Seoul: Hyean 혜안, 2004.

Kuhn, Thomas S. "History of Science." *The International Encyclopedia of the Social Sciences*, vol. 14. New York: Macmillan, 1968.

Kŭm Chang-t'ae 琴章泰. *Tasan Chŏng Yak-yong* 茶山 丁若鏞. Seoul: Sŏnggyun'gwan daehakkyo ch'ulp'anbu 成均館大學校出版部, 1999.

Kŭm Chang-t'ae 琴章泰. *Tasan sirhak t'amgu* 茶山實學探究. Seoul: Sohaksa 小學社, 2001.

Kŭm Chang-t'ae 琴章泰. "Tasan Kyŏnghak-ŭi t'al-Chujahakchŏk Segyegwan" 茶山 經學의 脫朱子學的 世界觀. *Tasanhak* 茶山學 1 (2000): 20-577.

Lam, Lay Yong. "Chu Shih-chieh's *Suan-hsueh ch'i-meng* (Introduction to Mathematical Studies)." *Archive for the History of Exact Sciences*, 21 (1979): 1-31.

Lam, Lay Yong. "The Geometrical Basis of the Ancient Chinese Square-Root Method." In *Science and Technology in East Asia*, edited by Nathan Sivin, 56-66. New York: Science History Publications, 1977.

Lam, Lay Yong. "The *Jih yung suan fa*: An Elementary Arithmetic Textbook of the Thirteenth Century." In *Science and Technology in East Asia*, edited by Nathan Sivin, 68-81. New York: Science History Publications, 1977.

Lau, D. C. "A Note on *Ke Wu*." *Bulletin of the School of Oriental and African Studies* 30 (1967): 353-57.

Lau, D. C. "On Mencius' Use of the Method of Analogy in Argument." In *Mencius*, translated by D.C. Lau, 235-63. Baltimore: Penguin Books, 1970.

Lau, D.C., trans. *Mencius.* Baltimore: Penguin Books, 1970.

Le, Aiguo 樂愛國. *Songdai de ruxue yu kexue* 宋代的儒學與科學. Beijing: Zhongguo kexuejishu chubanshe 中國科學技術出版社, 2007.

Lebra, Joyce C. *Japan's Greater East Asia Co-Prosperity Sphere in World War II: Selected Readings and Documents.* New York: Oxford University Press, 1975.

Lee, Thomas H.C. *Government Education and Examinations in Sung China.* Hong Kong: Chinese University Press, 1985.

Leslie, Donald. "Les theories de Wang Tch'ong sur la causalite." In *Melanges de sinology offerts a Monsieur Paul Demieville,* vol. 2, 179ff. Paris: Bibliothèque de l'institut des hautes études chinoises, 1974.

Leung, Angela Ki Che. "Medical Learning from the Song to the Ming." In *The Song-Yuan-Ming Transition in Chinese History*, edited by Paul J. Smith and Richard von Glahn, 374-398. Cambridge, Mass.: Harvard University Press, 2003.

Levenson, Joseph R. *Confucian China and Its Modern Fate.* Berkeley: University of California Press, 1968.

Lévy-Bruhl, L. *Les functions mentales dans les sociétés inférieures*, 2nd ed. Paris: Alcan, 1928.

Li, Yan and Du Shiran. *Chinese Mathematics: A Concise History*. Oxford: Clarendon Press, 1987.

Libbrecht, Ulrich. "Joseph Needham's Work in the Area of Chinese Mathematics." *Past and Present* 87 (1980): 30-39.

Libbrecht, Ulrich. *Chinese Mathematics in the Thirteenth Century*. Cambridge, Mass.; MIT Press, 1973.

Liji 禮記: *Liji zhushu* 禮記註疏. Taipei: Xinwenfeng 新文豊, reprint; 1977.

Lim, Jongtae. "Locating a Center on the Surface of a Globe: Negotiating China's Position on the Spherical Earth in Seventeenth and Eighteenth-Century China and Korea." *Historia Scientiarum* 17 (2008): 175-188.

Lindberg, David C. "Conceptions of the Scientific Revolution from Bacon to Butterfield: A Preliminary Sketch." In *Reappraisals of the Scientific Revolution*, edited by David C. Lindberg and Robert S. Westman, 1-26. Cambridge: Cambridge University Press, 1990.

Lindberg, David C., ed. *Science in the Middle Ages*. Chicago: University of Chicago Press, 1978.

Lindberg, David C. *John Pecham and the Science of Optics*. Madison: University of Wisconsin Press, 1970.

Lindberg, David C. and Robert S. Westman, eds. *Reappraisals of the Scientific Revolution*. Cambridge: Cambridge University Press, 1990.

Lindberg, David C. and Ronald L. Numbers, eds. *God and Nature: Historical Essays on the Encounter between Christianity and Science*. Berkeley: University of California Press, 1986.

Liu, Dun. "400 Years of the History of Mathematics in China: An Introduction to the Major Historians of Mathematics since 1592." *Historia Scientiarum* 4 (1994): 103-111.

Liu, James T. C. "How Did a Neo-Confucian School Become the State Orthodoxy?" *Philosophy East and West* 23 (1973): 483-505

Liu, James T. C. and P. Golas, eds. *Change in Sung China: Innovation or Renovation?* Boston: Heath, 1969.

Liu, Shu-hsien. "Theism from a Chinese Perspective." *Philosophy East and West* 28 (1978): 413-417.

Lloyd, Geoffrey E. R. *Adversaries and Authorities: Investigations into Ancient Greek and Chinese Science*. Cambridge: Cambridge University Press, 1996.

Lloyd, Geoffrey and Nathan Sivin. *The Way and the Word: Science and Medicine in Early China and Greece*. New Haven: Yale University Press, 2002.

Long, Pamela O. *Openness, Secrecy, Authorship: Technical Arts and the Culture of Knowledge from Antiquity to the Renaissance*. Baltimore: Johns Hopkins University Press, 2001.

Low, Morris. *Science and the Building of a New Japan.* New York: Palgrave Macmillan, 2005.

Maeyama, Y. "On the Astronomical Data of Ancient China (ca. -100~+200): A Numerical Analysis." *Archives internationales d'histoire des sciences* 25 (1975): 247-276; 26 (1976): 27-58.

Maier, Anneliese. *Studien zur Naturphilosophie der Spdtscholastik*, 5 vols. Rome: Edizioni di Storia e Letteratura, 1949-58.

Major, John S. "Myth, Cosmology and the Origins of Chinese Science." *Journal of Chinese Philosophy* 5 (1978): 1-20.

Major, John S. *Heaven and Earth in Early Han Thought: Chapters Three, Four, and Five of the* Huainanzi. Albany: State University of New York Press, 1993.

Maspero, Henri. *Taoism and Chinese Religion.* Amherst, Mass.: University of Massachusetts Press, 1981.

McKnight, Brian E. "Introduction." In *The Washing Away of Wrongs: Forensic Medicine in Thirteenth-Century China* by Sung T'zu, translated by Brian E. McKnight. Ann Arbor: University of Michigan Press, 1981.

Mencius (Mengzi 孟子).

Menzel, J. M., ed. *The Chinese Civil Service: Career Open to Talent?* Lexington, Mass.: Heath, 1963.

Merton, Robert K. "Science, Technology and Society in Seventeenth Century England." *Osiris* 4 (1938): 360-632.

Merton, Robert K. *Science, Technology & Society in Seventeenth Century England.* New York: Harper, 1970.

Meskill, John, ed. *The Pattern of Chinese History: Cycles, Development, or Stagnation?* Boston: Heath, 1965.

Metzger, Thomas A. *Escape from Predicament: Neo-Confucianism and China's Evolving Political Culture.* New York: Columbia University Press, 1977.

Min Yŏng-kyu 閔泳珪. "Sipchil segi Yijo hag'in-ŭi chidongsŏl – Kim Sŏk-mun-ŭi Yŏkhak isipsadohae" 十七世紀 李朝學人의 地動說 – 金錫文의 易學二十四圖解. *Tongbang hakchi* 東方學志16 (1975): 1-44.

Mingshi 明史.

Montgomery, Scott L. *Science in Translation: Movements of Knowledge through Cultures and Time.* Chicago: University of Chicago Press, 2000.

Moody, Ernest A. and Marshall Clagett. *The Medieval Science of Weights.* Madison: University of Wisconsin Press, 1952.

Moon, Joong-yang. "Traditional Cosmology Associated with the *I-ching* and Anti-Cosmological Discourses in 18th-Century Korea." *Seoul Journal of Korean Studies* 12 (1999): 177-227.

Moore, Charles A., ed. *The Chinese Mind.* Honolulu: University of Hawaii Press, 1967.

Morohashi Tetsuji 諸橋轍次. *Dai Kanwa jiten* 大漢和辭典. Tokyo: Taishukan, 1955-1960.

Morris-Suzuki, Tessa. *The Technological Transformation of Japan: From the Seventeenth to the Twenty-first Century.* Cambridge University Press, 1994.

Mote, Frederick W. "Confucian Eremitism in the Yuan Period." In *The Confucian Persuasion*, edited by Arthur F. Wright, 202-40. Stanford: Stanford University Press, 1960.

Mun Chung-yang 文重亮 (Joong-Yang Moon). "Sejongdae kwahakkisul-ŭi chajusŏng tasi pogi" 世宗代 科學技術의 '自主性' 다시 보기. *Yŏksahakpo* 歷史學報 189 (2006): 39-72.

Mun Chung-yang 文重亮. "Chŏnt'ongjŏk chayŏn insik ch'egye-ŭi sajŏk pyŏnhua: Sangsuhakchŏk, kironjŏk ujuron-ŭi simhua, chŏn'gae" 傳統的 自然認識 體系의 史的 變化: 象數學的, 氣論的 宇宙論의 深化, 展開. In *Han'guk sil'hak sasang yŏn'gu*, vol. 4, edited by Yŏnse taehakkyo Kukhak yŏn'guwon, 47-99. Seoul: Sinsŏwon, 1999.

Mun Chung-yang 文重亮. "Sipp'al segi huban Chosŏn kwahak kisul-ŭi ch'ui-wa sŏngkyŏk: Chŏngjodae chŏngbu bumun-ŭi ch'ŏnmun yŏksan hwaltong-ŭl chung-sim-ŭro" 18世紀 後半 朝鮮 科學技術의 推移와 性格 – 正祖代 政府 部門의 天文曆算 活動을 中心으로. *Yŏksa-wa Hyŏnsil* 歷史와 現實 39 (2001): 199-231.

Mun Chung-yang 文重亮. *Chosŏn hugi surihak-kwa suri tamnon* 朝鮮後期 水利學과 水利 談論. Seoul: Chimmundang 集文堂, 2000.

Murdoch, John E. "From Social into Intellectual Factors: An Aspect of the Unitary Character of Late Medieval Learning." In *The Cultural Context of Medieval Learning*, edited by John E. Murdoch and Edith D. Sylla, 271-339. Boston: Riedel, 1975.

Murdoch, John E. "Philosophy and the Enterprise of Science in the Later Middle Ages." In *The Interaction between Science and Philosophy*, edited by Yehuda Elkana, 51-74, 91-113. Atlantic Highlands, N.J.: Humanities Press, 1974.

Murdoch, John E. and Edith D. Sylla, eds. *The Cultural Context of Medieval Learning.* Boston: Riedel, 1975.

Murdoch, John E. and Edith D. Sylla. "The Science of Motion: The Role and Scope of Motion in Medieval Natural Philosophy." In *Science in the Middle Ages*, edited by David C. Lindberg, 206-264. Chicago: University of Chicago Press, 1978.

Nakayama, Shigeru and Nathan Sivin, eds. *Chinese Science: Explorations of an Ancient Tradition.* Cambridge, Mass.: MIT Press, 1973.

Nakayama, Shigeru. "History of East Asian Science: Needs and Opportunities." *Osiris* 10 (1995): 80-94.

Nakayama, Shigeru. "Characteristics of Chinese Astrology." *Isis* 57 (1966): 442-454.

Nakayama, Shigeru. "Joseph Needham, Organic Philosopher." In *Chinese Science: Explorations of an Ancient Tradition*, edited by Shigeru Nakayama and Nathan Sivin, 23-43. Cambridge, Mass.: MIT Press, 1973.

Nakayama, Shigeru. "Kyoto Group of the History of Science." *Japanese Studies in the History of Science* 9 (1970): 1-4.

Nakayama, Shigeru. "The Spread of Chinese Science into East Asia." .In *Current Perspectives in the History of Science in East Asia,* edited by Yung Sik Kim and Francesca Bray, 13-20. Seoul: Seoul National University Press, 1999.

Nakayama, Shigeru. *A History of Japanese Astronomy: Chinese Background and Western Impact.* Cambridge, Mass.: Harvard University Press, 1969.

Nappi, Carla. *The Monkey and the Inkpot: Natural History and Its Transformations in Early Modern China.* Cambridge, Massachusetts: Harvard University Press, 2009.

Needham, Joseph. *Clerks and Craftsmen in China and the West.* Cambridge: Cambridge University Press, 1970.

Needham, Joseph, et al. *Science and Civilisation in China*, 7 vols. Cambridge: Cambridge University Press, 1954- .

Needham, Joseph. "Social Position of Scientific Men and Physicians in Medieval China." *Proceedings, XIVth International Congress of the History of Science.* Tokyo and Kyoto: Science Council of Japan, 1974, IV:19-34.

Needham, Joseph. *The Grand Titration: Science and Society in East and West.* London: George Allen & Unwin, 1969.

Needham, Joseph. *Time and Eastern Man.* London: Royal Anthropological Institute, 1965.

Nelson, Benjamin. "Sciences and Civilizations, 'East' and 'West': Joseph Needham and Max Weber." In *Philosophical Foundations of Science*, edited by R. S. Cohen and M. Wartofsky, 445-493. Dordrecht: Reidel, 1974.

Ng, Wai-ming. "The *I Ching* and the Adaptation of Western Science in Tokugawa Japan." *Chinese Science* 15 (1998): 94-117.

No Tae-hwan 盧大煥. *Tongdo sŏki ron hyŏngsŏng kwajŏng yŏn'gu* 東道西器論 形成過程 研究. Seoul: Ilchisa 一志社, 2005.

Ogawa Haruhisa 小川晴久. "Chijŏn(dong)sŏl-esŏ uju muhannon-ŭro – Kim Sŏk-mun-kwa Hong Tae-yong-ŭi segye" 地轉(動)說에서 宇宙無限論으로 – 金錫文과 洪大容의 世界. *Tongbang hakchi* 東方學志 21 (1979): 55-90.

Ogawa Haruhisa 小川晴久. "Higashi Ajia ni okeru chiten(dō)setsu no seiritsu" 東アシアにおける地轉(動)說の成立. *Tongbang hakchi* 東方學志 23-24 (1980): 375-387.

Pak Kwon-su 朴權壽. "Sŏ Myŏng-ŭng-ŭi yokhakchŏk ch'ŏnmun'gwan" 徐命膺의 易學的 天文觀. *Han'guk Kwahaksa Hakhoeji* 韓國科學史學會誌 20 (1998): 57-101.

Pak Sŏng-rae 朴星來 (Seong-Rae Park). "Chosŏn sidae kwahaksa-rŭl ŏttŏk'e pol kŏt-in'ga" 朝鮮時代 科學技術을 어떻게 볼 것인가. *Han'guksa simin kangjwa* 韓國史市民講座 16 (1995): 145-166.

Pak Sŏng-rae 朴星來. "Chŏng Yak-yong-ŭi kwahak sasang 丁若鏞의 科學思想." In *Chŏng Yak-yong*, edited by Sa-sun Yun, 327-350. Seoul: Koryŏ daehakkyo chulp'anbu, 1990).

Pak Sŏng-rae 朴星來. *Han'guksa-edo kwahak-i innŭn'ga* 韓國史에도 科學이 있는가. Seoul: Kyobo mun'go 教保文庫, 1998.

Pan Jixing 潘吉星. *Song Yingxing pingzhuan* 宋應星評傳. Nanjing: Nanjing daxue chubanshe 南京大學出版社, 1990.

Pan Jixing 潘吉星. *Tiangong kaiwu jiaozhu ji yanjiu* 天工開物校注及研究. Chengdu: Bashu shushe 巴蜀書社, 1989.

Park, Seong-Rae. "Hong Tae-yong's Idea of the Rotating Earth." *Han'guk Kwahak-sa Hakhoe-ji* (Journal of the Korean History of Science Society) 1 (1979): 39-49.

Park, Seong-Rae. "Pride and Prejudice in the Historiography of Science in East Asia." In *Current Perspectives in the History of Science in East Asia,* edited by Yung Sik Kim and Francesca Bray, 3-12. Seoul: Seoul National University Press, 1999.

Peterson, Willard J. "Another Look at *Li*." *Bulletin of Sung-Yuan Studies*, no. 18 (1986): 13-32.

Peterson, Willard J. "'Chinese Scientific Philosophy' and Some Chinese Attitudes towards Knowledge about the Realm of Heaven-and-Earth." *Past and Present* 87 (1980): 20-30.

Peterson, Willard J. "Calendar Reform Prior to the Arrival of Missionaries at the Ming Court." *Ming Studies* 21 (1986): 45-61.

Peterson, Willard J. "Confucian Learning in Late Ming Thought." In *The Cambridge History of China*, vol. 8, edited by Denis Twitchett, and Frederick W. Mote, 708-788. Cambridge: Cambridge University Press, 1998.

Peterson, Willard J. "Fang I-chih: Western Learning and the 'Investigation of Things'." In *The Unfolding of Neo-Confucianism*, edited by William Theodore de Bary, 369-411. New York: Columbia University Press, 1975.

Peterson, Willard J. "Making Connections: The 'Commentary on the Attached Verbalizations' in the *Book of Changes*." *Harvard Journal of Asiatic Studies* 42 (1982): 67-116.

Peterson, Willard J. "Western Natural Philosophy Published in Late Ming China." *Proceedings of the American Philosophical Society* 117 (1973): 295-322.

Popkin, Richard H. *The History of Scepticism from Erasmus to Spinoza*, rev. ed. Berkeley: University of California Press, 1979.

Porkert, Manfred. *The Theoretical Foundations of Chinese Medicine: Systems of Correspondence.* Cambridge, MA: MIT Press, 1974.

Porter, Jonathan. "Bureaucracy and Science in Early Modern China: The Imperial Astronomical Bureau in the Ch'ing Period." *Journal of Oriental Studies* 18 (1980): 61-76.

Porter, Jonathan. "The Scientific Community in Early Modern China." *Isis* 73 (1982): 529-544.

Price, Derek J. de Solla. *Science since Babylon.* New Haven: Yale University Press, 1961.

Qian Mu 錢穆, *Zhuzi xin xuean* 朱子新學案, 5 vols. Taibei: Sanmin shuju 三民書局, 1971.

Ragep, F. Jamil and Sally P. Ragep, eds. *Tradition, Transmission, Transformation.* Leiden: Brill, 1996.

Reischauer, Edwin O. and John K. Fairbank. *East Asia: The Great Tradition.* Boston: Houghton Mifflin, 1958.

Restivo, Sal P. "Joseph Needham and the Comparative Sociology of Chinese Science and Modern Science." In *Research in Sociology of Knowledge, Science, and Art*, vol. 2., edited by Robert A. Jones and Henrika Kuklik, 25-51. Greenwich, Conn.: Aijai, 1979.

Robson, William A., ed. *The Civil Service in Britain and France.* London: Hogarth Press, 1956.

Rossi, Paolo. *Philosophy, Technology and the Arts in the Early Modern Era.* New York: Harper, 1970.

Ruan Yuan 阮元. *Chouren zhuan* 疇人傳.

Russell, C. A., ed. *Science and Religious Belief: A Selection of Recent Historical Studies.* London: Oxford University Press, 1973.

Sabra, A. I. "The Appropriation and Subsequent Naturalization of Greek Science in Medieval Islam: A Preliminary Statement." *History of Science* 25 (1987): 223-243.

Sabra, A. I. "Situating Arabic Science: Locality versus Essence." *Isis* 87 (1996): 654-670.

Sakai Tadao 酒井忠夫, ed. *Dōkyō no sōgōdeki skenkyū* 道教の綜合的研究. Tokyo, 1977.

Saso, M. "What is the Ho-t'u?" *History of Religions* 17 (1978): 399-416.

Schäfer, Dagmar. "The Congruence of Knowledge and Action: The *Tiangong kaiwu* and Its Author Song Yingxing." In *Chinese Handicrafts Regulations of the Qing Dynasty: Theory and Application*, edited by Hans-Ulrich Vogel, Christine Moll-Murata, and Jainze Song, 35-60. Munich: Indicum, 2005.

Schipper, K. "The Taoist Body." *History of Religions* 17 (1978): 355-386.

Schirokauer, Conrad. "Neo-Confucians under Attack: The Condemnation of *Wei-hsueh.*" In *Crisis and Prosperity in Sung*, edited by John W. Haeger, 163-198. Tucson: University of Arizona Press, 1975.

Schreck, Johann (Deng Yuhan 鄧玉函). *Cetian yueshuo* 測天約說.

Shapin, Steven. *The Scientific Revolution.* Chicago: University of Chicago Press, 1996.

Shen Gua 沈括. *Mengxi bitan* 夢溪筆談. Beijing: Wenwu chubanshe 文物出版社, reprint; 1975.

Shi Yunli 石云里. "Shiqi shiji Zhongguo de zhun Gebaini xueshuo" 十七世紀中國的準哥白尼學說. *Da ziran tansuo* 大自然探索 14 (1995): 122-127.

Shi Yunli 石云里. "*Tianbu zhenyuan* yu Gebaini tianwenxue zai Zhongguo de zaoqi chuanbo" 天步眞原與哥白尼天文學在中國的早期傳播. *Zhongguo keji shiliao* 中國科技史料 21 (2000): 83-91.

Shi, Yunli 石云里. "Cong Huang Daozhou dao Hong Darong: shiqi shiji Zhong Chao didongshuo de bijiao yanjiu" 從黃道周到洪大容 – 十七世紀中朝地動說的比較研究. *Ziran bianzhengfa tongxun* 自然辨證法通訊 19 (1997): 60-65.

Shiji 史記.

Shimao, Eikoh. "Some Aspects of Japanese Science." *Annals of Science* 46 (1989): 69-91.

Siku quanshu zongmu tiyao 四庫全書總目提要. Beijing: Zhonghua shuju 中華書局, 1965.

Sin Minchŏl 申敏澈. "Myŏngdae chŏnmun 'sasŭp'-ŭi kŭmjiryŏng-kwa ch'ŏnmun sŏjŏk-ŭi ch'ulp'an: kŭ yinyŏm-kwa silchae" 明代 天文 '私習'의 禁止令과 天文 書籍의 出版: 그 理念과 實在. *Han'guk Kwahaksa Hakhoeji* 韓國科學史學會誌, 29 (2007): 231-260.

Sin Minchŏl 申敏澈. Myŏngdae chŏnmun 'sasŭp'-ŭi kŭmji-wa yŏkpŏpkwan-ŭi chaejŏngnip 明代 天文 '私習'의 禁止와 曆法觀의 再定立. Master's thesis, Seoul National University, 2007.

Sivin, Nathan, ed. *Science and Technology in East Asia.* New York: Science History Publications, 1977.

Sivin, Nathan. "Copernicus in China." *Studia Copernicana* 6 (1973): 63-122.

Sivin, Nathan. "Why the Scientific Revolution Did Not Take Place in China – or Didn't It?" *Chinese Science* 5 (1982): 45-66.

Sivin, Nathan. "Chinese Alchemy and the Manipulation of Time." *Isis*, 67 (1976): 513-526.

Sivin, Nathan. "Cosmos and Computation in Early Chinese Mathematical Astronomy." *T'oung Pao*, Series 2, 6 (1969): 1-73.

Sivin, Nathan. "On the Word 'Taoist' as A Source of Perplexity. With Special Reference to the Relations of Science and Religion in Traditional China." *History of Religions* 17 (1978): 303-330.

Sivin, Nathan. "Shen Kua." In *Dictionary of Scientific Biography*, edited by Charles C. Gillispie, vol. 12, 369-393. New York: Charles Scribner's Sons, 1975.

Sivin, Nathan. "Social Relations of Curing in Traditional China." *Nihon ishigaku zasshi* 日本醫史學雜誌 (Japanese Journal of Medical History) 23 (1977): 505-532.

Sivin, Nathan. "Wang Hsi-shan." In *Dictionary of Scientific Biography*, edited by Charles C. Gillispie, vol. 14, 159-168. New York: Charles Scribner's Sons, 1976.

Sivin, Nathan. *Chinese Alchemy: Preliminary Studies.* Cambridge, Mass.: Harvard University Press, 1968.

Sleeswyk, Andre W. "Celestial River: A Reconstruction." *Technology and Culture* 19 (1978): 423-449.

Sleeswyk, Andre W. "Reconstruction of the South-pointing Chariots of the Sung Dynasty: Escapement and Differential Gearing in Eleventh Century China." *Chinese Science* 2 (1977): 4-36.

Smith, Paul J. and Richard von Glahn, eds. *The Song-Yuan-Ming Transition in Chinese History.* Cambridge, Mass.: Harvard University Press, 2003.

Smogulecki, Jean-Nicolas (Mu Nige 穆尼閣). *Tianbu zhenyuan* 天步眞原.

Snow, Charles P. *The Two Cultures and the Scientific Revolution.* Cambridge: Cambridge University Press, 1959.

Sŏ Myŏng-ŭng 徐命膺. *Pomanjae ch'ongsŏ* 保晚齋叢書 (Kyujanggak 奎章閣 ed.).

Standaert, Nicolas, ed. *Handbook of Christianity in China, Volume One: 635-1800.* Leiden: Brill, 2001.

Staudenmaier, John M. *Technology's Storytellers: Reweaving the Human Fabric.* Cambridge, Mass: MIT Press, 1985.

Stock, Brian. "Science, Technology, and Economic Progress in the Early Middle Ages." In *Science in the Middle Ages*, edited by David. C. Lindberg, 1-51. Chicago: University of Chicago Press, 1978.

Sudō, Yoshiyuki 周藤吉之. *Sōdai keizaishi kenkyū* 宋代經濟史研究. Tokyo: Tokyo University Press, 1962.

Sun, E. Z. and Sun, S.-c., trans. *T'ien-kung K'ai-wu: Chinese Technology in the Seventeenth Century.* University Park: Pennsylvania State University Press, 1966.

Szczesniak, Boleslaw. "Notes on the Penetration of the Copernican Theory into China (Seventeenth-Nineteenth Centuries)." *Journal of the Royal Asiatic Society* (April 1945): 30-38.

Taylor, Rodney L. *The Religious Dimensions of Confucianism.* Albany: State University of New York Press, 1990.

Tillman, Hoyt C. "The Idea and the Reality of the 'Thing' during the Sung." *Bulletin of Sung & Yuan Studies*, no. 14 (1978): 68-82.

Tu, Wei-ming, ed. *Confucian Traditions in East Asian Modernity: Moral Education and Economic culture in Japan and the Four Mini-Dragons.* Cambridge, Mass.: Harvard University Press, 1996.

Tu, Wei-ming. *Neo-Confucian Thought in Action: Wang Yang-ming's Youth* (1471-1509). Berkeley: University of California Press, 1976.

Twitchett, Denis and Frederick W. Mote, eds. *The Cambridge History of China*, vol. 8. Cambridge: Cambridge University Press, 1998.

Unschuld, Paul U. *Medical Ethics in Imperial China: A Study in Historical Anthropology.* Berkeley: University of California Press, 1979.

Vogel, Hans-Ulrich, Christine Moll-Murata, and Jainze Song, eds. *Chinese Handicrafts Regulations of the Qing Dynasty: Theory and Application.* Munich: Indicum, 2005.

Wagner, Donald B. "Liu Hui and Tsu Keng-chih on the Volume of a Sphere." *Chinese Science* 3 (1978): 1-26.

Wallace, William A. "The Philosophical Setting of Medieval Science." In *Science in the Middle Ages*, edited by David. C. Lindberg, 91-119. Chicago: University of Chicago Press, 1978.

Walle, Willy Wande. "Ferdinand Verbiest and the Chinese Bureaucracy." In *Ferdinand Verbiest (1623-1688): Jesuit Missionary, Scientist, Engineer and Diplomat*, edited by John W. Witek, 495-515. Nettetal: Steyler Verlag, 1994.

Wang Ermin 王爾民. *Wan Qing zhengzhi sixiangshilun* 晚淸政治思想史論. Taipei: Huashi chubanshe, 1969.

Wang Yangzong 王揚宗. "Xixue zhongyuan shuo zai MingQing zhi ji de youlai ji qi yanbian" '西學中源'說在明清之際的由來及其演變. *Dalu zazhi* 大陸雜誌 (Taibei 臺北) 90 (1995): 39-45.

Weber, Max. *The Religion of China, Confucianism and Taoism.* New York: Free Press, 1951.

Weisheipl, James A. "Classification of the Sciences in the Mediaeval Thought." *Mediaeval Studies* 27 (1965): 54-90.

Weisheipl, James A. "Curriculum of the Faculty of Arts at Oxford in the Early Fourteenth Century." *Mediaeval Studies* 26 (1964): 143-185.

Weisheipl, James A. "The Nature, Scope and Classification of the Sciences." In *Science in the Middle Ages*, edited by David. C. Lindberg, 461-482. Chicago: University of Chicago Press, 1978.

Welch, Holmes and Anna Seidel, eds. *Facets of Taoism: Essays in Chinese Religion.* New Haven: Yale University Press, 1979.

Welch, Holmes. *Taoism: The Parting of the Way*. Boston: Beacon, rev. ed.; 1965.

Westfall, Richard S. "Science and Patronage: Galileo and the Telescope." *Isis* 76 (1985): 11-30.

Wilson, Curtis. *William Heytesbury: Medieval Logic and the Rise of Mathematical Physics.* Madison: University of Wisconsin Press, 1960.

Witek, John W., ed. *Ferdinand Verbiest (1623-1688): Jesuit Missionary, Scientist, Engineer and Diplomat.* Nettetal: Steyler Verlag, 1994.

Wittenborn, Allen. "*Li* Revisited and Other Explorations." *Bulletin of Sung-Yuan Studies*, no. 17 (1981): 32-48.

Wright, Arthur F. *Buddhism in Chinese History.* Stanford: Stanford University Press, 1959.

Wright, Arthur F. "Values, Roles, and Personalities." In *Confucian Personalities*, edited by Arthur F. Wright and D. Twitchett, 3-23. Stanford: Stanford University Press, 1962.

Wright, Arthur F., ed. *The Confucian Persuasion.* Stanford: Stanford University Press, 1960.

Wright, Arthur F. and D. Twitchett, eds. *Confucian Personalities.* Stanford: Stanford University Press, 1962.

Xi, Zezong. "Chinese Studies in the History of Astronomy, 1949-1979." *Isis* 72 (1981): 456-470.

Xinfa suanshu 新法算書.

Xu Guangtai 徐光台. "Mingmo xixue yu Bailutong Shuyuan" 明末西學與白鹿洞書院. In *Zhongguo shuyuan luntan* 中國書院論壇 (Discussion Forum on the Chinese Academies), edited by Gao Feng 高峰, Hu Qing 胡青, Lai Gongou 賴功歐, and Ye Zunhong 葉存洪, 370-384. Beijing: Zuojia chubanshe 作家出版社, 2007.

Yabuuchi Kiyoshi, ed. 藪内清 編. *Tenkō kaibutsu no kenkyū* 天工開物の研究. Tokyo: Kōseisha 恒星社, 1954.

Yabuuchi Kiyoshi, trans. 藪内清 譯註. *Tenkō kaibutsu* 天工開物. Tokyo: Heibonsha 平凡社, 1969.

Yabuuchi Kiyoshi. "*Tenkō kaibutsu* ni tsuite" 天工開物について. In *Tenkō kaibutsu no kenkyū*, translated by Kiyoshi Yabuuchi, 1-24. Tokyo: Heibonsha, 1969.

Yabuuchi Kiyosi 藪内清. "Richo gakusha no chikyu kaitensetsu" 李朝科學者の地球回轉說. *Chōsen gakuhō* 朝鮮學報 49 (1968): 427-434.

Yabuuti, Kiyoshi. "The Study of the History of Chinese Science in Kyoto." *Acta Asiatica* 36 (1979): 1-6.

Yamada Keiji 山田慶兒. *Shushi no shizen-gaku* 朱子の自然學. Tokyo: Iwanami shoten 岩波書店, 1978.

Yamada, Keiji 山田慶兒. "Jujireki no michi. Genchō chika no tenmondai to tenmongakusha" 授時暦への道 – 元朝治下の天文臺と天文學者. In *Chūgoku no kagaku to kagakusha*, edited by Keiji Yamada, 1-207. Kyoto: Kyōto daigaku jinbunkagaku genkyūsho, 1978.

Yamada, Keiji, ed. 山田慶兒 編. *Chūgoku no kagaku to kagakusha* 中國の科學と科學者. Kyoto: Kyōto daigaku jinbunkagaku genkyūsho 京都大學 人文科學研究所, 1978.

Yang Xiaoming 楊小明. "Gebaini rixin didongshuo zai Zhongguo de zuizao jieshao" 哥白尼日心地動說在中國的最早介紹. *Zhongguo keji shiliao* 中國科技史料 20 (1999): 67-73.

Yi Ik 李瀷. *Sŏngoho sasŏl* 星湖僿說. Seoul: Kyŏnghi ch'ulp'ansa 慶熙出版社, 1967.

Yi Yong-bŏm 李龍範. "Kim Sŏk-mun-ŭi chijŏnnon-kwa kŭ sasangjŏk paegyŏng" 金錫文의 地轉論과 그 思想的 背景. *Chindan hakpo* 震檀學報 41 (1976): 81-107.

Yŏnse daehakkyo Kuk'hak yŏn'guwon, ed. 延世大學校 國學研究院 編. *Han'guk sil'hak sasang yŏn'gu* 4. *Kwahak kisul p'yŏn* 韓國實學思想研究, 4. 科學技術篇 Seoul: Hyean 慧眼, 2005.

Yŏnse taehakkyo Kukhak yŏn'guwon 延世大學校 國學研究院. Han'guk kŭndae yihaenggi chung'in yŏn'gu 韓國 近代移行期 中人研究. Seoul: Sinsŏwon 新書苑, 1999.

Yu Pong-hak 劉奉學. *Yŏn'am ilp'a Pukhak sasang yŏn'gu* 燕巖一派 北學思想 研究. Seoul: Ilchisa 一志社, 1995.

Yun Sa-sun, ed. 尹絲淳 編. *Chŏng Yak-yong* 丁若鏞. Seoul: Koryŏ daehakkyo chulp'anbu 高麗大學校 出版部, 1990.

Zeng, Xiongsheng. "Hiding in Doctoring: The Study of 1364 Medical Practitioners." In *Current Perspectives in the History of Science in East Asia,* edited by Yung Sik Kim and Francesca Bray, 438-448. Seoul: Seoul National University Press, 1999.

Zhang Hua 張華. *Bowuzhi* 博物志 (*Sibu beiyao* 四部備要 ed.).

Zhang Liwen 張立文. *Zhu Xi sixiang yanjiu* 朱熹思想研究. Beijing: Zhongguo shehui kexue chubanshe 中國社會科學出版社, 1981.

Zhang Yongtang 張永堂. *Mingmo Fangshi xuepai yanjiu chubian: Mingmo lixue yu kexue guanxi shilun* 明末方氏學派研究初編: 明末理學與科學關係試論. Taipei: Wenjing 文鏡, 1987.

Zhang Yongtang 張永堂. *Mingmo yu Qingchu lixue yu kexue guanxi zailun* 明末與清初理學與科學關係再論. Taipei: Xuesheng shuju 學生書局, 1994.

Zhang Zai 張載. *Zhangzi quanshu* 張子全書 (*Sibu beiyao* 四部備要 ed.).

Zhong Guangyan, ed. and trans. 鍾廣言 注釋. *Tiangong kaiwu* 天工開物. Hong Kong: Zhonghua Shuju 中華書局, 1978.

Zhongguo Nongye Bowuguan 中國農業博物館, ed. *Zhongguo gudai gengzhi tu* 中國古代耕織圖. Beijing: Xinhua shudian 新華書店, 1995.

Zhongguo tianwenxueshi zhengli yanjiu xiaozu 中國天文學史整理研究小組. *Zhongguo tianwenxueshi* 中國天文學史 [History of Chinese Astronomy]. Beijing: Kexue chubanshe 科學出版社, 1987.

Zhu Weijing, ed. 朱維靜 編. *Li Madou Zhongwen zhuyiji* 利瑪竇中文著譯集. Hong Kong: City University of Hong Kong Press香港城市大學出版社, 2001.

Zhu Xi 朱熹. *Huian xiansheng Zhuwengong wenji* 晦庵先生朱文公文集 (*Sibu beiyao* 四部備要 ed.).

Zhu Xi. *Sishu jizhu* 四書集註 (*Sibu beiyao* 四部備要 ed.).

Zhu Xi. *Zhuzi yulei* 朱子語類. Taipei: Zhengzhong shuju 正中書局, reprint; 1962.

Zhu, Pingyi 祝平一. "Song Ming zhi ji de yishi yu 'ruyi'" 宋明之際的醫史與儒醫. *Zhongyang yanjiuyuan Lishiyuyan yanjiusuo jikan* 中央研究院歷史語言研究所集刊 77 (2006): 401-449.

Zhuangzi 莊子.

Zürcher, Erik. *The Buddhist Conquest of China: the Spread and Adaptation of Buddhism in Early Medieval China.* Leiden: Brill, 1959.

Index

 | DOI 10.1163/9789004265318-016

www.ingramcontent.com/pod-product-compliance
Lightning Source LLC
LaVergne TN
LVHW010656300926
849161LV00005B/1143